Differential-Algebraic Equations Forum

Differential-Algebraic Equations Forum

The series *Differential-Algebraic Equations Forum* is concerned with analytical, algebraic, control theoretic and numerical aspects of differential algebraic equations, as well as their applications in science and engineering. It is aimed to contain survey and mathematically rigorous articles, research monographs and textbooks. Proposals are assigned to a Publishing Editor, who recommends publication on the basis of a detailed and careful evaluation by at least two referees. The appraisals will be based on the substance and quality of the exposition.

More information about this series at http://www.springer.com/series/11221

Stephen Campbell • Achim Ilchmann •
Volker Mehrmann • Timo Reis
Editors

Applications of Differential-Algebraic Equations: Examples and Benchmarks

Editors
Stephen Campbell
Department of Mathematics
North Carolina State University
Raleigh, NC, USA

Achim Ilchmann
Institut für Mathematik
Technische Universität Ilmenau
Ilmenau, Germany

Volker Mehrmann
Institut für Mathematik
Technische Universität Berlin
Berlin, Germany

Timo Reis
Fachbereich Mathematik
Universität Hamburg
Hamburg, Germany

ISSN 2199-7497 ISSN 2199-840X (electronic)
Differential-Algebraic Equations Forum
ISBN 978-3-030-03717-8 ISBN 978-3-030-03718-5 (eBook)
https://doi.org/10.1007/978-3-030-03718-5

Mathematics Subject Classification (2010): 34A09, 65L80, 00A71, 34B60

This Springer imprint is published by the registered company Springer Nature Switzerland AG.
The registered company address is: Gewerbestrasse 11, 6330 Cham, Switzerland

Preface of Applications of Differential-Algebraic Equations: Examples and Benchmarks

This volume in the series "DAE Forum" encompasses prototypical, innovative and emerging examples and benchmarks of differential-algebraic equations (DAEs) from application areas such as:

- Underactuated mechanical systems
- Computational electrodynamics
- Gas networks
- Coupled flow networks
- Vehicle dynamics
- Robotics
- Semi-discretized Navier-Stokes equations
- Tracking problems
- Nonsmooth, nonlinear DAEs

All articles have a modelling section, explaining whether the model is prototypical and for which applications it is used, followed by a mathematical analysis, and, if appropriate, a discussion of the numerical aspects including simulation. The volume may serve to achieve a deeper understanding of a broad spectrum of applications, to illustrate the many diverse areas of mathematics that are used in understanding and treatment of DAEs and to provide a reference for benchmarks in DAEs.

Raleigh, NC, USA
Ilmenau, Germany
Berlin, Germany
Hamburg, Germany
September 2018

Stephen Campbell
Achim Ilchmann
Volker Mehrmann
Timo Reis

Contents

General Nonlinear Differential Algebraic Equations and Tracking Problems: A Robotics Example

Stephen Campbell and Peter Kunkel

Abstract One of the ways that differential algebraic equations (DAEs) naturally arise is with tracking problems. This paper will discuss some of the tracking problems that occur, how they are interrelated, and how they relate to the theory of DAEs. This paper will focus on the theory and algorithms for unstructured tracking problems. These ideas will then be applied to a test problem involving a robot arm with a flexible joint. A variety of challenging test problems can be formulated from this model.

Keywords DAE · Differential Algebraic equation · Tracking control · Robotic arm

Mathematics Subject Classification (2010) 34A09, 34H05, 49J15, 65L80, 93C15, 93B40

1 Introduction

One of the ways that DAEs occur in applications from several areas are with tracking problems. We shall assume that the reader has some familiarity with DAEs and such concepts as the index of a DAE [8, 32]. There are generally three different types of tracking problems any of which can result in a DAE even if the original process is an ODE.

S. Campbell (✉)
Department of Mathematics, North Carolina State University, Raleigh, NC, USA
e-mail: slc@ncsu.edu

P. Kunkel
Mathematisches Institut, Universität Leipzig, Leipzig, Germany
e-mail: kunkel@math.uni-leipzig.de

S. Campbell et al. (eds.), *Applications of Differential-Algebraic Equations: Examples and Benchmarks*, Differential-Algebraic Equations Forum,
https://doi.org/10.1007/11221_2018_3

In its simplest formulation, a tracking problem consists of a system (1.1) and a reference trajectory $r(t)$,

$$F(\dot{x}, x, t, u) = 0 \tag{1.1a}$$

$$y = G(t, x). \tag{1.1b}$$

The goal is to have the control u cause the output y to track r in some sense. A number of examples come to mind where r can be the path of a vehicle, a desired temperature in a reactor, or the endpoint of a robotic arm to name just a few. However, the use of tracking is even more important than this suggests. The reason is that tracking is used in what we shall call index reduction in practice. In many applications the control u must be produced by another process and our real input is the input to that process. For example if u is a force, then there are inputs to a motor to produce the force. Thus we actually have a second system (1.2), called the actuator dynamics,

$$\hat{F}(\dot{z}, z, t, v) = 0 \tag{1.2a}$$

$$u = \tilde{G}(t, z). \tag{1.2b}$$

The combined system (1.1) and (1.2) viewed as a system in x, z, u, v along with a control objective on y will usually give a DAE of higher index than just (1.1). This is often approached by breaking the problem into two separate problems. One designs u to track r. Then one designs v to track u. This approach is often used in part because the actuator and the process may well be built by different groups and the same actuators used in a number of different applications. Not surprisingly there has been considerable research on tracking problems in the engineering literature. We shall reference just a few of these papers. In particular, we will not discuss such approaches as using observers [39].

The purpose of this paper is to present a challenging physical system that can be used to test different approaches to tracking problems involving DAEs. The particular problem is well known as a challenging problem from robotics that has been studied in the nonlinear control literature. Our intent here is different in that we are interested in using it as the basis of some different DAE tracking problems.

The organization of this paper is as follows. In the remainder of this section we discuss some preliminary material. Then in Sect. 2 we will discuss some of the different types of tracking and some of the issues that arise. Simple examples will be given as illustrations. Then in Sect. 3 we will discuss the needed mathematics behind tracking with general nonlinear DAEs. Section 4 will present the physical system that is our test problem. We shall point out several problem variations and then examine the techniques of the earlier sections applied to this test problem. We will see that sometimes one approach will be easier to apply than another and that the problem difficulty can be highly dependent on parameter values.

There are two sets of terminology that are used when working with DAE control systems such as (1.1a). In one of these, which we will refer to as the classical approach, the variable u is considered to be the control. Then the parts of x that are differentiated are called differential state variables and the rest of x are called algebraic state variables. Properties such as index refer to the dynamical response in x to a given u. In the other set of terminology, u is just considered to be additional algebraic state variables and one talks of the strangeness index. Having strangeness index one is closely related to having virtual index one. That is, there is a part of the algebraic state variables that can be chosen as the controls and the remaining DAE is index one [17, 24].

Here tracking can take three forms. One is exact tracking where we are given $r(t)$ and want u so that

$$y(t) - r(t) = 0. \tag{1.3}$$

This is also sometimes referred to as prescribed path control or systems inversion. That is, you want to go from a prescribed output r to an input u that generates that output.

The second type of tracking is where we want $y(t) - r(t)$ to be small in some appropriate function space norm. How small is interpreted can have a great impact on both numerical and analytic issues. Sometimes the goal is just to have $y(t) - r(t) \in \Sigma(t)$ for some time varying family of sets Σ. A parent watching their child just wants to keep the child in their field of view. This can also frequently arise in navigation problems. For example, say a robot has to move from one place to another around some obstacles. A path is designed off line that does this and which stays at least d away from the obstacles. Then one may have $\|y(t) - r(t)\| \leq d$ as a path inequality. The use of optimal control for approximate tracking of DAEs is considered in [1]. However, they restrict themselves to the case where the system is a differentiably flat system [25, 26], that is the planned trajectory and its derivatives completely determine the control. We do not make this assumption in our discussion although the test problem considered later is sometimes flat. The third is asymptotic tracking where $\|y(t) - r(t)\|$ goes to zero as $t \to \infty$. Often it is desirable to have some control over the rate the tracking error goes to zero. Note that with approximate tracking the tracking error does not need to go to zero which can be useful as shown later.

In addition, there may be restrictions on the available control, such as only so much control effort is available, and there may be endpoint or target conditions. All of these considerations can have an impact on the types of DAEs that arise and also how the DAE theory is interpreted.

There is a considerable amount of existing theory and experience with DAEs and tracking problems that comes into play with each of these scenarios. We will note only a tiny bit of this literature [29, 31, 37, 38]. Our goal is not to repeat this classical material here. Rather we will discuss some of what can be done with more general approaches. This is especially important with some classes of computer generated software models. It is important to note that even if one is using modeling software that is designed for index one and zero models, that higher index models readily occur.

The path $r(t)$ may or may not be known ahead of time. For path planning and in some other control applications one may consider that r is known. However, there are other scenarios where it is desired to estimate the value of $G(t, x(t))$. This is also sometimes referred to as tracking [27]. In this paper we consider r as known on the time interval of interest. The term path tracking is also used when following parameters using numerical homotopy methods [3]. Homotopy methods are also not considered here.

2 Introductory Discussion of Tracking

We begin first with elementary examples to illustrate some of the types of problems we are considering and issues that will be faced. The following examples in the introductory sections are all solved using either the direct transcription code GPOPS II [43] or boundary value solver `bvp4c` from Matlab.

2.1 *Exact Tracking*

Suppose that we have the ordinary differential equation (2.1) modeling a process

$$\dot{x}_1 = -x_1 + 0.2x_2 + u_1 \tag{2.1a}$$

$$\dot{x}_2 = -0.1x_1 - 2x_2 + u_2 \tag{2.1b}$$

$$\dot{x}_3 = x_2 \tag{2.1c}$$

$$\dot{x}_4 = x_3, \tag{2.1d}$$

and the goal is for x_4 to track a given r. Thus we have $y = G(t, x(t)) = x_4(t)$. If we add

$$x_4(t) - r(t) = 0 \tag{2.1e}$$

to (2.1a)–(2.1d) we get a DAE that is overdetermined in terms of x. It is not overdetermined in terms of x, u. In terms of x, u, it is differentiation index four. If r is sufficiently smooth, then clearly the DAE (2.1) is equivalent to

$$\dot{x}_1 = -x_1 + 0.2\ddot{r} + u_1 \tag{2.2a}$$

$$\dddot{r} = -0.1x_1 - 2\ddot{r} + u_2 \tag{2.2b}$$

$$x_2 = \ddot{r} \tag{2.2c}$$

$$x_3 = \dot{r} \tag{2.2d}$$

$$x_4 = r, \tag{2.2e}$$

which is an index one DAE in x_1, u_2 with u_1 left as an input that can be chosen depending on the control objectives. System (2.2) will be referred to as an index reduced version of (2.1). In this paper index reduction will always mean to an index one DAE with the same solutions as the original DAE.

Exact tracking was one of the first forms of tracking that were studied. The appearance of a DAE is immediate. It is important to note that since r in some of these applications is a known function, in fact it may be a designed path, and some of the derivatives of r may be available to high accuracy, even analytically. This is to be distinguished from cases where r is a measured output and hence is noisy which makes its differentiation much more problematic.

Note that the form of G is partially a design decision and partly restricted by physical considerations. However, it can have a major impact on the DAE structure. For example, a constrained mechanical system may have the form

$$\dot{x} = f(x, z, u) + G_z(z)^T \lambda \tag{2.3a}$$

$$\dot{z} = x \tag{2.3b}$$

$$r(t) = G(z) \tag{2.3c}$$

where z is position and x is velocity. This problem is index three in x, z, λ if G_z has full rank, otherwise it could be higher. However, this depends on the fact that (2.3c) is a physical constraint which generates a force so that its Jacobian appears in (2.3a). There is no reason a priori that the index be constant nor that it even be well defined if G is a desired output or trajectory that does not generate a force, that is, it is what is sometimes called a program constraint rather than a physical constraint. Then instead of (2.3) one has

$$\dot{x} = f(x, z, u) \tag{2.4a}$$

$$\dot{z} = x \tag{2.4b}$$

$$r(t) = G(z). \tag{2.4c}$$

Exact tracking for (2.3) for DAEs which are analytically transferable to semi-explicit index one DAEs is considered in several places. In [30, 31, 37] the approach is similar but the details differ. The goal is to get an index one system for numerical and application purposes. This involves both reduction, say by differentiating constraints, and by using feedback that reduces the index. A simple example of index reduction with feedback is

$$\dot{x}_1 = x_1 + 2x_2 + u \tag{2.5a}$$

$$0 = 4x_1 + u, \tag{2.5b}$$

which is index 2 in x for a given u. Letting $u = -2x_1 + x_2 + v$ gives

$$\dot{x} = -x_1 + 3x_2 + v \tag{2.6a}$$

$$0 = 2x_1 + x_2 + v \tag{2.6b}$$

which is now an index one DAE in x for a given v. In the case of an ODE plant the feedback can sometimes alter the index of the output tracking problem.

The other approach for exact tracking uses differential algebra and goes to a canonical form using nonlinear coordinate changes provided certain physical assumptions hold. The resulting form has a chain of integrators and a nonlinear equation much like (1.1) and one works with derivatives of r, u. From this form one may answer a number of control questions. The concept of the relative degree is closely related to the index when exact tracking is used. The application of this approach has been extensively studied in a number of papers. We note only [23, 37, 38]. The example (2.1) is similar to the canonical form used in [37, 38].

Exact tracking may not always be possible. Consider the following variant of (2.1) with new (2.1a)– (2.1b):

$$\dot{x}_1 = -x_1 + 0.2x_2 \tag{2.7a}$$

$$\dot{x}_2 = -0.1x_1 - 2x_2 + \sin(u_1) \tag{2.7b}$$

$$\dot{x}_3 = x_2 \tag{2.7c}$$

$$\dot{x}_4 = x_3. \tag{2.7d}$$

Controls of the form $\sin(u)$ appear in many vehicular problems where u is a steering angle. Clearly $x_4(t) = r(t)$ is not possible for all smooth $r(t)$. For example, if $r(t) = \sin(\alpha t)$ for a large α, then (2.7b) can be inconsistent and cannot be solved for u_1.

The role of initial conditions is important. With exact tracking they should be consistent for the tracking problem [9]. Since the tracking condition adds additional constraints the original initial conditions might not be consistent any longer. For this reason tracking problems are often broken up into two phases. A first alignment phase where, in essence the system gets to a consistent initial condition for the tracking problem solution and then a second phase which is the tracking.

2.2 *Approximate Tracking as Optimal Control*

Another approach to tracking is to attach a cost to the tracking error and try to reduce that cost. As we will see this has the effect of regularizing the problem numerically as long as the error tolerances are not too tight.

The first thing to think of is adding a quadratic cost of the form,

$$\begin{aligned} J = {} & (G(T, x(T)) - r(T))^T H(G(T, x(T)) - r(T)) \\ & + \int_0^T (G(t, x(t)) - r(t))^T Q(G(t, x(t)) - r(t)) + u^T Du \, dt, \end{aligned} \tag{2.8}$$

where $H \geq 0$, $Q \geq 0$, and $D \geq 0$. If there are no bounds on the control u, we will always take $D > 0$.

Suppose that we take (2.1) and desired path r and cost (2.8). If r is smooth, then the only issue is the need to interpolate for the higher derivatives of r. We take the time interval as [0 10]. For example if $r(t) = \sin(1.4t)$ and we take a consistent

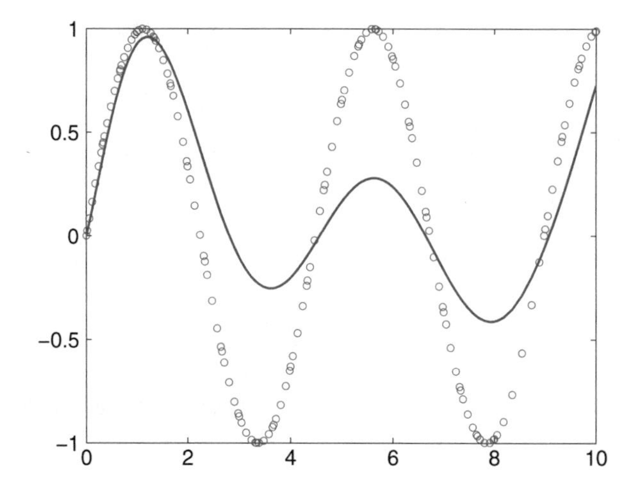

Fig. 1 State trajectory and target (circled line) for (2.2) and (2.8) on [0 10] with $D = 1$ and $Q = 10$

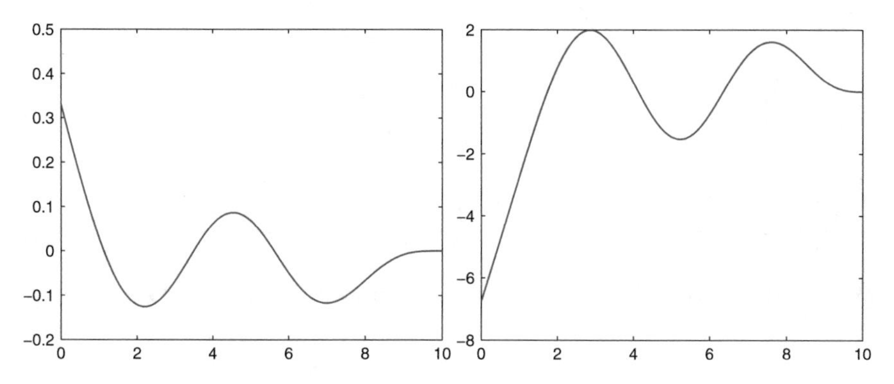

Fig. 2 Controls for (2.1) and (2.8) on [0 10] with $D = 1$ and $Q = 10$

initial condition of the constrained problem of $[3, 0, 1.4, 0]^T$ we get with $D = 1$, $Q = 10$ the results in Figs. 1 and 2.

If we set $Q = 10,000$ we get as expected much better tracking with larger control effort. In Fig. 3 we see that with high cost on the tracking error term and with smooth solutions, we quickly get a good answer even if the problem is theoretically index three for exact tracking.

But suppose that instead of a smooth reference trajectory we have one which is not smooth. For example, we could have a piecewise linear path generated by some route planning process. Figure 4 shows what happens with $Q = 10$.

Figure 4 was done with two iterations and it took several seconds for the iteration. Additional iterations were not helpful. This illustrates how a lack of smoothness can sometimes create problems. Suppose, however, we replace r with a smoother r that is very close to the original r. This was done by replacing the Matlab command `interp1` with `spline` and adding a few more points. The result is in Fig. 5. Not

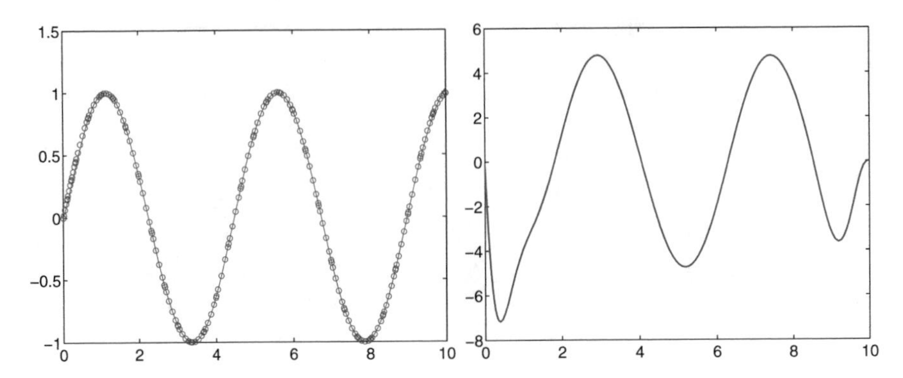

Fig. 3 State trajectory and target (circled line) for (2.1) and (2.8) on [0 10] with $D = 1$ and $Q = 10,000$

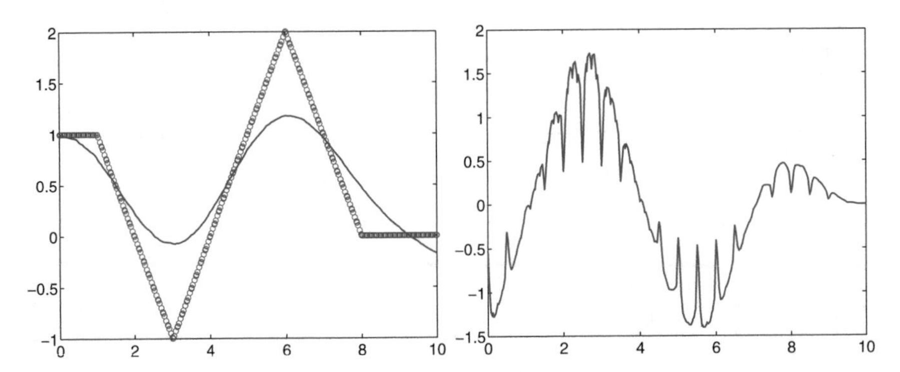

Fig. 4 State solution (solid) r (dotted) on the left and control u_2 (right) on [0 10] with $D = 1$ and $Q = 10$

only is the control much better with smoother r, but the iterations were much faster. Taking both more iterations and a higher Q, we get Fig. 6. Note that there is still a reasonable control even though tighter tracking is called for.

It is important to note that in these computations we are not explicitly differentiating r. Rather the optimizer is numerically working its way back through the equations to the control.

In the literature the type of target trajectories r is often restricted by assuming that r is the output of another dynamical process say $\dot{r} = h(t, r)$. This not only makes sure r is smooth but also gives some information on the derivatives of r if $h(t, r)$ is a known function.

2.3 Asymptotic Tracking

Asymptotic tracking is often treated on an infinite interval although, of course practically we are interested in doing so over a finite interval. Asymptotic tracking

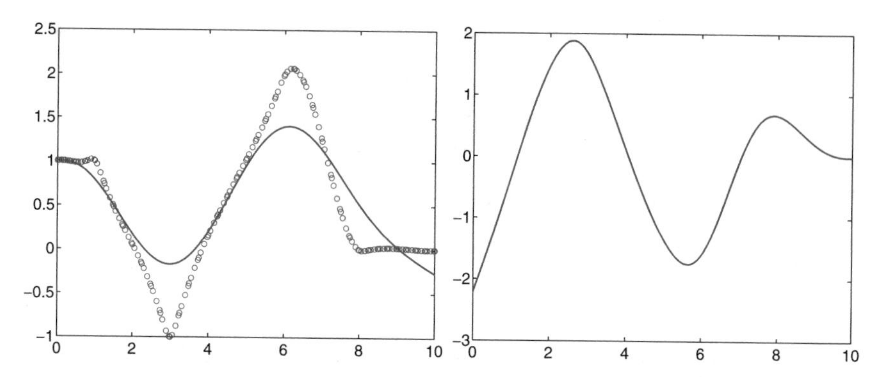

Fig. 5 State solution (solid) and r (dotted) on the left and control u_2 (right) on [0 10] with $D = 1$ and $Q = 10$ and smoothed piecewise linear r

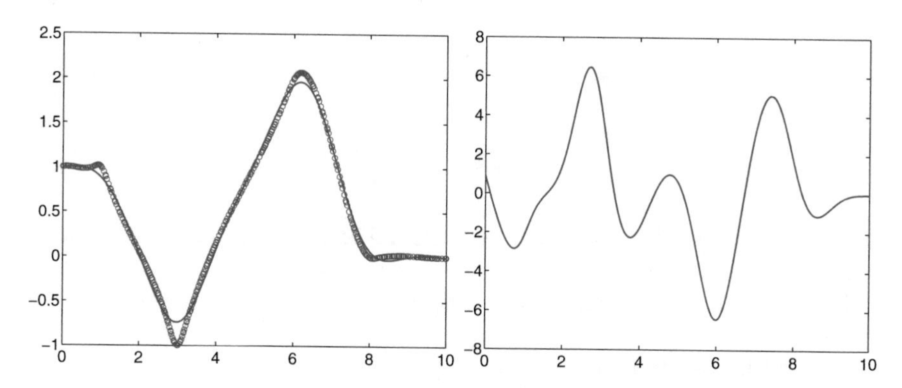

Fig. 6 State solution (left) and control u_2 (right) on [0 10] with $D = 1$ and $Q = 1000$, more iterations allowed, and smoothed piecewise linear r

has also been considered by a number of people. One way to get asymptotic tracking is to use optimal control formulations where the cost drives the tracking error very small. Another way is to utilize some of the theory of observers which is not discussed here. Suppose that we have the following system

$$\dot{x}_1 = -x_1 + 0.2x_2 + u_1 \tag{2.9a}$$

$$\dot{x}_2 = -0.1x_1 - 2x_2 + u_2 \tag{2.9b}$$

$$\dot{x}_3 = x_2 \tag{2.9c}$$

and we want x_3 to track r and we want this tracking to be independent of the initial conditions. Algebra shows that we get exact tracking if

$$x_3 = r \tag{2.10a}$$

$$x_2 = \dot{r} \tag{2.10b}$$

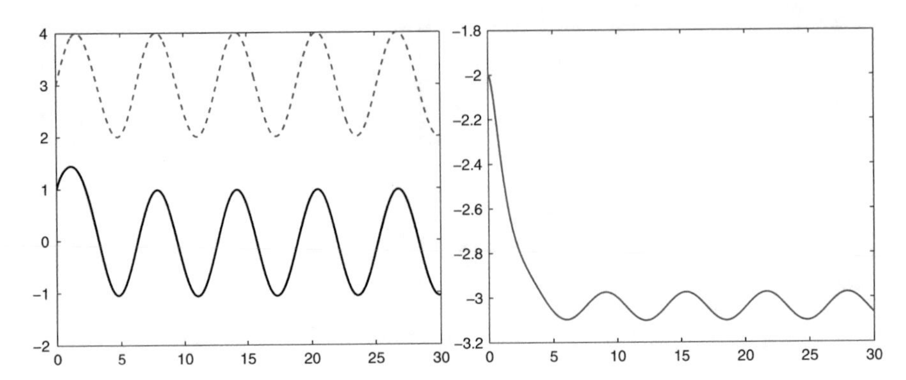

Fig. 7 State x_3 (solid) and r (dashed) on the left and tracking error (right) on [0 30] in original system using the open loop control (2.10d) and (2.10e) and initial condition $[1, 1, 1]^T$

$$x_1 = -10(2\dot{r} + \ddot{r}) \tag{2.10c}$$

$$u_2 = 0 \tag{2.10d}$$

$$u_1 = -(10r^{(3)} + 30\ddot{r} + 22.2\dot{r}) \tag{2.10e}$$

Suppose that we use the control in (2.10d) and (2.10e) but we do not have the correct initial conditions. To illustrate what happens, we take $r(t) = \sin(t)$ and our interval as [0 30]. The initial starting value is $x(0) = [1, 1, 1]^T$. The results are shown in Fig. 7.

This is not satisfactory. Let

$$\epsilon_3 = x_3 - r \tag{2.11}$$

$$\epsilon_2 = x_2 - \dot{r}. \tag{2.12}$$

Then we have the tracking error dynamics are

$$\dot{x}_1 = -x_1 + 0.2\epsilon_2 + u_1 + 0.2\dot{r} \tag{2.13a}$$

$$\dot{\epsilon}_2 = -0.1x_1 - 2\epsilon_2 - 2\dot{r} + u_2 - \ddot{r} \tag{2.13b}$$

$$\dot{\epsilon}_3 = \epsilon_2, \tag{2.13c}$$

which is an ODE in $x_1, \epsilon_2, \epsilon_3$. Since

$$\left\{ \begin{bmatrix} -1.0 & 0.2 & 0 \\ -0.1 & -2.0 & 0 \\ 0 & 1.0 & 0 \end{bmatrix}, \begin{bmatrix} 1 & 0 \\ 0 & 1 \\ 0 & 0 \end{bmatrix} \right\}$$

is a controllable pair, given any three desired eigenvalues, there is a feedback matrix F so that using the control

$$u = F \begin{bmatrix} x_1 \\ \epsilon_2 \\ \epsilon_3 \end{bmatrix} + \begin{bmatrix} -0.2\dot{r} \\ 2\dot{r} + \ddot{r} \end{bmatrix} \tag{2.14}$$

stabilizes the tracking error. For example, to place the eigenvalues at $\{-1, -2, -3\}$ we could use

$$F = -\begin{bmatrix} 1.0 & 0.1 & -0.1 \\ -0.1 & 2.0 & 3.0 \end{bmatrix},$$

which was found using the pole placement algorithm in Scilab [19], but any of a number of software packages could be used. This results in

$$\dot{x}_1 = -2x_1 + 0.1\epsilon_2 + 0.1\epsilon_3 \tag{2.15a}$$

$$\dot{\epsilon}_2 = -0.4\epsilon_2 - 3\epsilon_3 \tag{2.15b}$$

$$\dot{\epsilon}_3 = \epsilon_2, \tag{2.15c}$$

which is asymptotically stable with the desired eigenvalues. In terms of the original variables, we have our new control with feedback is

$$u = F \begin{bmatrix} x_1 \\ x_2 - \dot{r} \\ x_3 - r \end{bmatrix} + \begin{bmatrix} -0.2\dot{r} \\ 2\dot{r} + \ddot{r} \end{bmatrix}. \tag{2.16}$$

This choice of control presupposed that x_1, x_2, x_3 are available for feedback. If they are not, then an observer would have to be constructed for them.

Applying the tracking feedback (2.16) to the original system, and again using $x(0) = [1, 1, 1]^T$, we get the results in Fig. 8. Now the tracking error goes to zero.

3 Tracking with General Nonlinear DAEs

Given there has been such a large literature on tracking and DAEs it is natural to ask if there is anything that might be called new or worth discussing. However, most of this prior work considered the exploitation of some type of recognizable structure in the original equations. In large complex computer generated models there may not be an easily recognizable structure. The system models may be assembled from a number of other models. For the remainder of this section then we will assume that the starting DAE has a well defined solution manifold, a well defined index in the

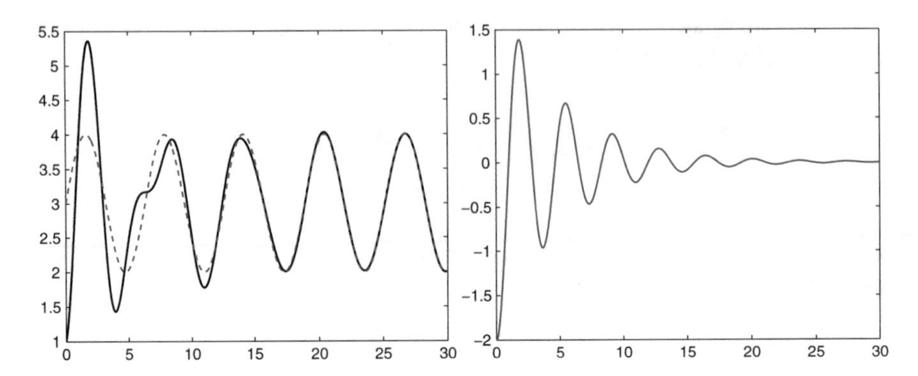

Fig. 8 State x_3 (solid) and r (dashed) on the left and tracking error (right) on [0 30] with feedback solution on original system

sense that follows, but we do not assume any other structure to the system. We do assume that the equations defining the system are smooth and can be differentiated the needed number of times. We do not assume this differentiation is symbolic. It may also be carried out with automatic differentiation software so that all of the derivatives are found numerically. The goal is to get algorithms that do not require differentiating any computed quantities and also avoid needing to perform complex algebraic manipulations other than numerical ones.

The algorithms given will compute an index one problem and carry out any needed simulation or optimization. But this transformation will be carried out numerically and locally at every time step and is thus invisible to the user.

3.1 Some Needed Mathematics

Some of the mathematics we shall need was developed in [14, 15, 20, 32], especially [13]. Since many readers will not be familiar with these results, we summarize some of them here. In a general DAE any control variables are viewed as additional algebraic state variables. Thus in this section we can absorb the u into the x and just write

$$F(t, x, \dot{x}) = 0. \tag{3.1}$$

Since by means of exact tracking and/or optimal control a unique solution is fixed, the final DAE to be treated consists of the same number n of equations and unknowns.

In the general case of unstructured nonlinear DAEs (3.1), the derivative array equations gotten by differentiating (3.1) ℓ times are given by

$$F_\ell(t, x, \dot{x}, \dots, x^{(\ell+1)}) = 0, \tag{3.2}$$

that is,

$$F_\ell(t, x, \dot{x}, \dots, x^{(\ell+1)}) = \begin{bmatrix} F(t, x, \dot{x}) \\ \frac{d}{dt} F(t, x, \dot{x}) \\ \vdots \\ \frac{d^\ell}{dt^\ell} F(t, x, \dot{x}) \end{bmatrix}. \tag{3.3}$$

The derivative of F_ℓ with respect to a variable v will be denoted $F_{\ell;v}$. The following hypothesis from [32] will be fundamental in what follows and generalizes what happens for linear time varying DAEs.

Hypothesis 1 *There exist integers μ, a, and d such that $\mathbb{L}_\mu = \{z_\mu \in \mathbb{I} \times \mathbb{R}^n \times \mathbb{R}^n \times \dots \times \mathbb{R}^n \mid F_\mu(z_\mu) = 0\}$ is not empty and for every point $(t_0, x_0, \dot{x}_0, \dots, x_0^{(\mu+1)}) \in \mathbb{L}_\mu$ there exists a (sufficiently small) neighborhood in which the following properties hold:*

1. *We have* $\operatorname{rank}(F_{\mu;\dot{x},\dots,x^{(\mu+1)}}) = (\mu+1)n - a$ *on $\mathbb{L}_\mu$. This implies that there exists a smooth full rank matrix function Z_2 of size $((\mu+1)n, a)$ satisfying*

$$Z_2^T F_{\mu;\dot{x},\dots,x^{(\mu+1)}} = 0$$

on $\mathbb{L}_\mu$.

2. *We have* $\operatorname{rank}(Z_2^T F_{\mu;x}) = a$ *on $\mathbb{L}_\mu$. This implies that there exists a smooth full rank matrix function T_2 of size $(n, n-a)$ satisfying*

$$Z_2^T F_{\mu;x} T_2 = 0.$$

3. *We have* $\operatorname{rank}(F_{\dot{x}} T_2) = d = n - a$. *This implies that there exists a smooth full rank matrix function Z_1 of size (n, d) satisfying*

$$\operatorname{rank} Z_1^T F_{\dot{x}} T_2 = d.$$

Again, alternative characterizations exist [12], but the preceding formulas fit our numerical procedures better. We may assume that μ is chosen minimally and set $\nu = \mu + 1$. For convenience, we use the shorthand notation $v = (\dot{x}, \dots, x^{(\mu+1)})$. Given $(t_0, x_0, v_0) \in \mathbb{L}_\mu$ we set

$$\hat{Z}_1 = Z_1(t_0, x_0, v_0), \quad \hat{Z}_2 = Z_2(t_0, x_0, v_0).$$

Moreover, due to Hypothesis 1 we can choose a $\hat{T}_1$ such that

$$\begin{bmatrix} F_{\mu;y}(t_0, x_0, v_0) & \hat{Z}_2 \\ \hat{T}_1^T & 0 \end{bmatrix}$$

is nonsingular. Defining

$$H(t, x, v, w) = \begin{bmatrix} F_\mu(t, x, v) + \hat{Z}_2 w \\ \hat{T}_1^T (v - v_0) \end{bmatrix},$$

we immediately see that $H(t_0, x_0, v_0, 0) = 0$ and that $H_{v,w}(t_0, x_0, v_0, 0)$ is nonsingular. Hence, the implicit function theorem shows that the equation $H(t, x, v, w) = 0$ can locally be solved for v, w, say according to

$$v = K(t, x), \quad w = L(t, x).$$

Obviously, every (t, x) with $L(t, x) = 0$ satisfies $F_\mu(t, x, K(t, x)) = 0$ and hence x is consistent at point t. But also the converse holds, that is, if x is consistent at point t, then (t, x) satisfies $L(t, x) = 0$. See [32, Ch. 4] for more details. It follows that the relation $L(t, x) = 0$ constitutes all constraints imposed by the given DAE. Moreover, the problem

$$\hat{Z}_1^T F(t, x, \dot{x}) = 0, \tag{3.4a}$$

$$L(t, x) = 0 \tag{3.4b}$$

is an index reduced DAE belonging to the original DAE. In particular, locally it possesses the same solutions as the original DAE (3.1) but is index one. Note that we are able to evaluate L (and K) numerically by means of Newton's method applied to $H(t, x, v, w) = 0$.

A completion of a DAE in x is an ODE in x that includes all the solutions of the original DAE. That is, the vector field defined by the DAE is completed to form a vector field of an ODE. There are a number of ways to compute a completion. The simplest is to just differentiate the constraints [2, 4, 16]. However, this can introduce undesirable effects such as drift off the solution manifold. Thus there has been considerable work on designing completions whose extra dynamics has desirable properties [10, 13, 40–42].

Let $\delta > 0$. A possible completion of (3.1) is then implicitly defined by performing the stabilized differentiation of the constraints

$$\hat{Z}_1^T F(t, x, \dot{x}) = 0, \tag{3.5a}$$

$$L_t(t, x) + L_x(t, x)\dot{x} + \delta L(t, x) = 0, \tag{3.5b}$$

which implies that now we have $L(t, x) = Ce^{-\delta t}$. Note that δ must be chosen sufficiently large to avoid drift off.

The derivatives L_t, L_x in (3.5b) can be obtained numerically due to the implicit function theorem by solving the system

$$H(t, x, v, w) = 0, \tag{3.6a}$$

$$\hat{Z}_1^T F(t, x, \dot{x}) = 0, \tag{3.6b}$$

$$L_1 + \beta L_2 \dot{x} = 0, \tag{3.6c}$$

$$\begin{bmatrix} F_{\mu;v}(t,x,v) & \hat{Z}_2 \\ \hat{T}_1^T & 0 \end{bmatrix} \begin{bmatrix} * & * \\ L_1 & L_2 \end{bmatrix} = - \begin{bmatrix} F_{\mu;t}(t,x,v) & F_{\mu;x}(t,x,v) \\ 0 & 0 \end{bmatrix}, \tag{3.6d}$$

where the last relation (3.6d) can be used to eliminate the unknowns L_1 and L_2. Note that utilizing in this way the structure of the nonlinear system (3.6), the computational costs compared with the standard integration of the DAE by the general purpose code GENDA [34, 35] are only slightly increased due to the additional bordering given by (3.6c).

3.2 *Exact Tracking of General DAEs*

Suppose that we have the DAE

$$F(t, x, \dot{x}, u) = 0, \tag{3.7a}$$

and the tracking condition

$$G(t, x(t)) - r(t) = 0. \tag{3.7b}$$

If we just want to know if tracking is possible in the given coordinates, then the condition is just whether or not (3.7) forms a well defined DAE. This can be verified by checking the Hypothesis and seeing if there are initial conditions that are consistent.

A related issue arises with tracking with a control versus behavioral tracking [28, 45]. Loosely speaking with behavioral tracking all algebraic variables are treated the same. In tracking with control some of the algebraic variables are labeled as controls and they are what are free to give tracking. As an example consider

$$\dot{x}_1 = x_1 + 2x_2 - x_3 \tag{3.8a}$$

$$0 = \sin t \; x_2 + \cos t \; x_3 \tag{3.8b}$$

and it is desired to have

$$\cos t \; x_2 - \sin t \; x_3 = r(t). \tag{3.8c}$$

If we wish to have x_3 be the control to give tracking we have difficulties when $t = n\pi$ and $r(t) \neq 0$. However, this is a well defined tracking problem in the behavioral sense since (3.8) implies that

$$\begin{bmatrix} x_2 \\ x_3 \end{bmatrix} = \begin{bmatrix} \sin t & \cos(t) \\ \cos t & -\sin t \end{bmatrix} \begin{bmatrix} 0 \\ r(t) \end{bmatrix}$$

and (3.8) is an index one DAE.

3.3 Approximate Tracking of General DAEs

Note that for (3.7a) if we add an output y, this does not change the DAE structure since y is an index one variable if it is not constrained in some manner. Suppose then that we have a general DAE and output

$$F(t, x, \dot{x}, u) = 0 \tag{3.9a}$$

$$y = G(t, x(t)) \tag{3.9b}$$

with cost

$$J = (G(T, x(T)) - r(T))^T H(G(T, x(T)) - r(T)) + \int_0^T (G(t, x(t)) - r(t))^T Q(G(t, x(t)) - r(t)) + u^T Du \, dt. \tag{3.9c}$$

In (3.9) y is an additional algebraic variable, and the index of (3.9) is the same as the index of (3.9a). There are a number of ways to try and solve this kind of optimal control problem. Some of these are discussed in [15].

Using the same techniques along the lines of the previous section, in particular using a suitable hypothesis for control problems, under further mild assumptions we can compute an index one formulation of (3.9a) of the form

$$F_1(t, x, \dot{x}, u) = 0 \tag{3.10a}$$

$$F_2(t, x, u) = 0, \tag{3.10b}$$

where $\partial F_1/\partial \dot{x}$ is full row rank and (3.10b) characterizes the solution manifold. For details we refer to [32]. Note that while (3.10) is index one it is not semi-explicit. It is possible to get a semi-explicit formulation, but in general that will only hold locally. The local formulation suffices for a simulation but it is not as desirable for a more global optimization problem.

If (3.10) is semi-explicit, then there are several options for finding the optimal control u and determining the tracking error. But for general DAEs where the $\dot{x}$ occurs implicitly more care is needed.

One option is to parameterize the controls, and then simulate (3.10) with a fully implicit index one DAE solver. However, for this to work easily it is important that the control not have hidden constraints. Note that (3.10b) allows for the possibility of control constraints if u appears in (3.10b). Control parametrization requires a decision about what are algebraic state variables and what are free control variables. While this is often possible it is not always the case that it can be used.

Another option, which can be useful if u may appear in (3.10b), but there are no separate control bounds, is to derive the necessary conditions. There are theoretical formulations of the necessary conditions and there are formulations designed for computation. The theoretical formulations use coordinate transformations based on the implicit function theorem and related concepts as we did earlier. We shall give a computational formulation from Section 4.2 of [33] for the special case that (3.10) is computationally available. Note that is the case throughout the present paper.

The necessary conditions are then the following boundary value problem on $[t_0, t_f]$.

$$F_1(t, x, \dot{x}, u) = 0, \tag{3.11a}$$

$$F_2(t, x, u) = 0 \tag{3.11b}$$

$$\frac{d}{dt}(E_1(t)^T\lambda_1) = \mathcal{K}_x(t, x, u)^T + (F_1)_x(t, x, \dot{x}, u)^T\lambda_1 + (F_2)_x(t, x, u)^T\lambda_2 \tag{3.11c}$$

$$0 = \mathcal{K}_u(t, x, u)^T + (F_1)_u(t, x, \dot{x}, u))^T\lambda_1 + (F_2)_u(t, x, u)^T\lambda_2 \tag{3.11d}$$

$$0 = E_1(t_0)(x(t_0) - x_0) \tag{3.11e}$$

$$0 = E_1(t_f)^+(\mathcal{M}_x(x(t_f)) + E_1(t_f)^T\lambda_1(t_f)), \tag{3.11f}$$

where $E_1(t) = (F_1)_{\dot{x}}(t, x(t), \dot{x}(t), u(t))$ and the subscripts $x, \dot{x}, u$ indicate differentiation with respect to the indicated variable. In addition,

$$\mathcal{M}(x) = (G(T, x) - r(T))^T H(G(T, x) - r(T)),$$

$$\mathcal{K}(t, x, u) = (G(t, x) - r(t))^T Q(G(t, x) - r(t)) + u^T Du$$

The boundary value problem may be solved with the techniques in [36].

The next problem is that the optimal control found can be noisy. This can occur on high index problems for example because the numerical approximations are working backwards through some differentiations. While good solutions can sometimes be found the more nonlinear the problem is and the higher the index is the more difficult this becomes. Examples are found in [11, 15]. One way to approach this issue is to make sure that r is smooth. This was illustrated earlier. However, even with smoother r the software still has to work back through the integrations.

An alternative is to try and make things nicer back where the control enters the problem. This was done in a different context in [7]. In this case the attempt is made to trade off some error in the tracking against restricting the optimization over a nicer class of u.

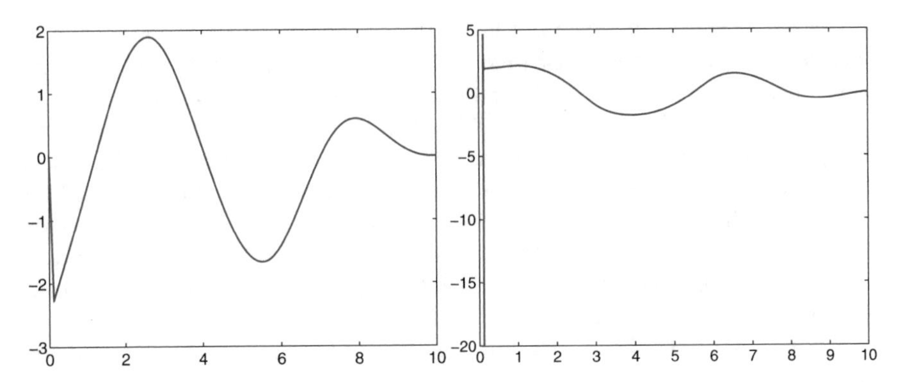

Fig. 9 Control u_2 (left) and auxiliary control v_2 (right) on [0 10] with $u' = v$

One option is to bound the derivative of the control. This is done by adding another differential equation to the DAE of the form $u' = v, u(0) = 0$. Then one can put a bound on v, or add a weight on v in the cost, or both.

To illustrate the effect that this can have, we return to the conditions that generate Fig. 4. We added two more dynamical equations $u' = v$ and a bound on v for $|v_i| \leq 20$. The result is now in Fig. 9.

Note that there is a spike in v_2 at the far left and a jump in u_2 at the left. That was due to our choice of $u(0) = 0$. This computation was just for illustration of how adding dynamics to the control can have a smoothing effect. In an actual application more attention could be paid to $u(0)$ or leaving $u(0)$ free.

3.4 Asymptotic Tracking of General DAEs

There are several different ways to incorporate DAEs and asymptotic tracking. One approach, and it is the most classical, is to design a feedback control that makes the error dynamics asmyptoticaly stable as illustrated earlier in Sect. 2.3. Alternatively using optimal control approaches can often generate controls that lead to asymptotic tracking. Here we shall first exploit an alternative approach using, in part, some ideas from [13]. We begin with (3.7). There are two types of constraints. One are the actual constraints implied by (3.7a). These we want to hold exactly so that we have a physically correct solution of the DAE. The other are (3.7b) which we would like to have hold but asymptotically is our best hope. We take the control u as an input and base the derivative array on just the x variables. We have then our problem can be computed to be

$$\hat{Z}_1^T F(t, x, \dot{x}, u) = 0 \tag{3.12a}$$

$$L(t, x, u) = 0 \tag{3.12b}$$

$$G(t, x) - r(t) = 0 \tag{3.12c}$$

where (3.12c) is a desired constraint. Note that u is free in (3.13b) since the reduction used only x. Performing a stabilized differentiation we get

$$\hat{Z}_1^T F(t, x, \dot{x}, u) = 0 \tag{3.13a}$$

$$L(t, x, u) = 0 \tag{3.13b}$$

$$\frac{d}{dt}(G(t, x) - r(t)) = \delta(G(t, x) - r(t)) \tag{3.13c}$$

where $\delta < 0$. If we treat a problem by inversion we find the control that gives exact tracking. But usually the system does not start on the track so the initial state value does not match the trajectory to be tracked. If we take a stabilized completion of the tracking problem then we can take any initial conditions and the solution will provide a control that will give us asymptotic tracking. We illustrate this later.

4 An Example from Robotics

A number of mechanical system models have natural connections to DAEs [44]. We will now apply the proceeding idea to the two link robotic arm shown in Fig. 10. Robot systems often use harmonic drives, belts, or long shafts as transmission elements between the motors and the links (arms) and typically display oscillations both in fast motion and after a sudden stop. Experimental tests and simulations have shown that the elasticity introduced at the joints by these transmission elements is the major reason for their vibrational behavior, so we must include elasticity in the dynamic model. As a result, the internal position of the motors does not determine the position of the driven arms. The dynamics of this displacement can be modeled by inserting a linear torsional spring at each elastic joint between the actuator and the link. The modeling process is described in full in [21, 22].

Note that in this problem the torques are the controls. One can make the problem even higher index as noted earlier by adding the motor dynamics that produce the torque. The more flexible the joint is the harder it is to control it since changes in the torque have to act through the spring. Here higher K in the model means greater flexibility. We will see that this parameter greatly affects the numerical solution of some tracking problems.

The mathematical model for the robot arm shown in Fig. 10 is given by (4.1). Here m_0, m_1, and m_p are masses, with m_p denoting the load or object being held and m_0 and m_1 the masses of the arms viewed as concentrated at the joints. l_1 and l_2 are the lengths of the arms, K is the coefficient of elasticity of joint 2, NT is the transmission ratio at the second joint, JR_i are rotor inertias, q_i are angular coordinates describing the robot's configuration, and τ_i are the rotational torques caused by the drive motors. In this model $x = [q_1, q_2, q_3, q_1', q_2', q_3']^T$, $u_1 = \tau_1$, and $u_2 = \tau_2$. If $K = 0$ the joint is inelastic and the elasticity increases as K

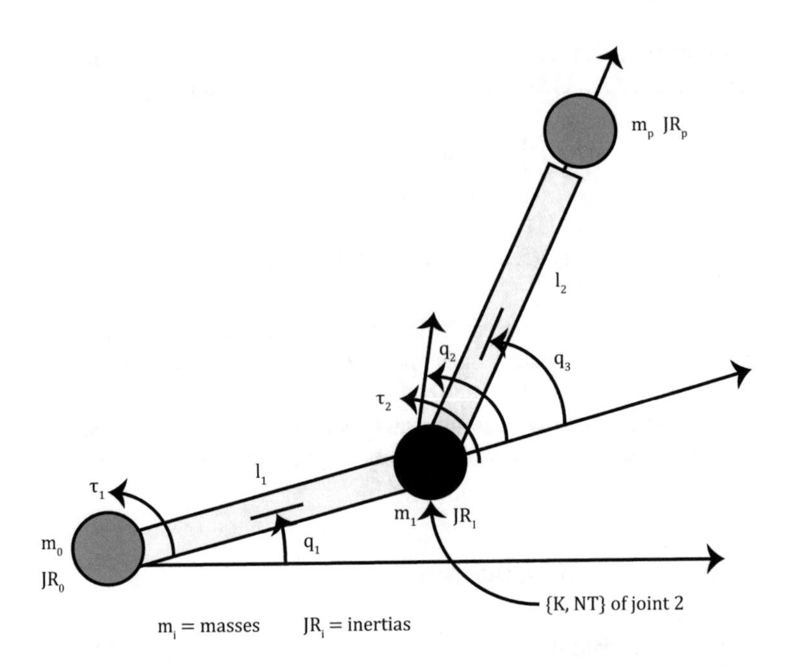

Fig. 10 Robotic arm with two flexible joints

increases. The dynamics equations are

$$x_1' = x_4 \tag{4.1a}$$

$$x_2' = x_5 \tag{4.1b}$$

$$x_3' = x_6 \tag{4.1c}$$

$$x_4' = f_4(x_2, x_3, x_4, x_6) + g_{41}(x_3)u_1 - g_{41}(x_3)u_2 \tag{4.1d}$$

$$x_5' = f_5(x_2, x_3, x_4, x_6) - g_{41}(x_3)u_1 + g_{52}(x_3)u_2 \tag{4.1e}$$

$$x_6' = f_6(x_2, x_3, x_4, x_6) + g_{61}(x_3)u_1 - g_{61}(x_3)u_2. \tag{4.1f}$$

The desired path is

$$0 = C(t, x_1, x_3). \tag{4.2}$$

The nonlinear functions in (4.1) are

$$g_{41}(x_3) = \frac{A_2}{A_3(A_4 - A_3 \cos^2 x_3)} \tag{4.3a}$$

$$g_{52}(x_3) = g_{41}(x_3) + \frac{1}{JR_1} \tag{4.3b}$$

$$g_{61}(x_3) = -g_{41}(x_3) - \frac{\cos x_3}{A_4 - A_3\cos^2 x_3}, \tag{4.3c}$$

$$\begin{aligned} f_4(x_2, x_3, x_4, x_6) &= \frac{A_2 \sin x_3 (x_4 + x_6)^2 + A_3 x_4^2 \sin x_3 \cos x_3}{A_4 - A_3 \cos^2 x_3} \\ &+ \frac{K\left(x_3 - \frac{x_2}{NT}\right)\left(\frac{A_2}{A_3}\left(\frac{NT-1}{NT}\right) + \cos x_3\right)}{A_4 - A_3\cos^2 x_3}, \end{aligned} \tag{4.3d}$$

$$\begin{aligned} f_5(x_2, x_3, x_4, x_6) &= -f_4(x_2, x_3, x_4, x_6) \\ &+ \frac{K}{NT}\left(x_3 - \frac{x_2}{NT}\right)\left(\frac{1}{JR_1} - 2g_{41}(x_3)\right), \end{aligned} \tag{4.3e}$$

$$\begin{aligned} f_6(x_2, x_3, x_4, x_6) &= -f_4(x_2, x_3, x_4, x_6) \\ &- \frac{K\left(x_3 - \frac{x_2}{NT}\right)\left(\frac{A_5}{A_3} - \left(\frac{3NT+1}{NT}\right)\cos x_3\right)}{A_4 - A_3\cos^2 x_3} \\ &- \frac{A_5 x_4^2 \sin x_3 + A_3 \sin x_3 \cos x_3 (x_4 + x_6)^2}{A_4 - A_3\cos^2 x_3}, \end{aligned} \tag{4.3f}$$

and the constants are

$$A_2 = JR_p + m_p l_2^2 \tag{4.4a}$$

$$A_3 = m_p l_1 l_2 \tag{4.4b}$$

$$A_4 = (m_1 + m_p) l_1 l_2 \tag{4.4c}$$

$$A_5 = (m_1 + m_p) l_1^2. \tag{4.4d}$$

Given a path in the work space, the path could be described in cartesian or polar coordinates. So the desired path is $\{p_1(t), p_2(t)\}$ or $\{r(t), \theta(t)\}$. It is also possible to be interested in controlling the arm while the endpoint lies on a particular surface. Note that a variant of this last problem was used in [18] where the motion was assumed vertical and a fault detection signal was being designed. Our interest here is quite different and is just in tracking problems which we will see changes the problem structure depending on the tracking desired.

Note that there are places where there are kinematic singularities. Suppose that this system is modeled in Cartesian coordinates and a path is prescribed for the endpoint mass. Note that $l_1 + l_2\cos(x_2)$ gives the distance of the mass from the base joint in the direction of the first link. As long as during the motion we have $l_2\cos(x_2) < l_2$ we have a well defined DAE and there are controls x_1, x_2 that can give exact tracking for consistent initial conditions and we can hope to find the needed control numerically. Suppose, however, that at some time the path requires $\cos(x_2) = 1$. That is, the arm is fully extended. Then there is a rank drop in a Jacobian and a corresponding drop in the dimension of the solution manifold. There may be no problem with the path. It is that there are singularities in the equations that must be solved. We will return to discussing this example later in the example section.

If $\{p_1(t), p_2(t)\}$ is the endpoint of the arm in Cartesian coordinates, then we have

$$p_1(t) = l_1 \cos(x_1) + l_2 \cos(x_1 + x_3) \tag{4.5a}$$

$$p_2(t) = l_1 \sin(x_1) + l_2 \sin(x_1 + x_3). \tag{4.5b}$$

In a global sense p does not uniquely determine x_1, x_3. If we reflect the arrangement across the line from the first joint to the endpoint we see that there is a second value of x_1, x_3 that give the same endpoint. Also x_1, x_3 can each be changed by any multiple of 2π. However, it is easy to show that the Jacobian of the map in (4.5) is nonsingular unless $x_3 = n\pi$ where n is an integer. When $x_3 = 2n\pi$, the arm is fully extended and the arm can move left or right or pull in but it cannot go out further. So there is a drop in the degrees of freedom. For now we assume $x_3 \neq 0$ $(x_3 \neq 2n\pi)$

The path completely determines the control, but we will see that the problem is index five in one of the control variables and that it does not simply decompose into a nicer problem without some nonlinear coordinate changes.

It is also important to note that this example is not trivially stable. That is, small perturbations can produce larger tracking errors. To illustrate this we took $u_1 = 0.1\sin(t)$, $u_2 = 0.1\sin(t)$, and looked at the difference in the simulation with $x_0 = [0, 0, \pi/4, 0, 0, 0]^T$ and x_0 perturbed by $x_{0p} = [0.01, 0.01, (\pi/4 - 0.01), 0.02, 0.03, -0.01]^T$. Simulation was done on [0 15] with RelTol set to 10^{-8} and AbsTol set to 10^{-9} using Matlab's ODE45. We took $K = 0.1$. The result is shown in Fig. 11. We see that the motion can be quite complex and parameter dependent.

4.1 *Stabilizing Feedback*

In what follows we will be finding open loop tracking controls but as noted above the system is not stable so that the open loop controls may not perform as well as hoped for. One option as noted earlier is to add a stabilzing feedback. For large scale motions this can be difficult and require some care. But for smaller motions one can often use linearizations. To illustrate we will take the robot arm and find its linearization around $x_1 = x_2 = x_4 = x_5 = x_6 = u_1 = u_2 = 0$, $x_3 = \pi/2$. This is at rest with the arm bent at 90° and the first link horizontal. The linearization takes the form

$$x' = Ax + Bu, \tag{4.6}$$

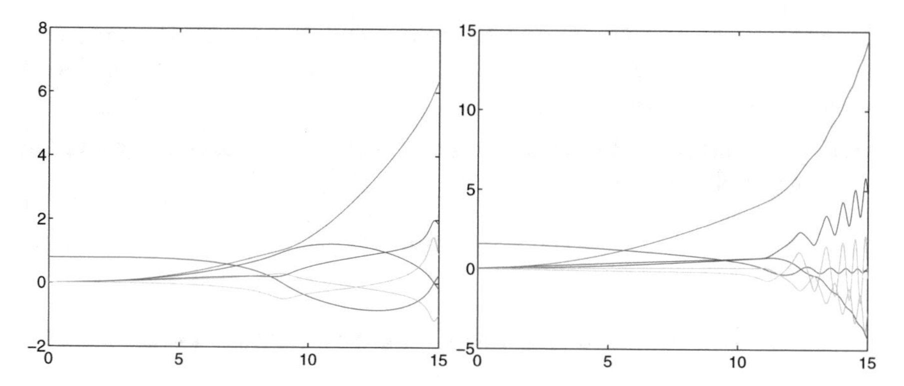

Fig. 11 Simulation of robot arm with $u = 0$ starting at x_0 (left) and $x_0 + x_{0p}$ (right)

where

$$A = \begin{bmatrix} 0 & 0 & 0 & 1 & 0 & 0 \\ 0 & 0 & 0 & 0 & 1 & 0 \\ 0 & 0 & 0 & 0 & 0 & 1 \\ 0 & a_{42} & a_{43} & 0 & 0 & 0 \\ 0 & a_{52} & a_{53} & 0 & 0 & 0 \\ 0 & a_{62} & a_{63} & 0 & 0 & 0 \end{bmatrix}, \quad B = \begin{bmatrix} 0 & 0 \\ 0 & 0 \\ 0 & 0 \\ b_{41} & b_{42} \\ b_{51} & b_{52} \\ b_{61} & b_{62} \end{bmatrix} \tag{4.7}$$

Here

$$a_{42} = -\frac{A_2K(NT-1)}{NT^2A_3A_4}, \ a_{43} = \frac{KA_2(NT-1)}{A_3NTA_4}, \tag{4.8}$$

$$a_{52} = -a_{42} - \frac{K}{NT}\left(\frac{1}{JR_1} - 2\frac{A_2}{A_3A_4}\right), \ a_{53} = -a_{43} + \frac{K}{NTJR_1}, \tag{4.9}$$

$$a_{62} = -a_{42} + \frac{KA_5}{NTA_3A_4}$$

$$a_{63} = -a_{43} + \frac{KA_5(3NT+1)}{A_3A_4NT}, \tag{4.10}$$

$$b_{41} = \frac{A_2}{A_3A_4}, \ b_{42} = -b_{41},$$

$$b_{51} = -b_{41}, \ b_{52} = b_{41} + \frac{1}{JR_1}, \tag{4.11}$$

$$b_{61} = -b_{41}, \quad b_{62} = b_{41}. \tag{4.12}$$

There are a number of tracking problems that can be built off of this example. We shall list four but several others are possible. We will solve some of these problems in the remaining sections of this paper.

Tracking Problem I: The dynamics are the robotic arm. The desired path is an arc of a small circle.

$$\begin{bmatrix} r_1 \\ r_2 \end{bmatrix} = \begin{bmatrix} 0.9 + 0.3\cos(\beta t) \\ 0.9 + 0.3\sin(\beta t) \end{bmatrix}, \tag{4.13}$$

where β is a parameter used to specify the speed of the target going around the circle.

Tracking Problem II: This is the same as Tracking Problem I (TPI) except that we add the hard physical constraint,

$$(p_1 - 0.9)^2 + (p_2 - 0.9)^2 - 0.09 = 0, \tag{4.14}$$

where p_1, p_2 are given by (4.5). Now the dynamics are a DAE but there is still one degree of freedom left for a control.

Tracking Problem III: This is similar to TPI except that we allow the target to go through a kinematic singularity by changing the target being tracked to the following. Let $\gamma = 2 - 1.2\sqrt{2}$ and the path be

$$\begin{bmatrix} r_1 \\ r_2 \end{bmatrix} = \begin{bmatrix} 1.2 + \gamma\cos(\beta t) \\ 1.2 + \gamma\sin(\beta t) \end{bmatrix}. \tag{4.15}$$

The target moves on the circle

$$(r_1 - 1.2)^2 + (r_2 - 1.2)^2 - \gamma^2 = 0, \tag{4.16}$$

which is tangent to the circle of radius 2 at $(\sqrt{2}, \sqrt{2})$.

Tracking Problem IV: This the same as TPIII except that we add (4.16) as a hard constraint so that we have DAE dynamics.

We now use our test problem to illustrate some of the ideas from the start of this paper.

4.2 *Tracking Problem I as Inversion*

If we approach this as an inversion problem, then it is an index five DAE and requires extensive index reduction which can be complicated to carry out. We shall solve this problem with an improved version of GENDA [34, 35]. GENDA takes the derivative array and locally computes an index one system with the same solutions at each time step. This is all done numerically and one does not actually compute

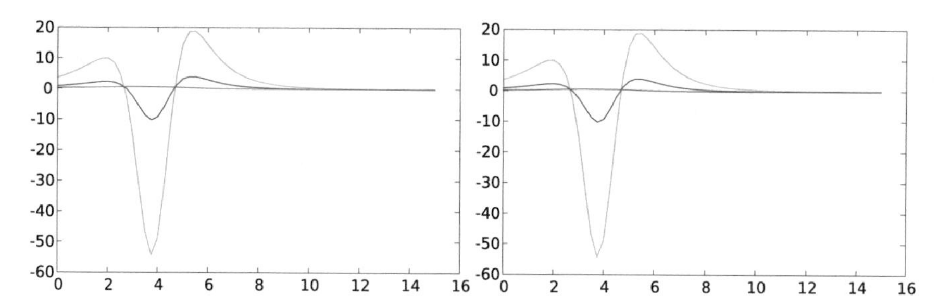

Fig. 12 Controls for exact tracking for different values of $K = 0.01, 0.05, 4.0$. u_1 on the left and u_2 on the right

the full reformulation. The original GENDA was a FORTRAN code. We use a C++ implementation that uses automatic differentiation at each time step both to compute the derivative array and also to compute any needed Jacobians.

An additional benefit of GENDA is that it also identifies a minimal set of equations in the derivative array that need to be differentiated in order to determine the dynamics. The number of differentiations of a given equation will be referred to as the differentiation index of a given equation. The choice of which equations are chosen is not unique but the indices and number of equations is.

For the example here see that there are: 2 equations of differentiation index 1, 2 equations of differentiation index 2, 2 equations of differentiation index 3, 1 equation of differentiation index 4, and 1 equation of differentiation index 5.

GENDA has several integrators implemented for integrating the underlying dynamics. For this simulation we used Gauss collocation of order six. Throughout this paper we take parameter values of $t_0 = 0$, $t_f = 15$; $\beta = 0.2$, $NT = 2$, $JR_1 = 1$, $JR_p = 1$, $l_1 = 1$, $l_2 = 1$, $m_1 = 1$, $m_2 = 10$, and $m_p = 10$. Since the initial condition must be consistent it is found by GENDA using a Guass-Newton method and the derivative array. For this example, the path uniquely determines the control and dynamics. We consider the three values of $K = 0.01, 0.05$, and 4.0. The software was able to solve the inverse problem for all values of K to high accuracy. The results are in Figs. 12 and 13. The controls u_1 and u_2 appear identical in Fig. 12, however, looking at the data graphed, one sees that the peaks differ by 1–2% and there are small differences in the curves.

4.3 Tracking Problem I as Stabilized Reduction

In this subsection we take the tracking problem and view it as high index DAE which we reduce to an index one problem using a stabilized reduction. This has the advantage that we can start with an initial value that is not on the track. Here we carried it out as described in [13]. Since adding the path gives a DAE which is completely algebraic, this corresponds to looking at the Gauss-Newton flow of

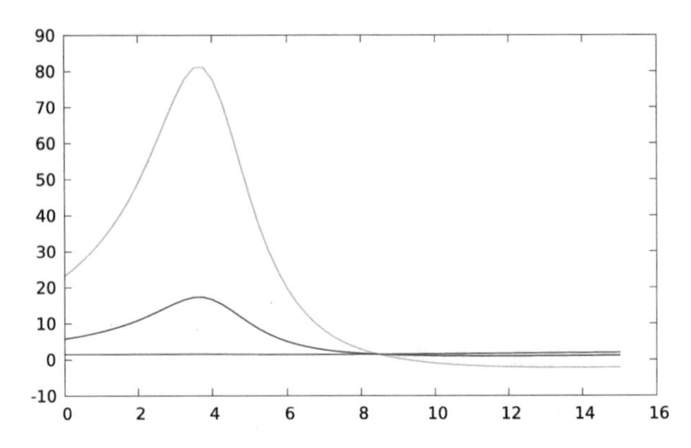

Fig. 13 x_3 angle of torsional spring for all three values of $K = 0.01, 0.05, 4.0$ with exact tracking

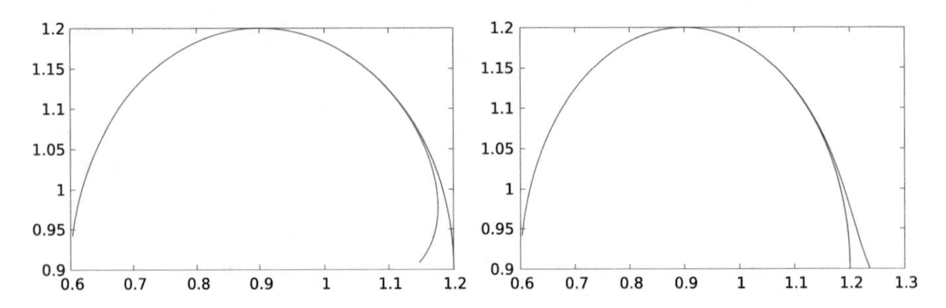

Fig. 14 Tracking for $K = 0.01$ using initial conditions (b) and (d) using stabilized reduction

$\hat{F}_2(t,x) = 0$ which is given by $\hat{F}_{2,t}(t,x) + \hat{F}_{2,x}(t,x)\dot{x} = 0$ where in our case $\hat{F}_{2,x}(t,x)$ is invertible. Hence, the ODE

$$\dot{x} = -\alpha \hat{F}_{2,x}(t,x)^{-1}\hat{F}_{2,t}(t,x)$$

is integrated with some appropriate stabilizing term α to control the attraction of the DAE solutions. In the computations reported here we used $\alpha = I$. We considered several different initial conditions.

$$\begin{aligned}
&\text{(a)}\ \ [0, 0, \pi/4, 0, 0, 0, 0, 0]^T \\
&\text{(b)}\ \ [-0.08, 20.0, 1.5, 0.06, 7.0, -0.06, 4.0, 4.0]^T \\
&\text{(c)}\ \ [-0.07, 25.0, 1.4, 0, 0, 0, 0, 0, 0]^T \\
&\text{(d)}\ \ [-0.07, 0, 1.4, 0, 0, 0, 0, 0, 0, 0, 0]^T
\end{aligned} \tag{4.17}$$

and $K = 0.01, 0.05, 4$ on $[0\ 15]$. It should be noted that we did not get a satisfactory answer with (a). It should also be noted that $25 - 8\pi$ is -0.1327 so that although (c) and (d) correspond to similar initial endpoints of the arm, they have different initial torque on the joint. Selected results are shown in Figs. 14 and 15. We see that for initial condition (b) we start inside the trajectory and with (d) we start outside.

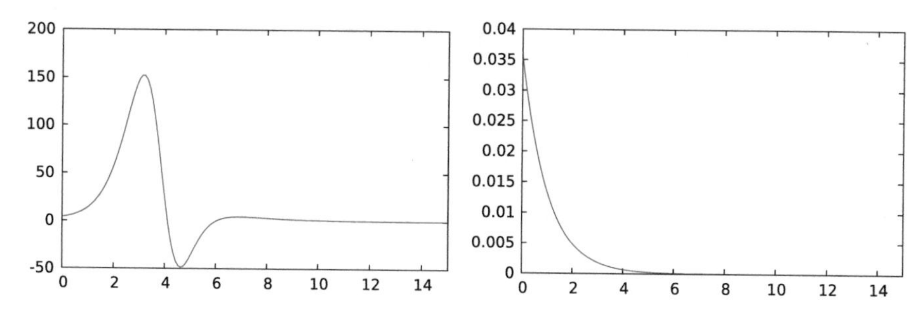

Fig. 15 For $K = 0.01$, computed control u_1 for (b) (left) and tracking error for (d) (right) using stabilized reduction

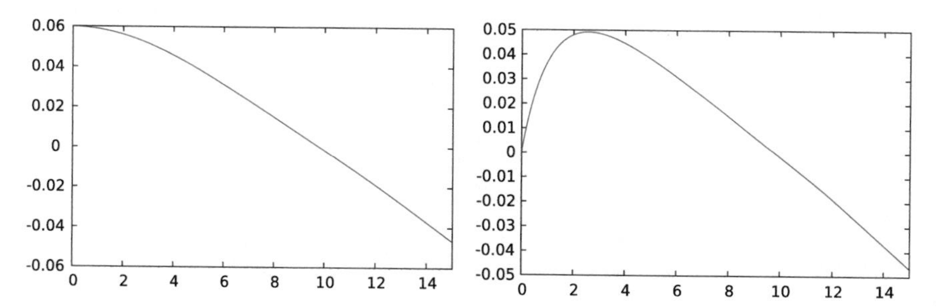

Fig. 16 Computed $\dot{x}_1$ for (b) (left) and (d) (right) for $K = 0.01$ using stabilized reduction

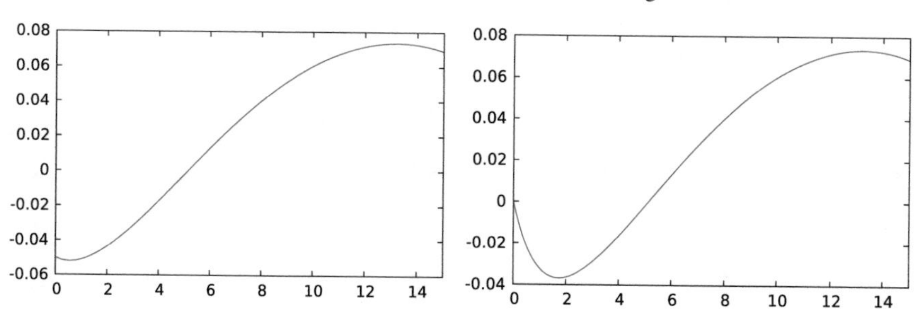

Fig. 17 Computed $\dot{x}_1$ for (b) (left) and (d) (right) for $K = 0.01$ using stabilized reduction

The controls were similar for all initial conditions and for both controls except that they were a bit smaller for initial condition (b). The computed values of x_1, x_2, x_3 were similar for all the initial conditions, but there were noticeable differences in the velocities.

When $K = 4$ the spring is much more elastic. This introduces a delay in the action of u_2 and makes it more difficult to control the arm. Several changes became apparent. Tracking was still like that shown in Fig. 14 and the tracking error was still like the right side of Fig. 15. But there were some major changes. First, as to be expected x_2 and x_3 were different. This is not surprising since the initial conditions on x_2 were not the same. But the difference is more dramatic then might be expected (Figs. 16 and 17). Note that in Fig. 18 there appears to be some difficulty in resolving x_4 numerically (Fig. 19).

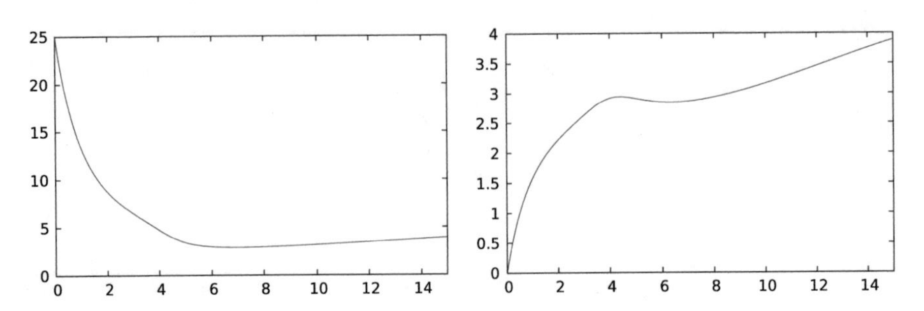

Fig. 18 Computed x_2 for (c) (left) and (d) (right) for $K = 4$ using stabilized reduction

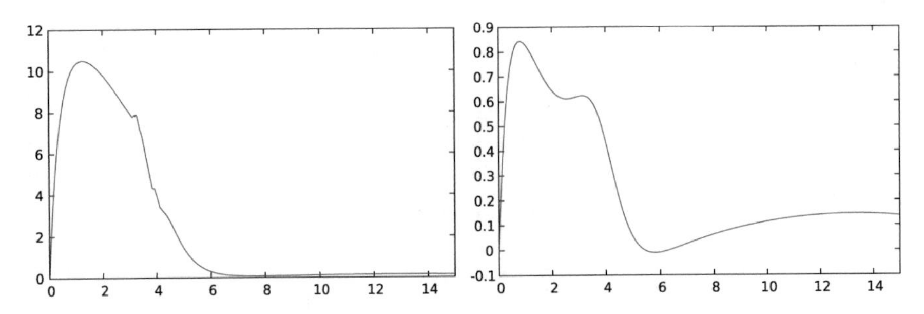

Fig. 19 Computed x_4 for c (left) and d (right) for $K = 4$ using stabilized reduction

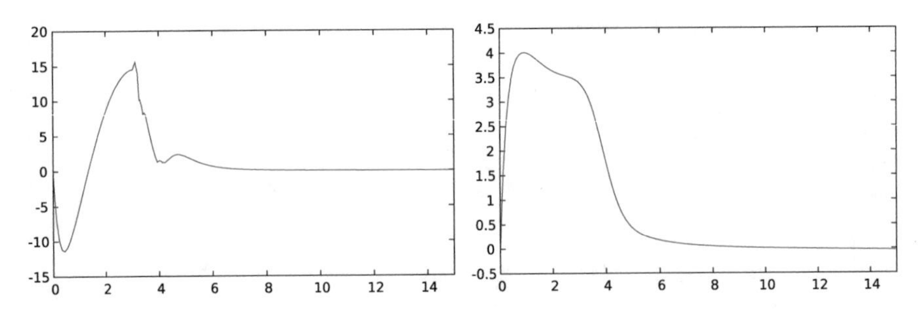

Fig. 20 Computed u_2 for (c) (left) and (d) (right) for $K = 4$ using stabilized reduction

The situation with the controls was even more difficult. Even though the tracking appeared excellent there was difficulty in determining u_2 for some initial conditions as shown on the left side of Fig. 20.

4.4 *Tracking Using Optimal Control*

We now turn to solving the problem with optimal control.

4.4.1 Tracking Problem I as Optimal Control

With an optimal control approach to TPI, in theory the situation is simpler since we discard (4.5) as a constraint and get an ODE process. We take the optimization problem as

$$J = \sum_{i=1,2} h_i(r_i(T) - p_i(T))^2 + \int_0^T q_i(r_i - p_i)^2 + d_i u_i^2 dt \tag{4.18}$$

We shall see that there are still computational challenges. We shall consider two different approaches. One is using direct transcription software and the other is working with the necessary conditions. Both have their challenges some of which will be mentioned. We used two different direct transcription codes. One is described in [15] and will be denote DT2. The other is GPOPS II which is denoted as DT1. Since it is more easily available we focused on using GPOPS II. DT2 was used to validate the answers computed using necessary conditions. It should be noted that for some problems with DT1 we would get good solutions if we limited the number of iterations but got long computer runs and bad solutions if we did not restrict the iteration count to a small number. The inability to get requested tolerances in higher index variables would drive the mesh too small. It should also be noted that even when the formulation is an optimal control problem with an ODE that the high index behavior of the tracking DAE comes into play because it means that the endpoint of the robotic arm is not directly linked to the control and thus the control problem can be sensitive.

We programmed this problem in GPOPS II and ran a number of cases using simplistic initial guesses using the formulation above without any index reduction. Depending on parameter values we either got good tracking or had a lot of difficulty. One factor is K the elasticity. If $K = 0$ or close to it, then we could often get a reasonable answer. Another factor is β. If β is small, then the target is moving slowly. The larger β is, the more challenging the tracking problem is. Recall that for exact tracking we know that several of the variables can involve up to fifth derivatives of $\sin \beta t$. Thus the larger β is, the greater amplitude we expect to see in solutions and the more rapidly they will oscillate. We assume $q_i = q, h_i = h$ and $d_i = d$. In this section we will consider the two initial conditions (a) and (d) from (4.17) and values of $K = 0.01, 4$. Many other computations were done that are not reported here. In some cases there were failures in getting feasible solutions.

For example, with $\beta = 0.2$, $K = 0.01$, $q = h = 100$, $d = 1$, starting at (a) which is at rest with $T = 15$, we observed the results in Fig. 21. In an optimal control solution the choice of necessary conditions plays a key role as we will see (Fig. 22).

Note that if we increase the elasticity by taking $K = 4$ we observe very different tracking results as shown in Fig. 23.

Increasing β with K small made the tracking difficult and at $\beta = 0.6$ we could not get reasonable tracking.

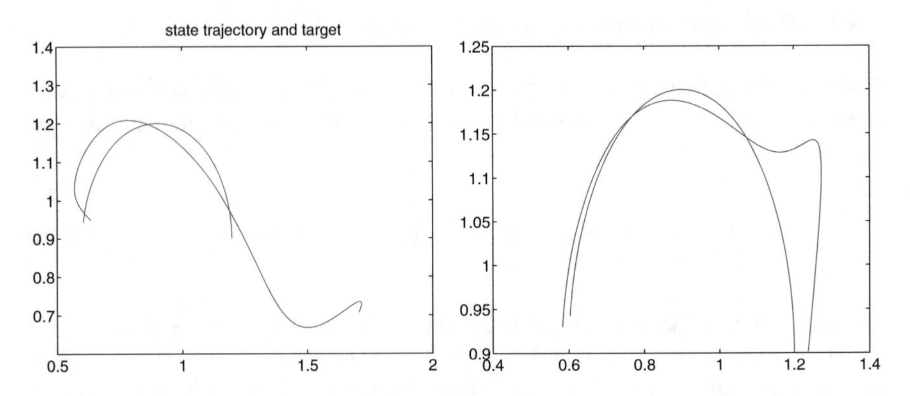

Fig. 21 State trajectory and target trajectories with $K = 0.01$, $\beta = 0.2$, from (a) (left) and (d) (right) using DT1

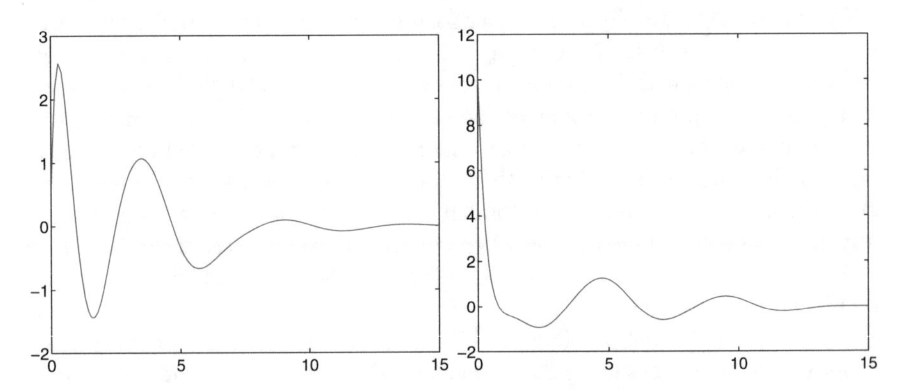

Fig. 22 Tracking controls with $K = 0.01$ and $\beta = 0.2$ from (a) u_1 (left) and u_2 (right) using DT1

4.4.2 Using Necessary Conditions

Direct transcription approaches have the advantage that they can attempt to solve very complex problems including those for which it is hard to form the necessary conditions. They have the disadvantage that sometimes it can be very challenging to get an initial guess that will lead to convergence to the correct solution.

As an alternative to using a direct transcription code for this version of the tracking problem we can write down the necessary conditions which are a boundary value problem (BVP). The problems then are twofold. For one, this is a nonlinear BVP and hence getting a good initial guess can still be hard. Secondly the necessary conditions are essentially saying the first derivative of the cost is zero in a function space sense. Thus when solving the BVP one could be finding a local maximum, a local minimum, or even a saddle point. For simple problems this is usually not an issue but for complex nonlinear problems it is an important issue.

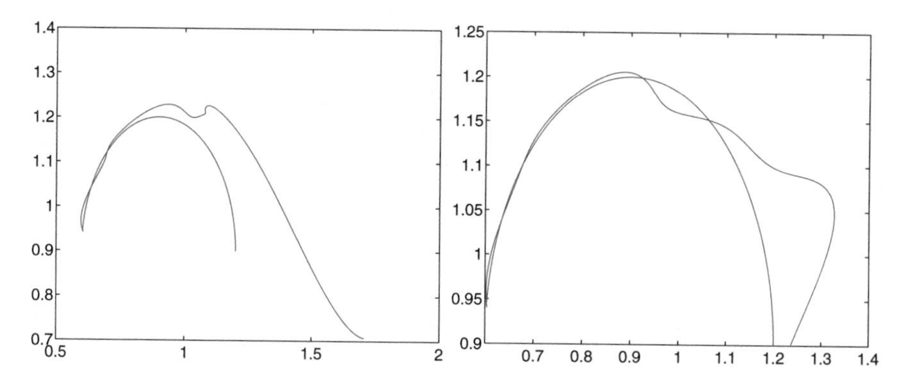

Fig. 23 Tracking with $K = 4$, $\beta = 0.2$, from initial conditions (a) on the left and (d) on the right using DT1

For a problem of the form

$$\dot{x} = F(t, x, u) \tag{4.19a}$$

$$J = \int_0^T \mathcal{K}(t, x, u)dt + \mathcal{M}(x(T)) \tag{4.19b}$$

$$x(0) = x_0, \tag{4.19c}$$

the necessary conditions (3.11) simplify to

$$\dot{x} = F(t, x, u) \tag{4.20a}$$

$$-\dot{\lambda} = F_x^T \lambda + \mathcal{K}_x^T \tag{4.20b}$$

$$0 = F_u^T \lambda + \mathcal{K}_u^T \tag{4.20c}$$

$$x(0) = x_0 \tag{4.20d}$$

$$0 = \lambda(T) - \mathcal{M}_x(x(T))^T. \tag{4.20e}$$

To find the solution of the BVP, it was first solved on a short interval and then iteratively solved on longer intervals. That is, we did homotopy in T. Note that starting with $T = 15$, $K = 4$, and doing homotopy on K we could get to $K = 0.01$ but starting at $K = 0.01$ and $T = 0.25$ we could not get to $T = 15$ with homotopy on T.

It should be pointed out that solving nonlinear control problems can be delicate. For example, we saw cases where GPOPS II produced a good answer when we limited the number of iterations but if we allowed for more iterations that results became overwhelmed and very noisy.

Comparing Fig. 23 with Fig. 24 and Fig. 21 with Fig. 27, we see that sometimes the solutions appeared somewhat different. It is informative to look at the optimal

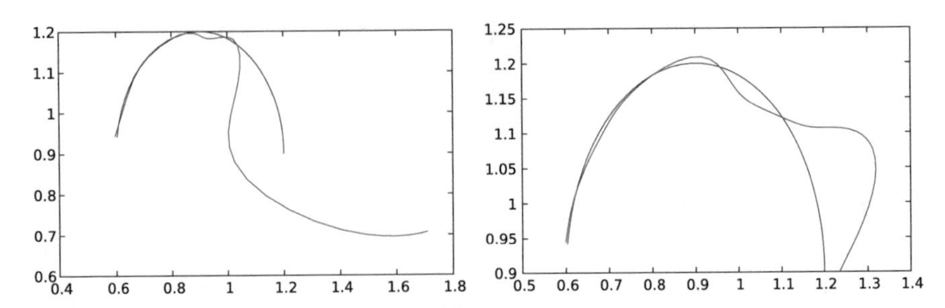

Fig. 24 Tracking on [0 15] with $K = 4$ and initial condition (a) (left) and (d) (right) using necessary conditions

Table 1 Optimal cost for several of the examples

K	Initial cond	DT1	DT2	Nec. conditions
4	(a)	5.3658434e+01		
4	(d)	2.0759187e+01	2.6993240e+01	2.6993192e+01
0.01	(a)	1.0874264e+02		
0.01	(d)	9.5560137e+00	2.6976860e+01	2.6976846e+01

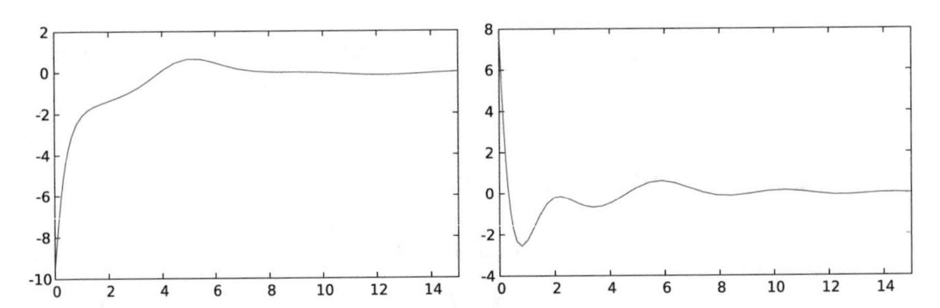

Fig. 25 Optimal control for tracking on [0 15] with $K = 4$ and initial condition (a) using necessary conditions

costs for some of these scenarios in Table 1. Looking at the table we see that DT2 and the necessary conditions obtained similar costs on the examples given which is to be expected since they worked off of similar reduced index formulations. DT1 got a smaller cost than DT2 when both solved the problem. This may be due to the very different type of initialization of the two methods and the existence of local minimums (Figs. 25, 26, 27, 28, 29).

5 Concluding Comments

While we were often able to find the tracking control in the previous section, there is a practical issue. The control is open loop and as noted the physical system is not long term stable so that actually applying the control might not give the desired

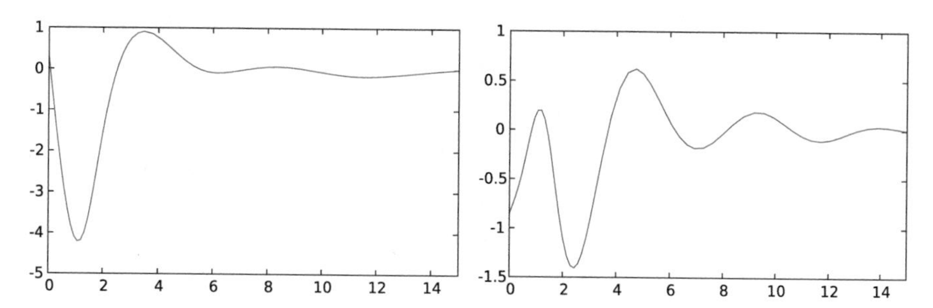

Fig. 26 Optimal control for tracking on [0 15] with $K = 4$ and initial condition (d) using necessary conditions

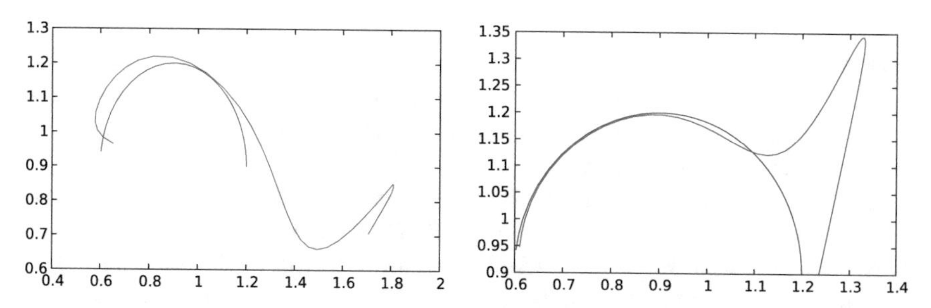

Fig. 27 Tracking on [0 15] with $K = 0.01$ and initial condition (a) (left) and (d) (right) using necessary conditions

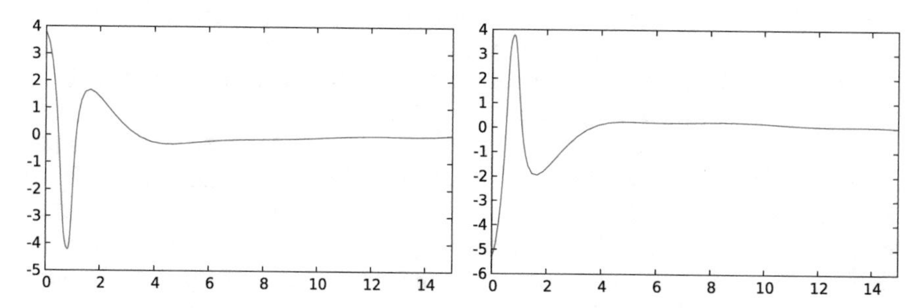

Fig. 28 Optimal control for tracking on [0 15] with $K = 0.01$ and initial condition (a) using necessary conditions

behavior. One way to overcome this problem is by using moving horizon control. Moving horizon control is used explicitly or implicitly in many systems including some biological ones.

Suppose that we have a physical system. We have the horizon length L and the control update frequency $\delta < L$. To begin we compute the optimal control over the horizon $[0\ L]$ and apply it open loop on $[0\ \delta]$. At time $t = \delta$ we see where our physical system actually is, $x(\delta)$ and compute a new optimal control on $[\delta\ \ L + \delta]$ starting at the actual $x(\delta)$ and apply it open loop on $[\delta\ \ 2\delta]$. This process is repeated over and over. While the control on each subinterval is open loop,

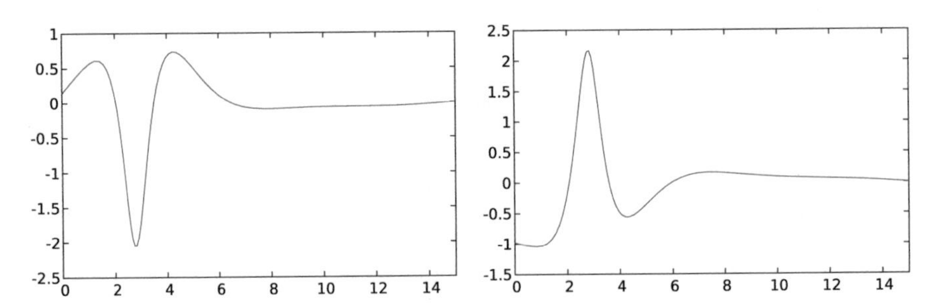

Fig. 29 Optimal control for tracking on [0 15] with $K = 0.01$ and initial condition (d) using necessary conditions

each of the optimization problems is computed from the actual initial state at the beginning of that interval so there is a feedback effect. This is particularly popular in chemical engineering and some areas of robotics including motion planning for robots operating on other planets.

Tracking is achieved by a feedback controller which might be static or dynamic. A noted earlier a central issue of the feedback is whether it depends on the output y only, or on the output and higher derivatives of the output, or on the full state x. This is an important issue but outside the focus of this paper. Another topic we have not addressed are the structural system properties which make tracking possible. This includes, for example, asymptotically stable zero dynamics, relative degree, and sign of the high-frequency gain. One approach is tracking with prescribed behavior, or "funnel control". There the behavior of the output evolves within a specified funnel. The approach is very robust but a number of structural system properties have to be satisfied. The interested reader is referred to [5, 6].

This paper has first reviewed some of the approaches for tracking problems when either the dynamics are a DAE or the tracking problem is formulated as a DAE. We then introduced a problem from robotics and applied some of these approaches to this tracking problem. The test problem presented, which is typical of many problems in robotics, has a number of formulations and can provide a challenging problem for both software and analysis. It is seen that which approaches are best depends on the specifics of the problem formulation chosen and often specialized DAE software may be needed.

References

1. Altmann, R., Heiland, J.: Simulation of multibody systems with servo constraints through optimal control. Multibody Sys. Dyn. **40**, 75–98 (2016)
2. Ascher, U.M., Chin, H., Reich, S.: Stabilization of DAEs and invariant manifolds. Numer. Math. **67**, 131–149 (1994)
3. Bates, D., Hauenstein, J., Sommese, A.: Efficient path tracking methods. Numer. Algorithms **58**, 451–459 (2011)

4. Baumgarte, J.: Stabilization of constraints and integrals of motion in dynamical systems. Comput. Methods Appl. Mech. Eng. **1**, 1–16 (1972)
5. Berger, T., Ilchmann, A., Reis, T.: Normal forms, high-gain, and funnel control for linear differential-algebraic system. In: Biegler, L.T., Campbell, S.L., Mehrmann, V. (eds.) Control and Optimization with Differential-Algebraic Constraints, pp. 127–164. SIAM, Philadelphia (2012)
6. Berger, T., Ilchmann, A., Reis, T.: Zero dynamics and funnel control of linear differential-algebraic systems. Math. Control Signals Systems **24**, 219–263 (2012)
7. Betts, J.T., Campbell, S.L., Thompson, K.: Solving optimal control problems with control delays using direct transcription. Appl. Numer. Math. **108**, 185–203 (2016)
8. Brenan, K.E., Campbell, S.L., Petzold, L.R.: Numerical Solution of Initial-Value Problems in Differential-Algebraic Equations. SIAM, Philadelphia (1996)
9. Brown, P.N., Hindmarsh, A.C., Petzold, L.R.: Consistent initial condition calculation for differential-algebraic systems. SIAM J. Sci. Comput. **19**, 1495–1512 (1998)
10. Campbell, S.L.: Uniqueness of completions for linear time varying differential algebraic equations. Linear Algebra Appl. **161**, 55–67 (1992)
11. Campbell, S.L., Betts, J.T.: Comments on direct transcription solution of DAE constrained optimal control problems with two discretization approaches. Numer. Algorithms **73**, 807–838 (2016)
12. Campbell, S.L., Griepentrog, E.: Solvability of general differential algebraic equations. SIAM J. Sci. Comput. **16**, 257–270 (1995)
13. Campbell, S.L., Kunkel, P.: Completions of nonlinear DAE flows based on index reduction techniques and their stabilization. J. Comput. Appl. Math. **233**, 1021–1034 (2009)
14. Campbell, S.L., Kunkel, P.: On the numerical treatment of linear-quadratic optimal control problems for general linear time-varying differential-algebraic equations. J. Comput. Appl. Math. **242**, 213–231 (2013)
15. Campbell, S.L., Kunkel, P.: Solving higher index DAE optimal control problems. Numer. Algebra Control Optim. **6**, 447–472 (2017)
16. Campbell, S.L., Leimkuhler, B.: Differentiation of constraints in differential-algebraic equations. Mech. Struct. Mach. **19**, 19–39 (1991)
17. Campbell, S.L., März, R.: Direct transcription solution of high index optimal control problems and regular Euler-Lagrange equations. J. Comput. Appl. Math. **202**, 186–202 (2007)
18. Campbell, S.L., Scott, J.R.: Active fault detection in nonlinear differential algebraic equations I: general systems. In: Martin, D. (ed.) Fault Detection: Methods, Applications and Technology, chap. 1, pp. 1–21. Nova Publishers, Hauppauge (2016). ISBN: 978-1-53610-359-5
19. Campbell, S.L., Chancelier, J.P., Nikoukhah, R.: Modeling and Simulation in Scilab/Scicos with ScicosLab 4.4. Springer, Berlin (2009)
20. Campbell, S.L., Kunkel, P., Mehrmann, V.: Regularization of Linear and Nonlinear Descriptor Systems. Control and Optimization with Differential-Algebraic Constraints, pp. 17–36. SIAM, Philadelphia (2012)
21. De Luca, A.: Control properties of robot arms with joint elasticity. In: Proceedings of International Symposium on the Mathematical Theory of Networks and Systems, Phoenix, AZ, June 1987
22. De Luca, A., Isidori, A.: Feedback linearization of invertible system. In: 2nd Duisburger Kolloquium Automation und Robotik, Duisburg, July 15–16, 1987
23. Devasia, S., Degang, C., Paden, B.: Nonlinear inversion-based output tracking. IEEE Trans. Autom. Control **41**, 930–942 (1996)
24. Engelsone, A., Campbell, S.L., Betts, J.T.: Direct transcription solution of higher-index optimal control problems and the virtual index. Appl. Numer. Math. **57**, 281–196 (2007)
25. Fliess, M., Lévine, J., Rouchon, P.: Index of an implicit time-varying linear differential equation: a noncommutative linear algebraic approach. Linear Algebra Appl. **186**, 59–71 (1993)
26. Fliess, M., Lévine, J., Martin, P., Rouchon, P.: Implicit differential equations and Lie-Bäcklund mappings. In: Proceedings of 34th IEEE Conference on Decision and Control, New Orleans, LA, pp. 2704–2709 (1995)

27. Friedland, B.: A nonlinear observer for estimating parameters in dynamic systems. Automatica **33**, 1525–1530 (1997)
28. Ilchmann, A., Mehrmann, V.: A behavioral approach to time-varying linear systems. part 2: descriptor systems. SIAM J. Control Optim. **44**, 1748–1765 (2005)
29. Krishnan, H., Mcclamroch, N.H.: Tracking in nonlinear differential-algebraic control systems with applications to constrained robot systems. Automatica **30**, 1885–1897 (1994)
30. Kumar, A., Daoutidis, P.: Feedback control of nonlinear differential-algebraic equation systems. AIChE J. **41**, 619–636 (1994)
31. Kumar, A., Daoutidis, P.: Control of Nonlinear Differential Algebraic Equation Systems. Chapman and Hall/CRC, New York (1999)
32. Kunkel, P., Mehrmann, V.: Differential-Algebraic Equations: Analysis and Numerical Solution. European Mathematical Society, Zürich (2006)
33. Kunkel, P., Mehrmann, V.: Optimal control for unstructured nonlinear differential algebraic equations of arbitrary index. Math. Control Signals Syst. **20**, 227–269 (2008)
34. Kunkel, P., Mehrmann, V.: Home page for GEneral Nonlinear Differential Algebraic equation solver. http://www.math.tu-berlin.de/numerik/mt/NumMat/Software/GENDA/info.shtml
35. Kunkel, P., Mehrmann, V., Seufer, I.: GENDA: a software package for the numerical solution of General Nonlinear Differential-Algebraic equations. Institut für Mathematik, TU Berlin Technical Report 730, Berlin (2002)
36. Kunkel, P., Mehrmann, V., Stöver, R.: Symmetric collocation for unstructured nonlinear differential-algebraic equations of arbitrary index. Numer. Math. **98**, 277–304 (2004)
37. Liu, X.P., Rohani, S., Jutan, A.: Tracking control of general nonlinear differential-algebraic equation systems. AICHE J. **49**, 1743–1760 (2003)
38. Lui, S.P.: Asymptotic output tracking of nonlinear differential-algebraic control systems. Automatica **34**, 393–397 (1998)
39. Martínez-Guerra, R., Suarez, R., de León-Morales, J.: Asymptotic output tracking of a class of nonlinear systems by means of an observer. Int. J. Robust Nonlinear Control **11**, 373–391 (2001)
40. Okay, I., Campbell, S.L., Kunkel, P.: The additional dynamics of least squares completions for linear differential algebraic equations. Linear Algebra Appl. **425**, 471–485 (2007)
41. Okay, I., Campbell, S.L., Kunkel, P.: Completions of implicitly defined vector fields and their applications. In: Proceedings of 18th International Symposium on Mathematical Theory of Networks and Systems (MTNS 08), Blacksburg (2008)
42. Okay, I., Campbell, S.L., Kunkel, P.: Completions of implicitly defined linear time varying vector fields. Linear Algebra Appl. **431**, 1422–1438 (2009)
43. Patterson, M.A., Rao, A.V.: GPOPS II: A MATLAB software for solving multiple-phase optimal control problems using hp-adaptive Gaussian quadrature collocation methods and sparse nonlinear programming. ACM Trans. Math Softw. **41**, 1–37 (2014)
44. Rabier, P.J., Rheinboldt, W.C.: Nonholonomic Motion of Rigid Mechanical Systems from a DAE Viewpoint. SIAM, Philadelphia (2000)
45. Willems, J.C.: The behavioral approach to open and interconnected systems. Control Syst. Mag. **27**, 46–99 (2007)

DAE Aspects in Vehicle Dynamics and Mobile Robotics

Michael Burger and Matthias Gerdts

Abstract The paper presents and discusses prototype applications occurring in path planning tasks for mobile robots and vehicle dynamics which involve differential-algebraic equations (DAEs). The focus is on modeling aspects and issues arising from the DAE formulation such as hidden constraints, determination of algebraic states, and consistency. The first part of the paper provides a general summary on modeling issues with DAEs while the second part discusses specific prototype applications in depth and presents numerical examples for selected examples arising in control tasks in robotics and vehicle dynamics.

Keywords Differential-algebraic equations · DAE models

Subject Classifications: 34A09, 34H05, 49N90, 93A30

1 Introduction

Differential-algebraic equations (DAEs) are frequently used to model the dynamic behavior of mechanical multibody systems, electric circuits, and systems in process engineering, see [17]. A common feature of these models is that they can

---

Electronic supplementary material The online version of this article (https://doi.org/10.1007/11221_2018_6) contains supplementary material, which is available to authorized users.

M. Burger
Fraunhofer-Institute for Industrial Mathematics ITWM, Department Mathematical Methods in Dynamics and Durability MDF, Kaiserslautern, Germany
e-mail: Michael.Burger@itwm.fraunhofer.de

M. Gerdts (✉)
Institute of Mathematics and Applied Computing, Department of Aerospace Engineering, Universität der Bundeswehr München, Neubiberg, Germany
e-mail: matthias.gerdts@unibw.de

S. Campbell et al. (eds.), *Applications of Differential-Algebraic Equations: Examples and Benchmarks*, Differential-Algebraic Equations Forum,
https://doi.org/10.1007/11221_2018_6

be automatically generated by software packages like SIMPACK, ADAMS, or MODELICA. This kind of automatism is convenient from a user's point of view, but it imposes high demands on numerical simulation or optimization methods. In this paper we will focus on modeling aspects rather than on theoretical properties or numerical algorithms. The latter topics are discussed in detail in recent survey papers, see, e.g., [14] for the treatment of initial value problems with DAEs, [2] for a discussion of mechanical multibody systems, [29] for optimal control techniques, and [50] for control theoretic results. In addition, theoretical and numerical properties are investigated in the monographs [11, 27, 30, 31, 33, 35].

All DAE models in this paper fit into the problem class semi-explicit DAEs of type

$$x'(t) = f(t, x(t), y(t), u(t)), \tag{1.1}$$

$$0 = g(t, x(t), y(t), u(t)), \tag{1.2}$$

where $x(\cdot)$ is referred to as *differential state*, $y(\cdot)$ is called *algebraic state*, and $u(\cdot)$ is an external *control input*. Correspondingly, (1.1) is called *differential equation* and (1.2) *algebraic equation*. An important subclass are mechanical multibody systems in descriptor form defined by

$$\begin{aligned} q'(t) &= v(t), \\ M(t, q(t))v'(t) &= f(t, q(t), v(t), u(t)) - g_q'(t, q(t))^\top \lambda(t), \\ 0 &= g(t, q(t)), \end{aligned} \tag{1.3}$$

where $q(\cdot)$ denotes the vector of generalized positions, $v(\cdot)$ the vector of generalized velocities, $\lambda(\cdot)$ are Lagrange multipliers, and $u(\cdot)$ is a control input. The mass matrix M is supposed to be symmetric and positive definite with a bounded inverse M^{-1} and thus, the second equation in (1.3) can be multiplied by $M(t, q(t))^{-1}$, in which case a semi-explicit DAE of type (1.1) and (1.2) with differential state $x = (q, v)^\top$, algebraic state $y = \lambda$, and control u occurs. The well-known *Grübler condition*, given by

$$\operatorname{rank} g_q'(t, q(t)) = m, \tag{1.4}$$

for all t guarantees unique solvability and excludes redundant constraints, where m is the dimension of g.

These classes of DAEs are well investigated with regard to theory and numerical treatment and common index definitions such as the differentiation index [26], the structural index [19], the strangeness index [33], the tractability index [35], and the perturbation index [31] coincide under some regularity assumptions. Despite its structural simplicity, semi-explicit DAEs arise in a high number of practically important applications, especially in robotics and vehicle dynamics, and we aim to discuss some of those applications in the following sections.

An outline of the paper is as follows. Section 2 is devoted to models in vehicle dynamics that naturally lead to a DAE formulation, such as connected (sub-)systems or rail systems. The purpose of this section is to provide a broad overview on known issues in DAE models. The presentation is kept brief in order to outline typical phenomena. Section 3 addresses control problems that lead to a DAE only by imposing additional path constraints. In Sect. 4 we discuss piecewise defined DAEs and their occurrence in docking maneuvers. These sections discuss in depth models in vehicle dynamics and mobile robotics and the occurrence of DAEs. In addition, numerical experiments are presented. Finally, Sect. 5 concludes the paper.

1.1 Notation

We use the following notation. The derivative w.r.t. time of a function $z(t)$ is denoted by $z'(t)$. The partial derivative of a function f with respect to a variable x will be denoted by $f'_x = \partial f/\partial x$. As an abbreviation of a function of type $f(t, x(t))$ we use the notation $f[t]$. For notional convenience we will often suppress the argument t in $x(t)$ and just write x instead (likewise for other functions depending on t).

2 Overview on Classical DAE Models

In the modeling process of mechanical systems and, in particular, in the building of full vehicle models or vehicular subsystems, like axles or suspensions, constraints and, thus, differential-algebraic equations, naturally arise. This section aims at providing a brief overview as well as an illustration of modeling steps leading to DAEs in the context of vehicle engineering. For a detailed discussion, we refer to the textbooks [44, 46, 49, 51] concerning general modeling aspects of multibody dynamics and we refer to [3, 20, 41, 42] for discussions and studies with a strong focus on vehicle system dynamics.

2.1 Modeling Kinematic Joints

First of all, the modeling or, to be more precise, the mathematical description of kinematic joints in terms of absolute coordinates is a natural source of algebraic constraint equations. To illustrate this, assume that there are two rigid bodies, body i and body j, which have absolute coordinates $q_i, q_j \in \mathbb{R}^{n_q}$, respectively (with $n_q = 6$ for 3D rigid bodies in 3D space) and which are coupled by a kinematic joint; the situation is sketched in Fig. 1. This coupling can be mathematically described by a nonlinear function $g : \mathbb{R}^{n_q} \times \mathbb{R}^{n_q} \longrightarrow \mathbb{R}^{n_J}$ and a set of equations of the following form

$$0 = g_{ij}(q_i, q_j). \tag{2.1}$$

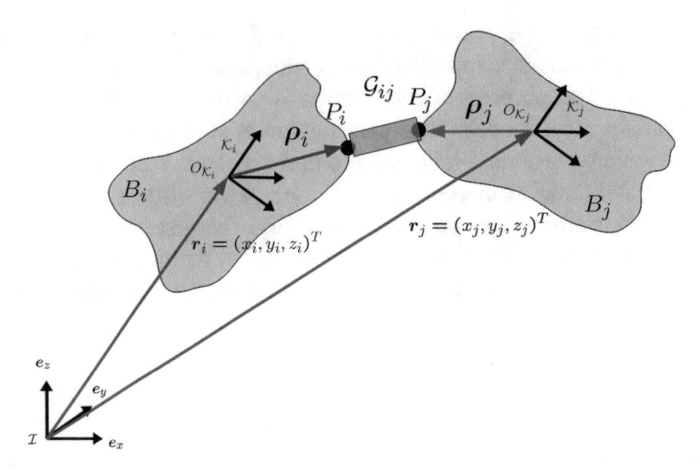

Fig. 1 Two rigid bodies coupled with a kinematic joint $\mathcal{G}_{ij}$

The function g characterizes the joint type and the number $0 < n_J \leq n_q$ denotes the number of restricted (or locked) degrees of freedom of the relative motion of the two involved bodies. This formulation is also called implicit joint description, see [51].

A typical example of such a joint is a so called spherical joint, which restricts relative translation of the two bodies and, accordingly, only allows relative rotation. Consider two rigid bodies with their absolute coordinates $q_{i/j} = (x_{i/j}, y_{i/j}, z_{i/j}, \alpha_{i/j}, \beta_{i/j}, \gamma_{i/j}) \in \mathbb{R}^6$, where $r_{i/j} := (x_{i/j}, y_{i/j}, z_{i/j})$ denote the Cartesian coordinates of a reference point on each body (typically the center-of-mass) and $(\alpha_{i/j}, \beta_{i/j}, \gamma_{i/j})$ denote, e.g., Cardan angles, which in turn specify the rotation-matrix of a body-fixed reference frame w.r.t. to a global frame at rest, see Fig. 1. Then, the positions of the two coupling points on each body can be represented as follows:

$$P_i = r_i + R(\alpha_i, \beta_i, \gamma_i)\rho^i, \quad P_j = r_j + R(\alpha_j, \beta_j, \gamma_j)\rho^j, \tag{2.2}$$

with a rotation matrix R and ρ^i and ρ^j being the positions of the coupling points expressed in the body-fixed reference-frame with origin r_i and r_j, respectively. In this setup, a spherical joint is simply described by the following set of equations

$$\begin{aligned} 0 &= g_{ij,spherical}(q_i, q_j) := P_i - P_j \\ &= r_i + R(\alpha_i, \beta_i, \gamma_i)\rho^i - \left(r_j + R(\alpha_j, \beta_j, \gamma_j)\rho^j\right). \end{aligned} \tag{2.3}$$

Consequently, defining a kinematic joint between two bodies in terms of absolute coordinates means adding an algebraic constraint equation (2.1). On the dynamic level, a kinematic constraint described in that way causes constraint forces on the involved bodies i and j, respectively, which have the form

$$f_{i/j} := -\left(\frac{\partial g_{ij}}{\partial q_{i/j}}\right)^{\top} \lambda_{ij}, \tag{2.4}$$

with unknown Lagrange multipliers λ_{ij} for each joint. These multipliers correspond to unknown algebraic variables in the DAE context, compare Eq. (1.3). Hence, for a multibody system consisting of N rigid bodies described in their absolute coordinates, collecting the dynamic equations for each body as well as all the joint constraint equations leads to the overall equations of motions, which are of the form as displayed in Eq. (1.3). In [20, 48], such an implicit joint modeling is studied for a small-size planar truck model.

An alternative way consists in choosing so called joint coordinates instead of absolute coordinates that describe explicitly the degrees of freedom in each joint, see [51]. Formally, if there is a kinematic joint between bodies i and j as before, one can choose a set of joint coordinates η_{ij} as well as an explicit joint function ϕ_{ij} such that

$$q_j = \phi_{ij}(q_i, \eta_{ij}), \tag{2.5}$$

describing in that sense the joint and the allowed motion explicitly and, thus, replacing the implicit description (2.1). For tree-structured multibody systems, cf. Fig. 2, this strategy completely avoids algebraic constraint equations resulting in an ODE as the system's overall equations of motion. If, however, the considered system has kinematically closed loops, cf. Fig. 2, this is no longer possible. For

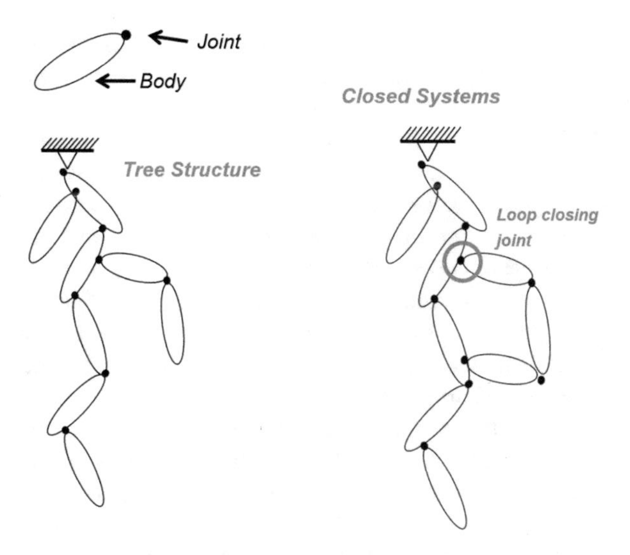

Fig. 2 Multibody system in tree structure (left) and a system with closed loop (right)

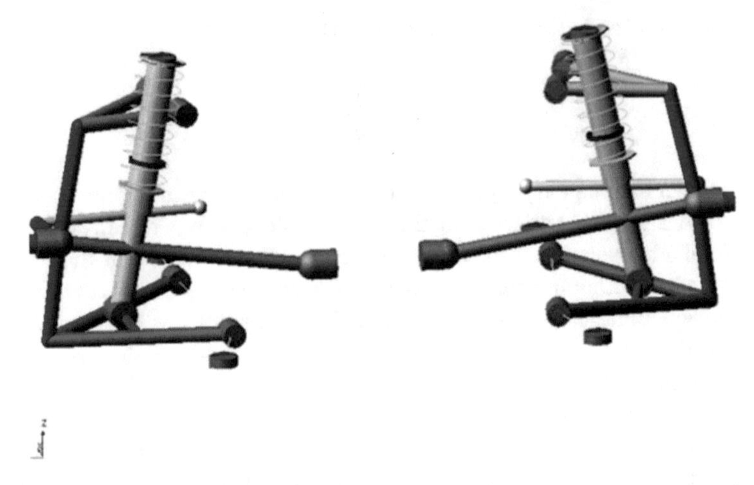

Fig. 3 Multibody system model of an axle with double wishbone suspensions, screenshot from the software tool ADAMS

such systems, the typical modeling procedure is divided into two steps: First, loop-closing joints (which are generally not unique, but must be chosen) are ignored leading to a tree-structured intermediate system, for which joint coordinates can be used avoiding constraints. Then, in a second step, the loop-closing joints are added and described implicitly as before in the form of Eq. (2.1).

A typical example in vehicle engineering, in which kinematically closed loops arise, is the modeling of suspension systems. Examplarily, in Fig. 3, a double wishbone suspension system with such a loop is shown as screenshot from MBS software tool ADAMS.

To summarize, if only absolute coordinates are used to describe rigid body positions, each kinematic joint leads to algebraic constraint equations and algebraic variables. In case of properly chosen joint coordinates, one can describe kinematic joints explicitly. For systems in tree-structure, compare the robotic arm in Sect. 3.1, constraint equations can be avoided and the system's equations of motion are in ODE form. If the system, however, has kinematically closed loops, cf. Fig. 2, also the joint coordinates are no longer independent and the loop closing joints are modeled implicitly using constraint equations again leading to a DAE system as equations of motion. Comparing both approaches, the description of kinematic joints (implicitly) by an constraint equation is rather straightforward. As in the case of the above mentioned spherical joint, the implicit joint functions g_{ij} can be stated more or less simply for a large class of different joint types, whereas the choice of joint coordinates and the derivation of the corresponding explicit joint function ϕ_{ij}, see Eq. (2.5), is often much more involved. In terms of the resulting equations of motion, for a tree-structured system, the implicit approach using absolute coordinates leads to a large DAE system of maximum dimension, which is, however, highly structured and sparse. In contrast, using joint coordinates gives rise to a complicated and nonlinear ODE system of minimum dimension with

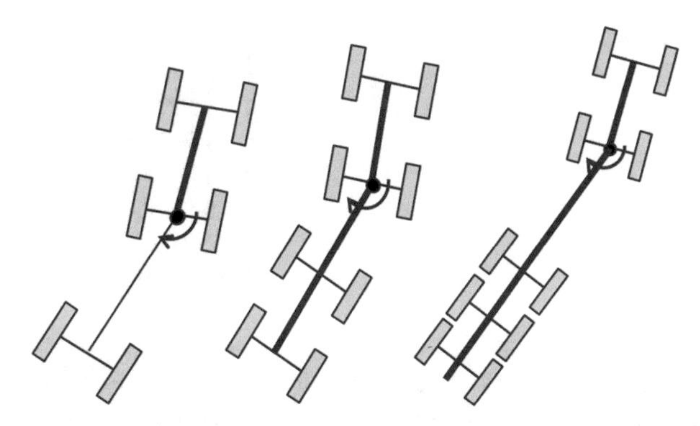

Fig. 4 Car-/tractor-trailer combinations

dense system matrices. An example for the use of a kinematic joint is the vehicle-trailer coupling in Example 2.1.

Example 2.1 (Vehicle-Trailer Coupling) In the field of modeling vehicular systems, there is a further strongly related scenario, in which the modeling of a kinematic joint coupling by an algebraic constraint equation is favorable: namely, the coupling of a vehicle and a trailer. Again, a main benefit of the coupling via constraints is that it can be established in a simple and straightforward manner. Both systems, the car, truck or tractor on the one hand-side and the trailer on the other hand-side can be modeled independently from each other and, if needed, combined afterwards by imposing a constraint equation describing the coupling joint, e.g., a revolute joint, i.e., a single-degree of freedom joint only allowing relative rotation around one common rotation axis, compare [23, 43].

In the context of vehicle-trailer systems, one can consider both planar models of medium complexity, cf. Fig. 4, for instance, in order to derive optimized controls for parking or other steering maneuvers with vehicle-trailer combinations using optimal control approaches. Coupling using joints and constraint equations can be used as well in connection with more complex 3D MBS vehicle models. The latter is offered, e.g., within the commercial software tool TESIS DYNAware, see [21–23]. An extension of the vehicle-trailer coupling towards docking maneuvers is discussed in Sect. 4.

2.2 Contact Modeling

In the context of vehicle dynamics, the DAE framework provides a simple and efficient approach to model contacts and, in particular, the most prominent scenario here is the wheel-ground contact. Specific applications in this direction are discussed in Sects. 3.3 and 3.4.

Concerning road vehicles, this wheel-ground contact is typically realized within the tire model, in which dynamic contact forces (using point-contacts or distributed

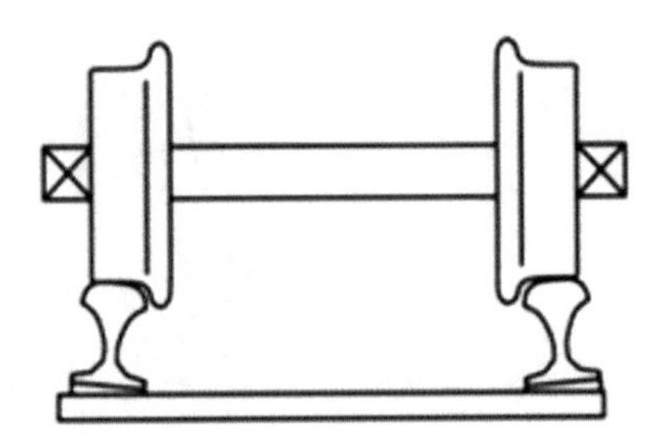

Fig. 5 Wheel set model for rail vehicle dynamics, see [47]

contact patch approaches) are calculated via force elements. In the field of rail vehicle, however, the contact between wheel and rail is typically not considered dynamically, but as contact of two rigid bodies with smooth contact surfaces, see Fig. 5. Here, the DAE modeling approach has been used very successfully.

To illustrate the approach, consider two bodies i and j with smooth surfaces being in permanent contact. As this notion already indicates, the contact can be regarded as a joint and, thus, permanent contact can be modeled as constraint equation in terms of the following contact condition, see [3],

$$0 = g_{ij}(q) = (n^{(i)})^\top \left(P^{(i)}(q, s^{(i)}(q)) - P^{(j)}(q, s^{(j)}(q)) \right). \tag{2.6}$$

Herein, $P^{(i/j)}$ denote the position of the contact points, which depend on the body coordinates q as well as on (also unknown) coordinates $s^{(i/j)}$ that parameterize the surfaces of bodies i and j. Last not least, $n^{(i)}$ denotes the normal vector on the surface of body i at the point $P^{(i)}$. The surface coordinates have to satisfy a set of geometrical conditions, which can be denoted as

$$0 = h_{ij}(q, s^{(i)}, s^{(j)}). \tag{2.7}$$

These equations are auxiliary equations that are needed to find the contact point. Classically, first, Eq. (2.7) has to be solved internally to obtain $s^{(i/j)}$, which then allows to evaluate and to solve Eq. (2.6). In contrast, making use of a DAE modeling approach, Eqs. (2.6) and (2.7) can just be added to the remaining system equations and combined with other algebraic equations, e.g., due to other kinematic joints. The overall system equations can be solved as a monolithic system by one solver. For a more detailed discussion, we refer to [3, 44], in [47], the wheelset model from Fig. 5 is studied and all equations including the contact conditions are explicitly stated.

A last important issue concerning contact modeling worth mentioning here is the fact that in case of modeled contact dynamics, often also friction has to be taken into account, compare [3]. Typically, friction forces, or, to be more precise, the friction force law, depend on contact forces, and, thus on constraint forces. In the context of the formulation in Eq. (1.3) and provided that the contact is modeled within the algebraic constraint equations, there are friction forces that depend on $-g_q'(t, q)^\top \lambda$, and, thus, the dependencies of the right-hand side force vector f is

$f = f(t, q, v, u, \lambda)$. With the modified force vector Eq. (1.3) becomes

$$\begin{aligned} q' &= v, \\ M(t,q)v' &= f(t,q,v,u,\lambda) - g_q'(t,q)^\top \lambda, \\ 0 &= g(t,q), \end{aligned}$$

where we suppressed the argument t in q, v, λ for notational convenience. Differentiating the algebraic constraint twice w.r.t. time yields the constraint on acceleration level

$$\begin{aligned} 0 = {} & g_{tt}''(t,q) + g_{tq}''(t,q)v + g_{qt}''(t,q)v + g_{qq}''(t,q)(v,v) \\ & + g_q'(t,q)M(t,q)^{-1}(f(t,q,v,u,\lambda) - g_q'(t,q)^\top \lambda). \end{aligned} \tag{2.8}$$

Now, if the modified Grübler condition

$$\operatorname{rank}\left(g_q'(t,q)M(t,q)^{-1}\left(\frac{\partial f}{\partial \lambda} - (g_q')^\top \right) \right) = m$$

holds in a solution, then the Eq. (2.8) can be solved for λ by the implicit function theorem and the differentiation index is three.

Remark 2.1 We like to point out that the Grübler condition will be violated, if redundant algebraic constraints are introduced. This happens, for instance, in the presence of redundant joint restrictions. As an example consider a door with two hinges, which are modeled as revolute joints. Herein, joint forces can be distributed in different ways on the two joints. We refer the reader to [52–54] for a more detailed analysis on the treatment of redundant constraints.

3 Path Constraints and Dynamic Inversion Control

Often, the dynamics of a system do not directly lead to a DAE. Instead the DAE arises from the task to follow a prescribed trajectory, compare, e.g., [10], [11, Chapter 6.3, pp. 157] for an application from flight trajectory control. To this end, consider the control system

$$x'(t) = f(t, x(t), u(t)) \tag{3.1}$$

with $x(t) \in \mathbb{R}^{n_x}$, $u(t) \in \mathbb{R}^{n_u}$ and prescribed path constraints

$$0 = g(t, x(t), u(t)), \qquad g : \mathbb{R} \times \mathbb{R}^{n_x} \times \mathbb{R}^{n_u} \longrightarrow \mathbb{R}^{n_y}, \tag{3.2}$$

where the control vector u may or may not appear explicitly in the constraints. The combined system leads to a semi-explicit DAE where some components of the control vector u serve as algebraic variables. Typically the number of constraints has to be less than or equal to the number of controls, i.e. $n_y \leq n_u$, since otherwise – except in degenerate cases – not enough degrees of freedom remain to satisfy the constraints by an appropriate choice of the controls. Suppose the DAE possesses a solution. Let u be partitioned into two components $u = (y, w)^\top$ with $y \in \mathbb{R}^{n_y}$ and $w \in \mathbb{R}^{n_u - n_y}$. We intend to interpret the component y as an algebraic state, whereas w can be seen as the "true" degrees of freedom in the control vector u. Using this partition, the algebraic constraint reads

$$0 = g(t, x(t), y(t), w(t)).$$

Now, if the Jacobian g'_y is non-singular and essentially bounded along the solution, then the DAE has index one in the sense of the differentiation index. By the implicit function theorem, the component y is implicitly determined by $(t, x(t), w(t))$. Likewise if g does not depend on $u = (y, w)^\top$ explicitly and if $g'_x(t, x) f'_y(t, x, u)$ is non-singular and essentially bounded along the solution, then the DAE has index two (in the sense of the differentiation index) and the derivative of the constraint, i.e. $0 = g_t(t, x(t)) + g_x(t, x(t)) f(t, x(t), u(t))$, determines the component y of u implicitly. This type of reasoning can be applied more generally if the DAE exhibits a Hessenberg structure with respect to the component y of u, compare [14]. Note that most of the DAEs in this paper have Hessenberg structure and the determination of a suitable y is straightforward by choosing y such that the Jacobian of the ℓ-th derivative of the algebraic constraints has full rank, where ℓ is minimal. For general unstructured DAEs it is much more involved to identify suitable components y of u in a systematic way, compare the concept of the strangeness index in [33]. A way to achieve this is excellently described in the upcoming paper [16]. It uses the so-called derivative array, which contains the DAE itself and its time derivatives up to a certain order. Then a completion of the DAE is determined, compare [15], and solved numerically using GENDA, see [34] and [32].

Please note, that high index DAEs may actually occur, compare the index five problem in [8, Example 2].

The outlined procedure is sometimes called dynamic inversion control, see [4], or servo constraints, see [1, 9, 40]. A detailed analysis for mechanical multibody systems can be found in [5–9]. Optimal control techniques were used in [1, 4] and in [12, Chapter 4, pp. 67–82], [13] to find road input data for measured quantities like chassis accelerations. A similar dynamic inversion approach was used in flight trajectory optimization scenarios, see [18]. In this paper dynamic inversion (also for kinematic models) means the process of identifying components of u in the combined system (3.1) and (3.2) and identifying these components with algebraic states y in (1.1) and (1.2) such that the DAE is well-defined in the sense of the differentiation index. We illustrate the approach for specific applications from mobile robotics and vehicle control. We like to demonstrate with these specific examples that, depending on the respective task, a DAE with index three, index two, or a DAE of mixed index arises.

Fig. 6 Configuration of a KUKA youBot with an omnidirectional platform and a five link robotic arm

3.1 A Mobile Robot

We apply the dynamic inversion control approach in order to control a robot on a prescribed path. To this end we consider a simplified model of the KUKA youBot in Fig. 6.

The youBot consists of a platform with omnidirectional wheels, i.e. the platform is able to move in any direction and to rotate simultaneously. A 5-link robotic arm with 5 degrees of freedom is mounted on the platform (one base and four links). We use the following simplified model which decouples the motion of the platform and the motion of the robotic arm. This is justified since the platform is much heavier than the arm and the joints of the links are such that electric motors inside keep the requested angle and compensate gravity effects. The motion of the platform in the (x, y)-plane is described by the ODEs

$$x'(t) = v_x(t), \tag{3.3}$$

$$y'(t) = v_y(t), \tag{3.4}$$

$$\psi'(t) = \omega(t), \tag{3.5}$$

$$v_x'(t) = u_x(t)\cos\psi(t) - u_y(t)\sin\psi(t), \tag{3.6}$$

$$v_y'(t) = u_x(t)\sin\psi(t) + u_y(t)\cos\psi(t), \tag{3.7}$$

where (x, y) denotes the center of gravity of the platform, ψ it's yaw angle, and (v_x, v_y) the velocity vector. The platform can be controlled by the yaw rate ω and the acceleration vector (u_x, u_y), which is given in the platform's local coordinate system.

The joint angles of the $N = 5$ links of the robotic arm are denoted by q_i and its velocities by v_i, $i = 1, \ldots, N$. Technically it is possible to control the velocities directly within the bounds $v_{min} = -90$ degrees/s and $v_{max} = +90$ degrees/s. This leads to the simple kinematic equations

$$q'(t) = v(t) \tag{3.8}$$

with $q = (q_1, \ldots, q_N)^\top$ and $v = (v_1, \ldots, v_N)^\top$. Owing to physical constraints, the joint angle vector q is bounded by

$$q_{min} \leq q \leq q_{max} \tag{3.9}$$

with $q_{min} = (-169, -65, -151, -102, -167)^\top$ and $q_{max} = (169, 90, 146, 102, 167)^\top$ (the values are in degree). The lengths of the arms 1–4, see Fig. 6 are $\ell_1 = 0.155\,[\mathrm{m}]$, $\ell_2 = 0.135\,[\mathrm{m}]$, $\ell_3 = 0.081\,[\mathrm{m}]$, $\ell_4 = 0.09\,[\mathrm{m}]$. The offset vector from the center of gravity of the platform to the mount point of the base in the platform's coordinate system is denoted by $a = (0.153, 0, 0)^\top\,[\mathrm{m}]$ and the vector from the mounting point to the first joint in base's coordinate system is $b = (0.035, 0, 0.147)^\top\,[\mathrm{m}]$.

Let the rotation matrices be defined as

$$R_z(\alpha) = \begin{pmatrix} \cos\alpha & -\sin\alpha & 0 \\ \sin\alpha & \cos\alpha & 0 \\ 0 & 0 & 1 \end{pmatrix}, \quad R_y(\beta) = \begin{pmatrix} \cos\beta & 0 & \sin\beta \\ 0 & 1 & 0 \\ -\sin\beta & 0 & \cos\beta \end{pmatrix},$$
$$R_x(\gamma) = \begin{pmatrix} 1 & 0 & 0 \\ 0 & \cos\gamma & -\sin\gamma \\ 0 & \sin\gamma & \cos\gamma \end{pmatrix}, \quad R_{zy}(\alpha, \beta) = R_z(\alpha) R_y(\beta),$$
$$R_{zyy}(\alpha, \beta, \gamma) = R_z(\alpha) R_y(\beta + \gamma),$$
$$R_{zyyy}(\alpha, \beta, \gamma, \delta) = R_z(\alpha) R_y(\beta + \gamma + \delta),$$
$$R_{zyyyx}(\alpha, \beta, \gamma, \delta, \eta) = R_z(\alpha) R_y(\beta + \gamma + \delta) R_x(\eta),$$

where the subscripts indicate the rotation axes for consecutive rotations. Let $z = (x, y, \psi, v_x, v_y, q_1, \ldots, q_5)^\top$. The center of gravity of the platform is given by $r_0(z) = (x, y, h)^\top$, where $h = 0.084\,[\mathrm{m}]$ is the height of the platform. Moreover, the mount points r_i, $i = 1, \ldots, 5$, of the links and the gripper position are given by

$$r_1(z) = r_0(z) + R_z(\psi)(a + R_z(q_1) b),$$
$$r_2(z) = r_1(z) + R_{zy}(\psi + q_1, q_2 - \frac{\pi}{2}) \begin{pmatrix} \ell_1 \\ 0 \\ 0 \end{pmatrix},$$
$$r_3(z) = r_2(z) + R_{zyy}(\psi + q_1, q_2 - \frac{\pi}{2}, q_3) \begin{pmatrix} \ell_2 \\ 0 \\ 0 \end{pmatrix},$$

Table 1 Parameters of the youBot model

Parameter	Value	Unit
v_{min}	-90	[degree/s]
v_{max}	90	[degree/s]
q_{min}	$(-169, -65, -151, -102, -167)^\top$	[degree]
q_{max}	$(169, 90, 146, 102, 167)^\top$	[degree]
ℓ_1	0.155	[m]
ℓ_2	0.135	[m]
ℓ_3	0.081	[m]
ℓ_4	0.09	[m]
a	$(0.153, 0, 0)^\top$	[m]
b	$(0.035, 0, 0.147)^\top$	[m]
h	0.084	[m]

$$r_4(z) = r_3(z) + R_{zyyy}(\psi + q_1, q_2 - \frac{\pi}{2}, q_3, q_4) \begin{pmatrix} \ell_3 \\ 0 \\ 0 \end{pmatrix},$$

$$r_5(z) = r_4(z) + R_{zyyyx}(\psi + q_1, q_2 - \frac{\pi}{2}, q_3, q_4, q_5) \begin{pmatrix} \ell_4 \\ 0 \\ 0 \end{pmatrix}.$$

Summarizing, the motion of the robot is described by the ODE (3.3)–(3.7) and (3.8), where $z = (x, y, \psi, v_x, v_y, q)^\top$ denotes the 10-dimensional state vector and $u = (u_x, u_y, \omega, v)^\top$ is the 8-dimensional control vector.

A DAE arises, if parts of the motion of the robot are fixed to prescribed paths. We investigate three cases. For notational convenience we use the generic names z, y, w, g, and γ for the differential state vector, the algebraic state vector, the (free) control vector, the algebraic constraints, and a curve, respectively. Table 1 summarizes all parameters of the model for the reader's convenience.

Remark 3.1

(a) Note, that $q = 0$ corresponds to the upright configuration of the robotic arm.
(b) The angle q_5 does not influence the above positions. The angle q_5 becomes relevant if the orientations of the bodies (and the gripper) have to be considered, e.g., in collision avoidance scenarios or in situations where the gripper has to be positioned in a particular way.
(c) A full mechanical multibody system describing the dynamics of the robot more accurately can be derived using the Lagrangian equations of motion. This leads to a highly nonlinear system with 8 degrees of freedom (9 if the gripper is modeled as well). However, for many path planning tasks the presented kinematic model (called simplified model) is sufficient.
(d) Please note that the choice of controls in u in the previous model is mainly a modeling decision. Likewise we could have chosen to control the accelerations

of the joint angles and the yaw angle. However, as we shall see later in Sect. 3.1.2, our choice of controls will lead to a DAE with a mixed index. This is what we intend to illustrate.

3.1.1 Prescribed Path for the Platform

The center of gravity (x, y) of the platform is restricted to follow a given curve

$$\gamma(t) = \begin{pmatrix} x_d(t) \\ y_d(t) \end{pmatrix}, \qquad t \in [0, T],$$

which is parameterized with respect to time t. This leads to the path constraint

$$0 = g(t, z(t)) := \begin{pmatrix} x(t) - x_d(t) \\ y(t) - y_d(t) \end{pmatrix}. \tag{3.10}$$

If γ is twice continuously differentiable, then twofold differentiation with respect to time and exploitation of (3.3) and (3.4) leads to the relations

$$\begin{aligned} 0 &= \frac{d^2}{dt^2} g(t, z(t)) =: g^{(2)}(t, z(t), u(t)) \\ &= \begin{pmatrix} u_x(t) \cos\psi(t) - u_y(t) \sin\psi(t) - x_d''(t) \\ u_x(t) \sin\psi(t) + u_y(t) \cos\psi(t) - y_d''(t) \end{pmatrix}. \end{aligned} \tag{3.11}$$

The Jacobian $\frac{\partial g^{(2)}}{\partial u}$ reads

$$\frac{\partial g^{(2)}}{\partial u} = \begin{pmatrix} \cos\psi & -\sin\psi & 0\,0\,0\,0\,0\,0 \\ \sin\psi & \cos\psi & 0\,0\,0\,0\,0\,0 \end{pmatrix}.$$

This matrix is of rank two since

$$\det \begin{pmatrix} \cos\psi & -\sin\psi \\ \sin\psi & \cos\psi \end{pmatrix} = \cos^2\psi + \sin^2\psi = 1.$$

Hence, it is natural to consider the controls u_x and u_y as algebraic state $y = (u_x, u_y)^\top$. The combined system of (3.3)–(3.4), (3.8), and (3.10) is a Hessenberg-DAE of index three with differential state z, algebraic state y, and control $w = (\omega, v)^\top$. The initial value $z(0) = z_0 = (x_0, y_0, \psi_0, v_{x,0}, v_{y,0}, q_0)^\top$ has to be consistent, that is, it has to satisfy the equations

$$0 = g(0, z_0) = \begin{pmatrix} x_0 - x_d(0) \\ y_0 - y_d(0) \end{pmatrix}, \tag{3.12}$$

and

$$0 = \frac{d}{dt} g(t, z(t))\bigg|_{t=0} = \begin{pmatrix} v_{x,0} - x_d'(0) \\ v_{y,0} - y_d'(0) \end{pmatrix}. \tag{3.13}$$

The algebraic state y is determined by (3.11), that is,

$$y(t) = \begin{pmatrix} u_x(t) \\ u_y(t) \end{pmatrix} = \begin{pmatrix} \cos\psi(t) & -\sin\psi(t) \\ \sin\psi(t) & \cos\psi(t) \end{pmatrix}^\top \begin{pmatrix} x_d''(t) \\ y_d''(t) \end{pmatrix}. \tag{3.14}$$

The remaining controls in w can be used to optimize the motion of the mobile robot on the curve γ in view of a given performance criterion. Please note that in this context it is not meaningful in general to impose constraints on y since y is already determined by x_d'', y_d'', and ψ.

Example 3.1 Let the curve $\gamma(t) = (x_d(t), y_d(t))^\top$ for $t \in [0, T]$ be given by

$$x_d(t) = r \sin t, \qquad y_d(t) = r \sin t \cos t$$

with $r = 2$ and $T = 2\pi$. Then

$$x_d'(t) = r \cos t, \qquad y_d'(t) = r(\cos^2 t - \sin^2 t)$$

and

$$x_d''(t) = -r \sin t, \qquad y_d''(t) = -4r \cos t \sin t.$$

Let the initial state be given by $z_0 = (0, 0, 0, r, r, q_0)^\top$ with $q_0 = (0, 0, 0, 0, 0)^\top$. The controls $\omega \in [-\pi/2, \pi/2]$ and $v_i \in [-\pi/2, \pi/2]$, $i = 1, \ldots, 5$, are chosen such that the robotic arm stays in the upright position $q_1 = (-\psi(t_1), 0, 0, 0, 0)^\top$ until time $t_1 = \pi/2$, picks up some object at time $t_2 = \pi$ from the left, moves back into the upright position $q_3 = (-\psi(t_3), 0, 0, 0, 0)^\top$ until time $t_3 = 3\pi/2$, and puts down the object at time $T = 2\pi$ to the right. The angular positions at time t_2 and T are given by $q_2 = (1.5708 - \psi(t_2), 1.5254, 0.661786, 0.95445, 0)^\top$ and $q_T = (-1.5708 - \psi(T), 1.5254, 0.661786, 0.95445, 0)^\top$.

This can be achieved by solving the following optimal control problem:

Minimize

$$\alpha_0 \sum_{i=1}^{3} \|q(t_i) - q_i\|^2 + \alpha_1 \int_0^T \omega^2(t)dt + \alpha_2 \int_0^T \|v(t)\|^2 dt$$

subject to (3.3)–(3.7), (3.8), (3.9), $\omega \in [-\pi/2, \pi/2]$, $v_i \in [-\pi/2, \pi/2]$, $i = 1, \ldots, 5$, *initial state* $z_0 = (0, 0, 0, r, r, q_0)^\top$ *with* $q_0 = (0, 0, 0, 0, 0)^\top$, *and*

$$q(T) = q_T.$$

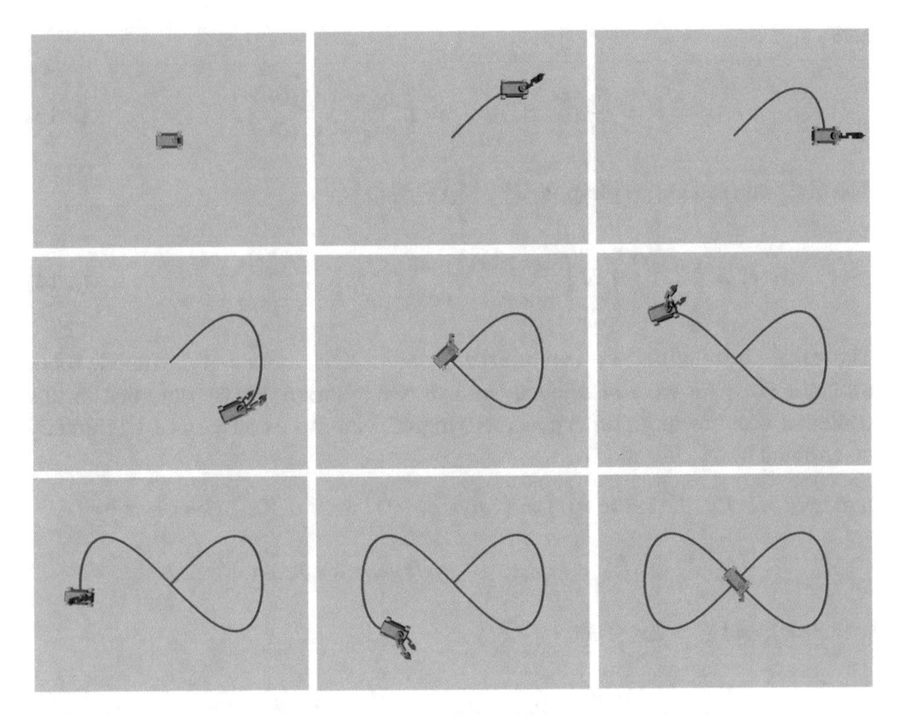

Fig. 7 Snapshots of the motion of the mobile robot (Example 3.1): Tracking of a prescribed path for the platform from $t = 0$ (top left) to $t = 2\pi$ (bottom right) and simultaneous relocation of an object by the robotic arm. Please note that the wheels are omnidirectional (although they appear as standard wheels in the graphics)

Figure 7 shows the motion of the robot for $\alpha_0 = 50$, $\alpha_1 = \alpha_2 = 10^{-2}$. Note, that the weights are chosen such that a good compromise between reaching the intermediate configurations q_i, $i = 1, 2, 3$, and minimizing control effort is obtained. The positive weights α_1 and α_2 have a regularizing effect in the Hamiltonian of the system. A detailed discussion of optimal control problems is beyond the scope of this paper and we refer the reader to [27, 29] for details.

Figure 8 shows selected states and controls. Please note that the controls are discontinuous, i.e., the velocities for the joints jump. It is not possible to realize this in a practical implementation, but we assume that an internal controller is able to approximate this discontinuous behavior nearly instantaneously. The numerical results have been obtained with the software OCPID-DAE1 [28]. OCPID-DAE1 is a software package for solving optimal control problems subject to DAEs. It uses a direct shooting discretization and a sequential-quadratic programming (SQP) method with Armijo linesearch to solve the discretized optimal control problem. Derivatives required by the SQP method are computed by solving a sensitivity DAE. The package offers various options regarding control approximations and integrators. In addition, it has a simulation mode, which can be used to simulate a DAE for a given control input. For the optimal control problem we used an

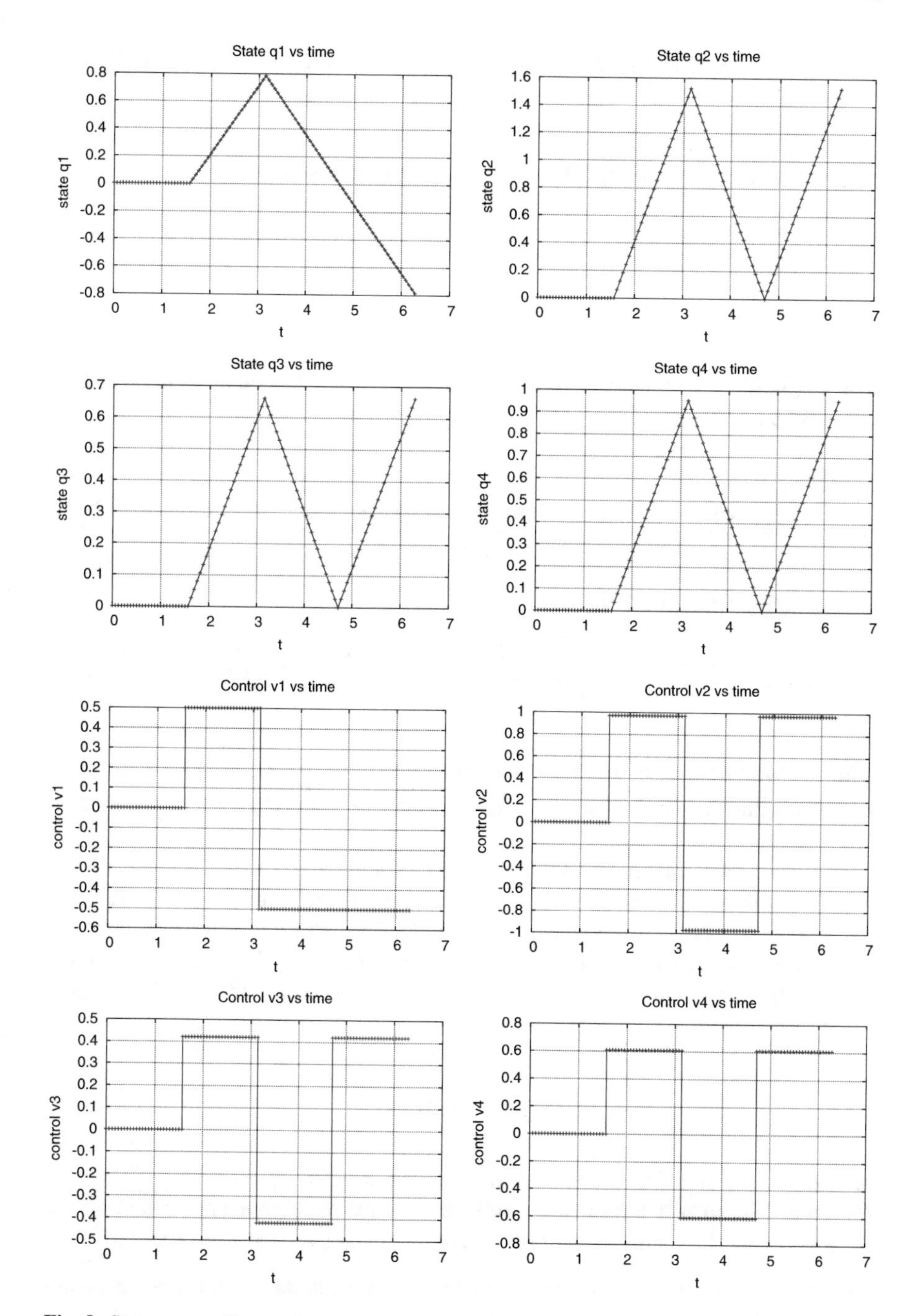

Fig. 8 States q_1, q_2 (first row), states q_3, q_4 (second row), controls v_1, v_2 (third row), and controls v_3, v_4 (fourth row) in Example 3.1

equidistant discretization with 101 grid points, a feasibility tolerance of 10^{-10} and an optimality tolerance of 10^{-7} for the KKT conditions.

3.1.2 Prescribed Path and Orientation for the Platform

We modify the problem in Sect. 3.1.1 by adding the algebraic constraint

$$0 = \psi(t) - \arctan\left(\frac{y_d'(t)}{x_d'(t)}\right) \tag{3.15}$$

to g in (3.10). This constraint forces the robot to keep its orientation tangential to the curve γ. Differentiating (3.15) with respect to time yields

$$\omega(t) = \frac{v_x(t)(u_x(t)\sin\psi(t)+u_y(t)\cos\psi(t))-v_y(t)(u_x(t)\cos\psi(t)-u_y(t)\sin\psi(t))}{v_x(t)^2+v_y(t)^2}.$$

Introducing y from (3.14) yields

$$\omega(t) = \frac{v_x(t)y_d''(t) - v_y(t)x_d''(t)}{v_x(t)^2+v_y(t)^2} = \frac{x_d'(t)y_d''(t) - y_d'(t)x_d''(t)}{v_x(t)^2+v_y(t)^2}, \tag{3.16}$$

which is well-defined unless the total velocity $\bar{v}(t) = \sqrt{v_x(t)^2+v_y(t)^2}$ becomes zero. Note that the relation $\omega(t) = \bar{v}(t)\kappa(t)$ holds, where $\kappa(t)$ denotes the curvature of the curve γ.

We have thus shown that the combined system of (3.3), (3.4), (3.8), (3.10), and (3.15) is a DAE with differential state z, algebraic state $y = (u_x, u_y, \omega)^\top$, and control v. It is not a Hessenberg-DAE, but a DAE with a mixed index where u_x and u_y are index-3 algebraic states while ω is an index-2 algebraic state. The initial value $z(0) = z_0 = (x_0, y_0, \psi_0, v_{x,0}, v_{y,0}, q_0)^\top$ in addition to (3.12), (3.13) has to satisfy the condition

$$0 = \psi_0 - \arctan\left(\frac{y_d'(0)}{x_d'(0)}\right).$$

As before, the remaining controls in v can be used to optimize a performance criterion. Likewise it is not meaningful in general to impose constraints on the algebraic states in y since those a fully determined by (3.14) and (3.16).

Example 3.2 We consider again Example 3.1 with the additional algebraic constraint (3.15).

Figure 9 shows the motion of the robot for $\alpha_0 = 50, \alpha_1 = \alpha_2 = 10^{-2}$. Figure 10 shows selected states and controls. Again, the velocities for the joints are discontinuous and we tacitly assume that the internal controller is able to approximate this discontinuity almost instantaneously. The numerical results have been obtained

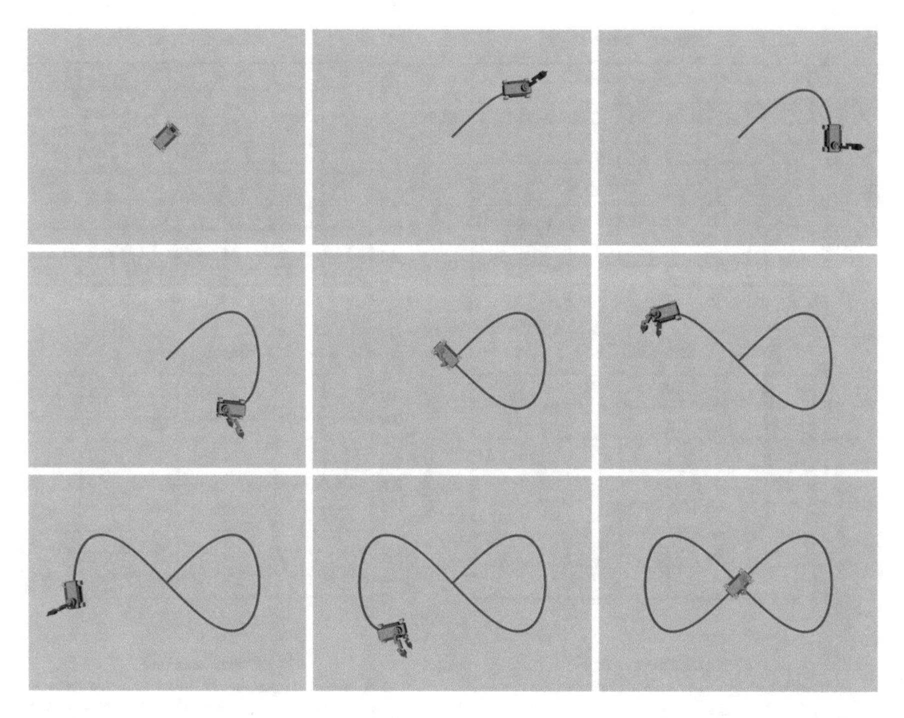

Fig. 9 Snapshots of the motion of the mobile robot (Example 3.2): Tracking of a prescribed path with tangential orientation of the platform from $t = 0$ (top left) to $t = 2\pi$ (bottom right) and simultaneous relocation of an object by the robotic arm. Please note that the wheels are omnidirectional (although they appear as standard wheels in the graphics)

with the software OCPID-DAE1 [28] with an equidistant discretization with 101 grid points, a feasibility tolerance of 10^{-10} and an optimality tolerance of 10^{-7} for the KKT conditions. Comparing the results in Figs. 10 and 8 we notice that only the velocity v_1 (and consequently the joint angle q_1) of the first robot joint was affected by introducing the additional algebraic constraint (3.15) for the yaw angle. Of course, the yaw angles in Examples 3.1 and 3.2 differ as well owing to the prescribed yaw angle in (3.15).

3.1.3 Prescribed Path for the End Affector

A situation more complicated arises if the end affector position r_5 has to follow a prescribed path. In order to simplify the analysis, we additionally restrict the motion of the platform to a straight line at a constant velocity and postulate that the end affector is oriented horizontally at all times, e.g. to perform a welding task on a vertical wall. This leads to the algebraic constraints

$$0 = g(t, z(t)) := \begin{pmatrix} g_1(t, z(t)) \\ g_2(t, z(t)) \end{pmatrix} \tag{3.17}$$

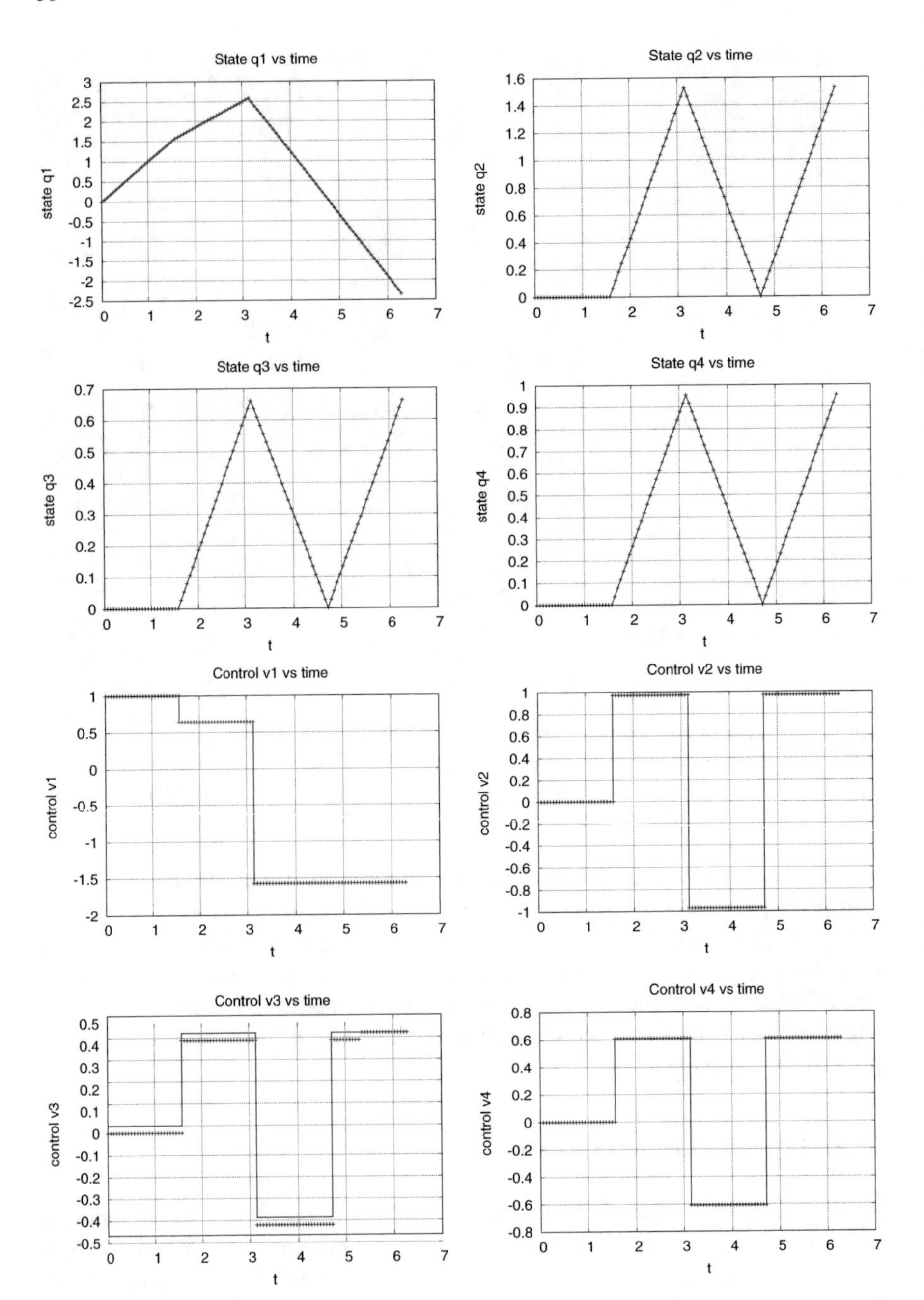

Fig. 10 States q_1, q_2 (first row), states q_3, q_4 (second row), controls v_1, v_2 (third row), and controls v_3, v_4 (fourth row) in Example 3.2

with

$$g_1(t,z(t)) := \begin{pmatrix} v_x(t) - v_d \\ v_y(t) \\ \psi(t) \end{pmatrix}, \quad g_2(t,z(t)) := \begin{pmatrix} q_2(t) + q_3(t) + q_4(t) - \pi/2 \\ r_5(z(t)) - \gamma(t) \end{pmatrix},$$

where v_d is the desired velocity in x-direction and the curve $\gamma : [0,T] \to \mathbb{R}^3$, $\gamma(t) := (x_d(t), y_d(t), z_d(t))^\top$ is the desired path of the end affector position r_5. Differentiation of the algebraic constraints in g leads to the conditions

$$0 = \frac{d}{dt} g_1(t,z(t)) = \begin{pmatrix} v_x'(t) \\ v_y'(t) \\ \psi'(t) \end{pmatrix} = \begin{pmatrix} u_x(t)\cos\psi(t) - u_y(t)\sin\psi(t) \\ u_x(t)\sin\psi(t) + u_y(t)\cos\psi(t) \\ \omega(t) \end{pmatrix}$$

and

$$0 = g_2^{(1)}(t,z(t)) := \frac{d}{dt} g_2(t,z(t)) = \begin{pmatrix} v_2(t) + v_3(t) + v_4(t) \\ r_5'(z(t))z'(t) - \gamma'(t) \end{pmatrix}.$$

Thus, the first set of constraints immediately yields

$$u_x(t) = u_y(t) = \omega(t) = 0.$$

The explicit evaluation of the remaining equations and exploitation of the constraints yields

$$r_5'(z)z' - \gamma' = \begin{pmatrix} (\ell_2(v_2+v_3)\cos(q_2+q_3) + \ell_1 v_2 \cos q_2)\cos q_1 \\ -v_1((\zeta_1+b_1)\sin q_1 + b_2\cos q_1) + v_x \\ \hline (\ell_2(v_2+v_3)\cos(q_2+q_3) + \ell_1 v_2 \cos q_2)\sin q_1 \\ +v_1((\zeta_1+b_1)\cos(q_1) - b_2\sin q_1) \\ \hline -\ell_2(v_2+v_3)\sin(q_2+q_3) - \ell_1 v_2 \sin q_2 \end{pmatrix} - \gamma',$$

where $\zeta_1 := \ell_1 \sin q_2 + \ell_2 \sin(q_2+q_3)$, $\zeta_2 := \ell_1 \cos q_2 + \ell_2 \cos(q_2+q_3)$. It turns out that the Jacobian of $g_2^{(1)}$ with respect to $\tilde{v} = (v_1, v_2, v_3, v_4)^\top$, i.e. the matrix

$$\frac{\partial g_2^{(1)}}{\partial \tilde{v}} = \begin{pmatrix} 0 & 1 & 1 & 1 \\ -(\zeta_1+b_1)\sin q_1 - b_2\cos q_1 & \zeta_2\cos q_1 & \ell_2\cos(q_2+q_3)\cos q_1 & 0 \\ (\zeta_1+b_1)\cos q_1 - b_2\sin q_1 & \zeta_2\sin q_1 & \ell_2\cos(q_2+q_3)\sin q_1 & 0 \\ 0 & -\zeta_1 & -\ell_2\sin(q_2+q_3) & 0 \end{pmatrix},$$

is non-singular if and only if

$$\begin{aligned} 0 \neq & -\ell_2 \cos q_2 \cos^2(q_2+q_3) \\ & +(\ell_1\cos^2 q_2 - \ell_2 \sin q_2 \sin(q_2+q_3) - b_1 \sin q_2 - \ell_1)\cos(q_2+q_3) \\ & +((\ell_1 \sin q_2 + b_1)\sin(q_2+q_3) + \ell_2)\cos q_2. \end{aligned} \tag{3.18}$$

Fig. 11 Snapshots of the motion of the mobile robot (Example 3.3): Tracking of a prescribed path for the end affector from $t = 0$ (top left) to $t = 2\pi$ (bottom right). Please note that the wheels are omnidirectional (although they appear as standard wheels in the graphics)

Hence, subject to the regularity condition in (3.18) we have shown that the DAE, which consists of (3.3), (3.4), (3.8), and (3.17) is an index-2 DAE with differential state z, algebraic state $y = (u_x, u_y, \omega, v_1, v_2, v_3, v_4)^\top$, and control v_5. The latter does not have any influence in the present model and hence it can be set to zero.

The initial value $z(0) = z_0 = (x_0, y_0, \psi_0, v_{x,0}, v_{y,0}, q_0)^\top$ has to be consistent, that is, it has to satisfy the equation $0 = g(0, z_0)$ with g in (3.17).

Example 3.3 Figure 11 shows the motion of the robot for a desired velocity $v_d = 1$, $v_5 \equiv 0$, and

$$\gamma(t) = \begin{pmatrix} t + 0.143 - 0.02\sin(2t) \\ 0.341 \\ 0.397 + 0.05(1 - \cos(2t)) \end{pmatrix}, \qquad t \in [0, 2\pi].$$

The initial state is given by $z_0 = (0, 0, 0, 1, 0, \pi/2, 0, \pi/2, 0, 0)^\top$.

Figure 12 shows selected states and algebraic states. Again, the numerical results have been obtained with the software OCPID-DAE1 [28] with an equidistant discretization with 201 grid points. In this example, no degrees of freedom are left for optimization (v_5 is set to zero) and hence we used the simulation mode of

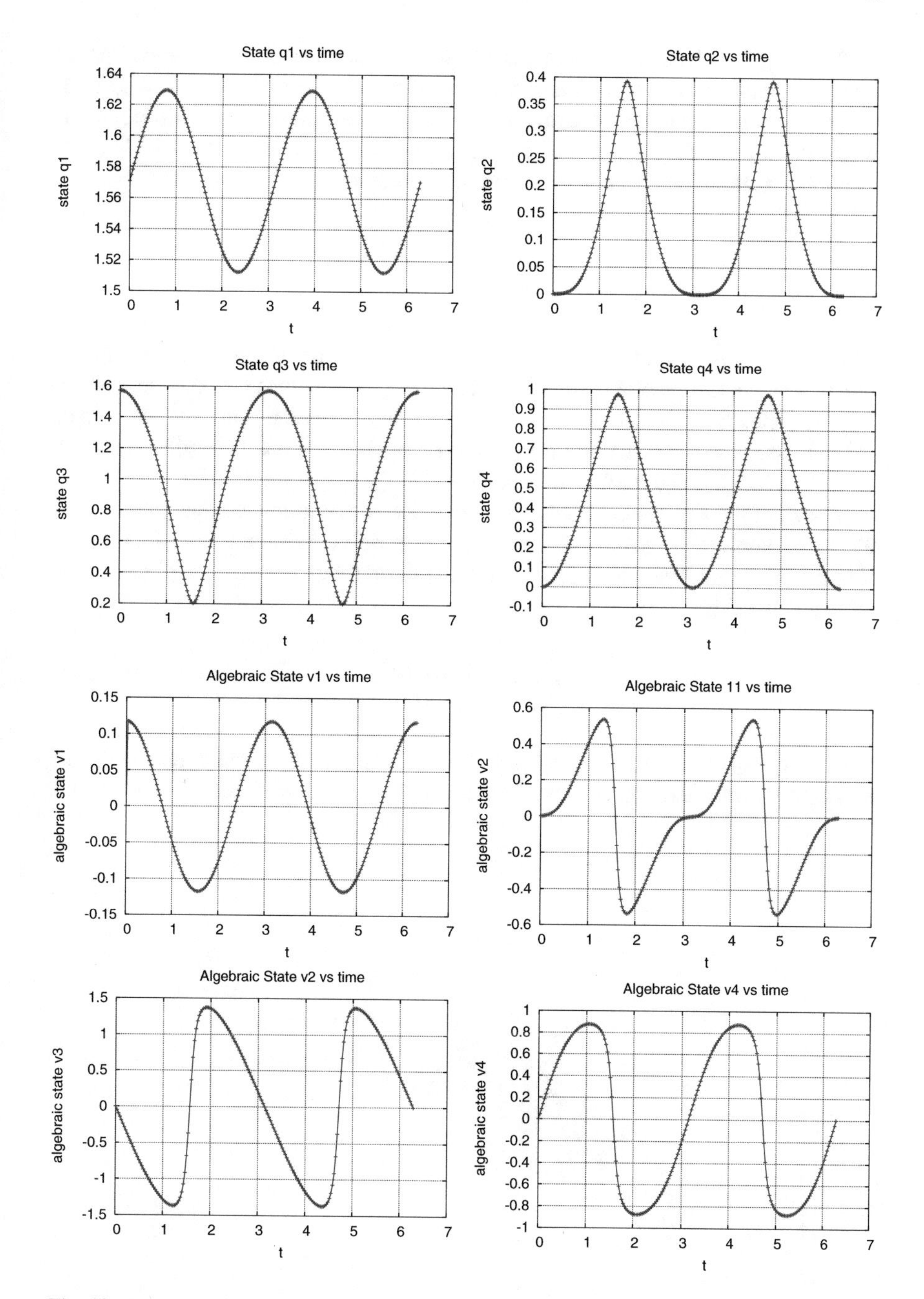

Fig. 12 States q_1, q_2 (first row), states q_3, q_4 (second row), algebraic states v_1, v_2 (third row), and algebraic states v_3, v_4 (fourth row) in Example 3.3

OCPID-DAE1 with an adaption of DASSL with automatic step-size selection as the integrator with an absolute and relative integration tolerance of 10^{-6}. In contrast to the previous examples, the algebraic states, i.e., the joint velocities, are differentiable in this example. This is a direct consequence of the prescribed path in (3.17) being a smooth function (note that γ is chosen to be smooth).

3.2 Vehicle Moving on Prescribed Path

Consider a vehicle moving in the plane. The position of the vehicle and its orientation in a Cartesian coordinate system (x_I, y_I) are given by (x, y, ψ), where (x, y) denotes the center of gravity (or some other reference point) of the vehicle and ψ is the yaw angle, i.e. the rotation angle of the car's local coordinate system (x_C, y_C) relative to the inertial coordinate system (x_I, y_I), see Fig. 13.

Instead of the Cartesian coordinate system it is often more convenient to use an alternative coordinate system – the curvilinear coordinate system, see, e.g., [36]. To this end let the midline of the track (or some other reference line) be given by the curve $\gamma_m : [0, L] \longrightarrow \mathbb{R}^2$ with

$$\gamma_m(s) := \begin{pmatrix} x_m(s) \\ y_m(s) \end{pmatrix}.$$

The curve γ_m of length L is supposed to be parameterized with respect to its arclength s and thus, $\|\gamma_m'(s)\| = 1$ and $\langle \gamma_m''(s), \gamma_m'(s) \rangle = 0$ for all $s \in [0, L]$.

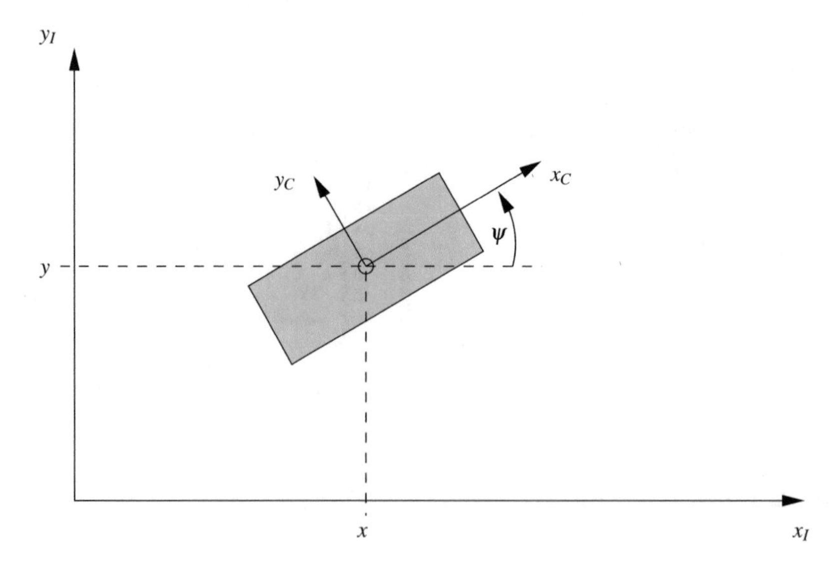

Fig. 13 Cartesian coordinates and orientation of vehicle

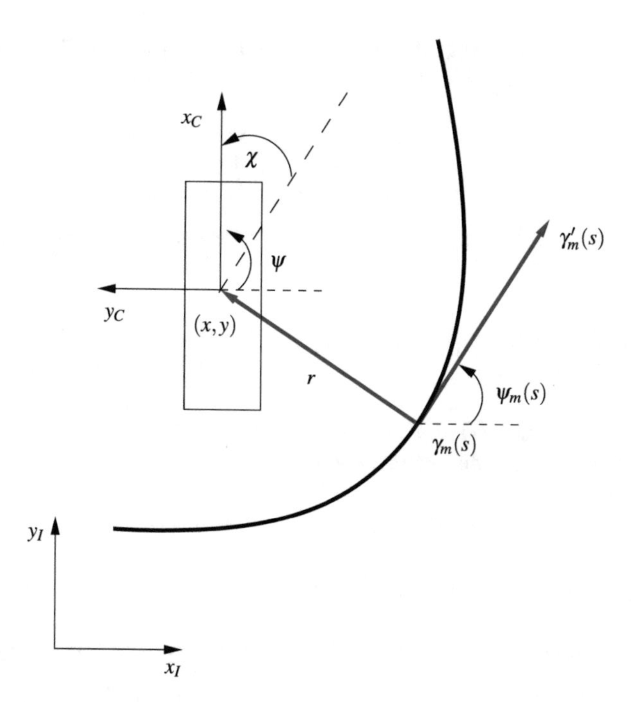

Fig. 14 Curvilinear coordinates

Moreover, we have

$$\gamma_m'(s) = \begin{pmatrix} x_m'(s) \\ y_m'(s) \end{pmatrix} = \begin{pmatrix} \cos \psi_m(s) \\ \sin \psi_m(s) \end{pmatrix} \quad \text{with} \quad \psi_m(s) = \arctan\left(\frac{y_m'(s)}{x_m'(s)}\right)$$

and the curvature of γ_m is given by

$$\kappa_m(s) = \psi_m'(s) = x_m'(s)y_m''(s) - y_m'(s)x_m''(s).$$

We now represent any point (x, y), which is given in the Cartesian coordinate system, by

$$\begin{pmatrix} x \\ y \end{pmatrix} = \begin{pmatrix} x_m(s) \\ y_m(s) \end{pmatrix} + n_m(s)r$$

using coordinates $s \in [0, L]$ and $r \in \mathbb{R}$, compare Fig. 14.

The vector $n_m(s)$ denotes the normal (to the left) vector

$$n_m(s) = \gamma_m''(s) = \begin{pmatrix} -y_m'(s) \\ x_m'(s) \end{pmatrix} = \begin{pmatrix} -\sin \psi_m(s) \\ \cos \psi_m(s) \end{pmatrix},$$

which is perpendicular to the curve's velocity vector γ_m' at s.

Suppose the car is driving in the (s, r) coordinate system with velocity v. The car's position is given by the curve $\gamma_c : [0, T] \longrightarrow \mathbb{R}^2$ with

$$\gamma_c(t) := \begin{pmatrix} x(t) \\ y(t) \end{pmatrix}.$$

The curve γ_c is supposed to be parameterized with respect to time t. At time t we then have the relation

$$\begin{pmatrix} x(t) \\ y(t) \end{pmatrix} = \begin{pmatrix} x_m(s(t)) \\ y_m(s(t)) \end{pmatrix} + n_m(s(t))r(t) = \begin{pmatrix} x_m(s(t)) - r(t)y_m'(s(t)) \\ y_m(s(t)) + r(t)x_m'(s(t)) \end{pmatrix} \tag{3.19}$$

with $s(t) \in [0, L]$ and $r(t) \in \mathbb{R}$. Let ψ denote the yaw angle of the vehicle in the Cartesian (x, y)-coordinate system. Then

$$x'(t) = v(t)\cos\psi(t) \qquad \text{and} \qquad y'(\ell) = v(t)\sin\psi(t). \tag{3.20}$$

Differentiation of (3.19) and exploiting (3.20) yields

$$v(t)\cos\psi(t) = \left(x_m'(s(t)) - r(t)y_m''(s(t))\right)s'(t) - r'(t)y_m'(s(t)), \tag{3.21}$$

$$v(t)\sin\psi(t) = \left(y_m'(s(t)) + r(t)x_m''(s(t))\right)s'(t) + r'(t)x_m'(s(t)). \tag{3.22}$$

Multiplication of the first equation by $x_m'(s(t))$ and the second by $y_m'(s(t))$, adding them, and solving for $s'(t)$ yields the differential equation

$$s'(t) = \frac{v(t)\left(x_m'(s(t))\cos\psi(t) + y_m'(s(t))\sin\psi(t)\right)}{1 - r(t)\kappa_m(s(t))}.$$

Likewise, we find

$$r'(t) = v(t)\left(-y_m'(s(t))\cos\psi(t) + x_m'(s(t))\sin\psi(t)\right).$$

Setting $\chi(t) := \psi(t) - \psi_m(s(t))$, exploiting $x_m'(s(t)) = \cos\psi_m(s(t))$ and $y_m'(s(t)) = \sin\psi_m(s(t))$ yields the system of differential equations

$$\begin{aligned}
s'(t) &= \frac{v(t)\cos\chi(t)}{1 - r(t)\kappa_m(s(t))}, \\
r'(t) &= v(t)\sin\chi(t), \\
\chi'(t) &= \psi'(t) - \kappa_m(s(t))s'(t) \\
&= v(t)\left(u(t) - \frac{\kappa_m(s(t))\cos\chi(t)}{1 - r(t)\kappa_m(s(t))}\right),
\end{aligned}$$

where the control $u(t)$ is the curvature of the curve γ_c at time t.

Remark 3.2 Please note that there is a one-to-one relation between the curvature u of the driving path and the steering angle δ of the car. In a kinematic car model the yaw angle ψ satisfies the differential equation $\psi'(t) = \frac{v(t)}{\ell} \tan \delta(t)$, where ℓ is the distance between front and rear axle. Since we also have $\psi'(t) = v(t)u(t)$, we find $\delta(t) = \arctan(\ell \cdot u(t))$.

Now, we prescribe a sufficiently smooth path that the car is supposed to follow. For simplicity we choose the midline as the reference path. This leads to the algebraic constraint

$$r(t) = 0.$$

Differentiating twice yields

$$\begin{aligned} 0 &= r'(t) = v(t) \sin \chi(t), \\ 0 &= r''(t) = v'(t) \sin \chi(t) + v(t)\chi'(t) \cos \chi(t) \\ &= v'(t) \sin \chi(t) + v(t)^2 \left(u(t) - \frac{\kappa_m(s(t)) \cos \chi(t)}{1 - r(t)\kappa_m(s(t))} \right) \cos \chi(t). \end{aligned}$$

If $v \neq 0$ then the former yields $\sin \chi(t) = 0$ and the latter can be solved for u with

$$u(t) = \frac{\kappa_m(s(t)) \cos \chi(t)}{1 - r(t)\kappa_m(s(t))} = \kappa_m(s(t)) \cos \chi(t). \tag{3.23}$$

Hence, u can be considered an algebraic state and the (differentiation) index of the DAE is three. A consistent initial value for the DAE has to satisfy $r(t_0) = 0$, $\chi(t_0) = k\pi$, $k \in \mathbb{Z}$, if $v(t_0) \neq 0$.

Remark 3.3 Note that this dynamic inversion control might not work in the presence of additional control constraints of type $u(t) \in U$, since the control is fixed by the dynamic inversion and it depends on the prescribed trajectory whether or not it can be controlled subject to control constraints.

In essence the above DAE approach (or dynamic inversion approach) can be exploited to design a controller for systems that are called differentially flat in the control community, compare [24, 37, 45]. Herein, the task is to create a feedback control law that moves the system back to the desired reference trajectory if perturbations occur. Let the desired trajectory be given by $r_d \equiv 0$, $\chi_d \equiv 0$, $s_d' = v_d$.

Let $y := r$ be the observed quantity. Then, $y' = r' = v \sin \chi$ and $y'' = r'' = v' \sin \chi + v^2 \left(u - \frac{\kappa_m(s) \cos \chi}{1 - r\kappa_m(s)} \right) \cos \chi$. Thus

$$u = \frac{1}{v^2 \cos \chi} \left(r'' - v' \sin \chi \right) + \frac{\kappa_m(s) \cos \chi}{1 - r\kappa_m(s)}$$

The feedback-law is then given by

$$u = \frac{1}{v_d^2 \cos \chi_d} \left(r_d'' - k_1(r' - r_d') - k_2(r - r_d) - v_d' \sin \chi_d \right) + \frac{\kappa_m(s_d) \cos \chi_d}{1 - r_d \kappa_m(s_d)},$$

$$= \frac{1}{v_d^2} \left(-k_1 r' - k_2 r \right) + \kappa_m(s_d),$$

compare [45]. Note that u coincides with the algebraic state u in (3.23) if no perturbations in r and s occur. The constants k_1 and k_2 have to be chosen such that the closed-loop system is asymptotically stable.

Example 3.4 We apply the controller for the racing track of Hockenheim and modify the above controller by adding control bounds, i.e. we use the feedback controller

$$u = \max \left\{ -u_{max}, \min \left\{ u_{max}, \frac{1}{v_d^2} \left(-k_1 r' - k_2 r \right) + \kappa_m(s_d) \right\} \right\}$$

with $k_1 = 2$, $k_2 = 0.5$, $u_{max} = \tan(\delta_{max})/\ell$, where $\delta_{max} = 0.3$ [rad] denotes the maximum steering angle and $\ell = 4$ [m] the length of the car. In order to simulate measurement errors, the values of r and r' are perturbed by equally distributed errors $\epsilon_r \in [-0.05, 0.05]$ and $\epsilon_{r'} \in [-0.1, 0.1]$, respectively.

Let the desired velocity v_d at arclength s be

$$v_d(s) = \min \left\{ v_{max}, \sqrt{\frac{a_{max}}{|\kappa(s+h)|}} \right\},$$

where $a_{max} = 9.81$ [m/s^2] denotes the maximal lateral acceleration, $v_{max} = 40$ [m/s] the maximal velocity, and $\kappa(s+h)$ is the curvature of the track at arclength $s + h$. Herein, $h = 20$ [m] is a look-ahead distance and s denotes the current arclength position on the track. Moreover we introduce the differential equation

$$v'(t) = \frac{v_d(s(t)) - v(t)}{T}, \qquad v(0) = v_0,$$

in order to model a delay in the tracking of the desired velocity with a constant $T = 1.5$.

Figure 15 shows the controlled drive along the racing track of Hockenheim with initial value $s(0) = 0$, $r(0) = 1$, $\chi(0) = 0.05$, $v(0) = 10$ and time interval [0, 300] with a control frequency of 10 [Hz]. One can see nicely that the velocity controller slows down the car before the bends owing to the lookahead parameter h.

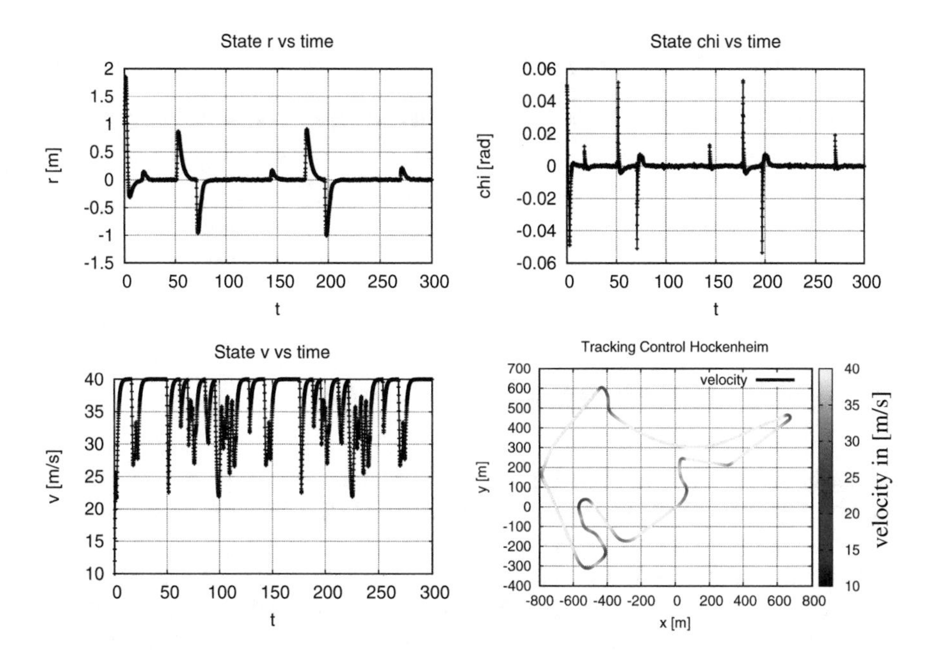

Fig. 15 Controlled motion: offset r from reference line (top left), deviation χ from the track's yaw angle (top right), velocity (bottom left), velocity on the track (bottom right). The colors indicate the velocity in [m/s]

3.3 *Identification of Road Profiles*

In the sequel, we discuss an additional application field: the usage of path constraints for (dynamic) inversion, compare [1, 12]. We consider the case $n_y = n_u$ and system dynamics modeled as an ODE,

$$x'(t) = f(t, x(t), u(t)), \tag{3.24}$$

and system outputs,

$$z(t) := h(t, x(t), u(t)) \in \mathbb{R}^{n_y}. \tag{3.25}$$

Assume additionally that there are measured reference quantities, $z_{REF} : \mathbb{R} \to \mathbb{R}^{n_y}$ and the task is to find control inputs u that lead as an excitation to an exact tracking, i.e.,

$$0 = g(t, x(t), u(t)) := h(t, x(t), u(t)) - z_{REF}(t), \tag{3.26}$$

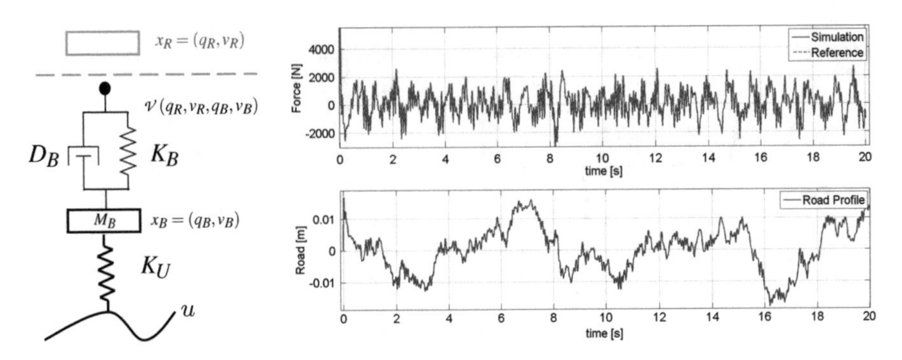

Fig. 16 Left: vertical tire-surrogate model, right: simulation results

with x being the solution of (3.24) given u. The solution strategy followed here is simply adding Eq. (3.26) and solving the resulting DAE in the differential variables x and in the algebraic variables u.

We consider the special case of identifying road profiles using that approach. To that end, assume that there are vehicle measurements available, among others, vertical wheel forces and rim displacements. It is of certain relevance to derive road profiles based on these quantities and information of the measurement vehicle and the tire, since road profiles characterize the road roughness of the traveled route and can serve as (invariant) input quantity, which can be used to excite other vehicle models, possibly different from the model of the measurement vehicle, see [12] for further details and discussions. To solve the task, we consider a tire-surrogate model as depicted in Fig. 16 on the left.

We assume given (measured) rim motion q_R, v_R and consider the vertical dynamic equations

$$\begin{aligned} \dot{q}_B &= v_B, \\ M_B \dot{v}_B &= -F(q_R, v_R, q_B, v_B) - K_U \cdot (q_B - u), \end{aligned} \tag{3.27}$$

together with the control constraint equation (tracking a given vertical force),

$$0 = F(q_R, v_R, q_B, v_B) - F_{REF}(t). \tag{3.28}$$

With a linear force law $F(q_R, v_R, q_B, v_B) = -K_B(q_R - q_B) - D_B(v_R - v_B)$, it is straightforward to verify that the resulting DAE has index 2. In Fig. 16, some results are presented, the DAE has been solved numerically by the implicit Euler method. In the upper plot on the right, the simulation quantity and reference quantity are displayed, they coincide perfectly up to integration tolerance. In the lower plot on the right, the corresponding road profile is shown. For further and more detailed discussions, we refer to our work in [12, 13].

3.4 *Identification of Tyre Loads*

A problem related to the identification of road profiles is the identification of tyre loads. To this end we consider a simplified full vehicle model, see Fig. 17, where vertical and horizontal motion are decoupled, drag forces and rolling resistance forces are neglected.

The equations of motion read as follows, compare [25, 39] for related models:

$$
\begin{aligned}
x' &= v_x \cos\psi - v_y \sin\psi, \qquad y' = v_x \sin\psi + v_y \sin\psi, \qquad z' = v_z,\\
\phi' &= w_\phi, \qquad \kappa' = w_\kappa, \qquad \psi' = w_\psi, \qquad \delta' = w_\delta,\\
mv_x' &= mv_y w_\psi + L_{rl} + L_{rr} + F_{fl}^x + F_{fr}^x,\\
mv_y' &= -mv_x w_\psi + S_{rl} + S_{rr} + F_{fl}^y + F_{fr}^y,\\
mv_z' &= -mg + F_{fl}^{sd} + F_{fr}^{sd} + F_{rl}^{sd} + F_{rr}^{sd},\\
z_*' &= v_*^z, \qquad m_*(v_*^z)' = -m_* g - F_*^{sd} + F_*^r, \qquad * \in \{fl, fr, rl, rr\},\\
J_\phi w_\phi' &= M_{sd}^\phi + h_{rl} S_{rl} + h_{rr} S_{rr} + h_{fl} F_{fl}^y + h_{fr} F_{fr}^y,\\
J_\kappa w_\kappa' &= M_{sd}^\kappa - h_{rl} L_{rl} - h_{rr} L_{rr} - h_{fl} F_{fl}^x - h_{fr} F_{fr}^x,\\
J_\psi w_\psi' &= \ell_f\left(F_{fl}^y + F_{fr}^y\right) - \ell_r\left(S_{rl} + S_{rr}\right) + w_r\left(L_{rr} + F_{fr}^x\right) - w_\ell\left(L_{rl} + F_{fl}^x\right),
\end{aligned}
$$

where

$$
\begin{aligned}
F_{fl}^x &= L_{fl}\cos\delta - S_{fl}\sin\delta, \qquad F_{fr}^x = L_{fr}\cos\delta - S_{fr}\sin\delta,\\
F_{fl}^y &= L_{fl}\sin\delta + S_{fl}\cos\delta, \qquad F_{fr}^y = L_{fr}\sin\delta + S_{fr}\cos\delta,\\
M_{sd}^\phi &= -w_r\left(F_{fr}^{sd} + F_{rr}^{sd}\right) + w_\ell\left(F_{fl}^{sd} + F_{rl}^{sd}\right),\\
M_{sd}^\kappa &= -\ell_f\left(F_{fl}^{sd} + F_{fr}^{sd}\right) + \ell_r\left(F_{rl}^{sd} + F_{rr}^{sd}\right),\\
h_{fl} &= z - \ell_f\sin\kappa + w_\ell\cos\kappa\sin\phi, \qquad h_{fr} = z - \ell_f\sin\kappa - w_r\cos\kappa\sin\phi,\\
h_{rl} &= z + \ell_r\sin\kappa + w_\ell\cos\kappa\sin\phi, \qquad h_{rr} = z + \ell_r\sin\kappa - w_r\cos\kappa\sin\phi,\\
F_*^{sd} &= c_*^w(z_* - h_*) + k_*^w(z_*' - h_*'), \qquad * \in \{fl, fr, rl, rr\}.
\end{aligned}
$$

The meaning of the occurring quantities is summarized in Table 2 and parameters can be found in Table 3. The subscripts fr, fl, rr, rl indicate the location on the car. To this end, f and r in the first position mean front and rear, respectively, and l and r in the second position refer to left and right, respectively.

We are now interested in identifying the tyre loads F_*^r, $* \in \{fl, fr, rl, rr\}$, from measured data ξ_* of the z-positions z_*, $* \in \{fr, fl, rr, rl\}$, of the wheels, i.e. we

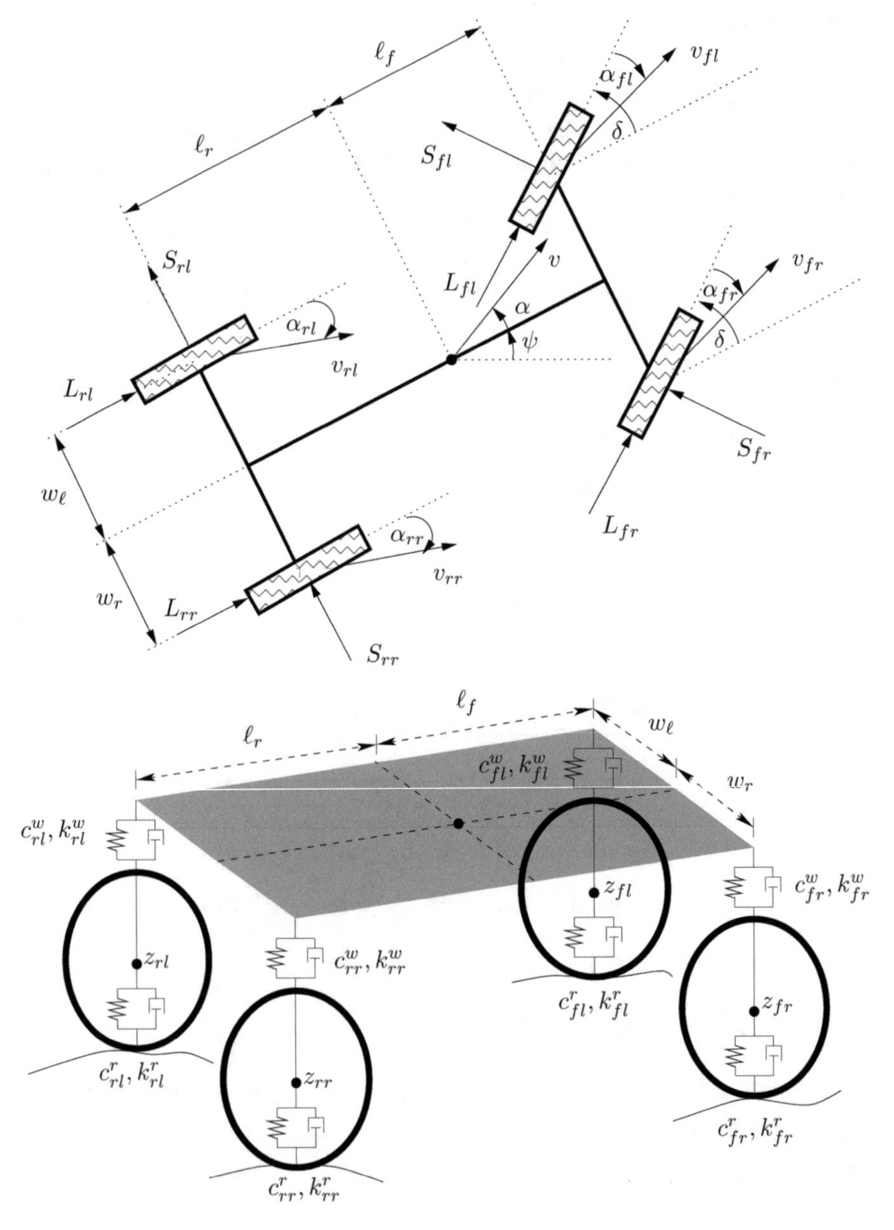

Fig. 17 Simplified full car model: horizontal and vertical geometry

have the algebraic constraints

$$0 = z_* - \xi_*(t), \qquad * \in \{fl, fr, rl, rr\}.$$

Table 2 Variables of the simplified fullcar model

Name	Description
(x, y, z)	Center of gravity of chassis
(ϕ, κ, ψ)	Roll angle, pitch angle, yaw angle of chassis
$z_*^w, * \in \{fr, fl, rr, rl\}$	z-Position of center of gravity of wheels
$v, v_*, * \in \{fr, fl, rr, rl\}$	Velocities at center of gravity and at tyre-road contact
$v_*^z, * \in \{fr, fl, rr, rl\}$	Velocities in z-direction of wheels
$\alpha, \alpha_*, * \in \{fr, fl, rr, rl\}$	Side-slip angles at center of gravity and at tyre-road contact
δ, w_δ	Steering angle and steering angle velocity
$F_*^{sd}, * \in \{fr, fl, rr, rl\}$	Spring/damper forces between wheels and chassis
$F_*^r, * \in \{fr, fl, rr, rl\}$	Tyre forces between road and wheel
$S_*, * \in \{fr, fl, rr, rl\}$	Lateral tyre forces
$L_*, * \in \{fr, fl, rr, rl\}$	Longitudinal tyre forces

Table 3 Parameters of the simplified fullcar model (parameter values are rounded to five decimals)

Name	Value	Description
ℓ_f, ℓ_r	1.0212, 1.4788 [m]	Length from center of gravity to front axle and rear axle
w_ℓ, w_r	0.7182, 0.7818 [m]	Distance from center of gravity to left and right side of vehicle
m	1510 [kg]	Mass of chassis (and passengers)
$m_*, * \in \{fl, fr, rl, rr\}$	50 [kg]	Masses of wheel and brake assemblies
J_ϕ, J_κ, J_ψ	371.95, 887.38, 1047.6 [kg m^2]	Moments of inertia roll, pitch, yaw
$c_*^w, * \in \{fl, fr, rl, rr\}$	25,000 [N/m]	Spring coefficients of spring suspension at wheels
$k_*^w, * \in \{fl, fr, rl, rr\}$	3500 [Ns/m]	Damper coefficients of spring suspension at wheels

For simplicity we assume that the car is driving on a straight line at constant velocity, that is $y \equiv 0\,[\text{m}]$, $\delta \equiv 0$, $v_x \equiv 20\,[\text{m/s}]$, $v_y \equiv 0\,[\text{m/s}]$. Moreover, in this situation the longitudinal and lateral tyre forces vanish, i.e. $L_* \equiv 0$, $S_* \equiv 0$, $* \in \{fl, fr, rl, rr\}$. With these simplifications we obtain the following mechanical multibody system with generalized forces λ_*, $* \in \{fl, fr, rl, rr\}$, which relate to the tyre loads according to $F_*^r = -\lambda_*$, $* \in \{fl, fr, rl, rr\}$:

$$
\begin{aligned}
z' &= v_z, \qquad \phi' = w_\phi, \qquad \kappa' = w_\kappa, \\
mv_z' &= -mg + F_{fl}^{sd} + F_{fr}^{sd} + F_{rl}^{sd} + F_{rr}^{sd}, \quad J_\phi w_\phi' = M_{sd}^\phi, \quad J_\kappa w_\kappa' = M_{sd}^\kappa, \\
z_*' &= v_*^z, \qquad m_*(v_*^z)' = -m_* g - F_*^{sd} - \lambda_*, \qquad * \in \{fl, fr, rl, rr\}, \\
0 &= z_* - \xi_*, \qquad * \in \{fl, fr, rl, rr\}.
\end{aligned}
\tag{3.29}
$$

This DAE is of index three and twofold differentiation of the algebraic constraints (3.29) yields

$$\lambda_* = -m_* g - F_*^{sd} - m_* \xi_*'', \qquad * \in \{fl, fr, rl, rr\}.$$

Herein, we assumed that the measurements ξ_* (or a smooth interpolation or approximation thereof) are twice continuously differentiable. Figure 18 shows the results for the drive with a duration of 10 [s] on a bumpy road segment (bumps on the left and right with offset) with initial values $z(0) = 0.46571$, $v_z(0) = 0$, $\phi(0) = -0.00934$, $w_\phi(0) = 0$, $\kappa(0) = 0.0242$, $w_\kappa(0) = 0$, $z_*(0) = 0.27$, $v_*^z(0) = 0$, $* \in \{fl, fr, rl, rr\}$. The wheel measurements with a maximum absolute excitation of 0.02 [m] are modeled by

$$\xi_{fl}(t) = z(0) + \xi(t), \qquad \xi_{fr}(t) = z(0) - \xi(t-1),$$

$$\xi_{rl}(t) = z(0) + \xi(t - (\ell_f + \ell_r)/v_x), \quad \xi_{rr}(t) = z(0) - \xi(t - 1 - (\ell_f + \ell_r)/v_x),$$

with

$$\xi(t) = \begin{cases} 0.01 \left(\cos\left(\frac{2\pi}{0.1}(t - 2.5 \cdot k)\right) - 1\right), & \text{if } t \in [2.5 \cdot k, 2.5 \cdot k + 0.1], k = 1, 2, 3, \\ 0, & \text{otherwise.} \end{cases}$$

The generalized forces λ_* are related to the tyre loads F_*^r by $F_*^r = -\lambda_*$ for $* \in \{fl, fr, rl, rr\}$ and Fig. 18 shows that the generalized forces are negative and consequently the tyre loads are positive. Thus, there is road contact at all times. If the excitation ξ_* is increased then it may happen that the tyre loads become negative, which is not meaningful physically. Thus, the presented model is restricted to non-negative tyre loads.

4 Piecewise Defined DAEs and Dockings

Consider two independent multibody systems with generalized coordinates q_1 and q_2, respectively. The equations of motion are given by

$$M_j(q_j) q_j'' = f_j(q_j, q_j', u_j), \qquad j = 1, 2. \tag{4.1}$$

For simplicity, we neglect algebraic constraints, but emphasize that adding them is straightforward. Let us assume that the two systems move independently in the time interval $[t_0, t_c]$. The time point t_c is supposed to be a docking or coupling time point, where both systems turn into a (physically or virtually) coupled system with permanent contact described by the coupling constraints

$$0 = g(q_1(t), q_2(t)) \qquad t \geq t_c.$$

Throughout we assume that at least one of the Jacobians g_{q_j}', $j = 1, 2$, has full rank.

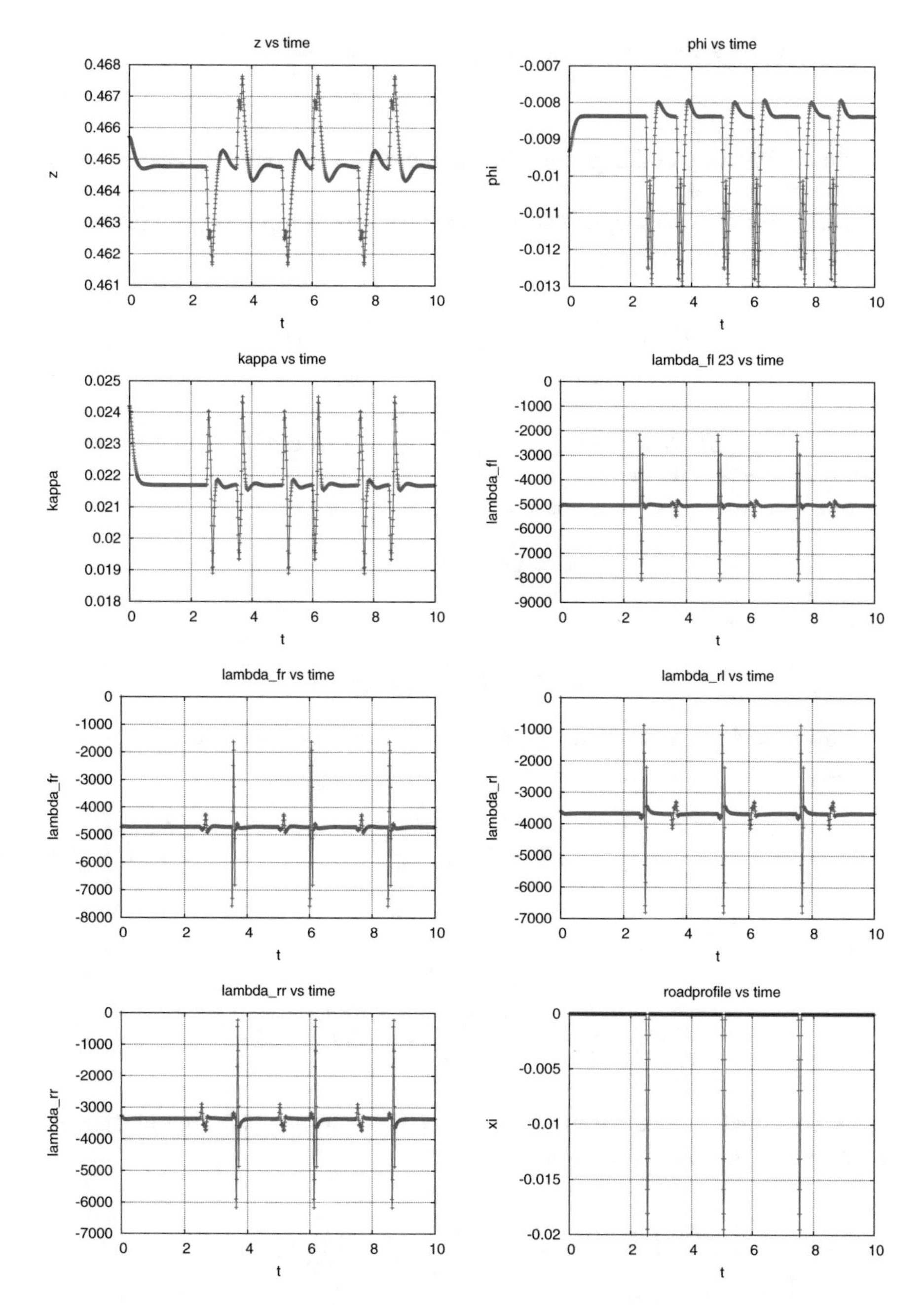

Fig. 18 Selection of differential and algebraic states: z, roll angle ϕ, pitch angle κ, λ_*, $* \in \{fl, fr, rl, rr\}$, $\xi(t)$ (from top left to bottom right). The plot at bottom right shows the road profile $\xi(t)$ with a maximum absolute excitation of 0.02 [m]

Often, the coupling constraints exhibit a separated structure

$$g(q_1, q_2) := g_1(q_1) - g_2(q_2),$$

where, for instance, g_1 and g_2 describe the coordinates of contact points on system 1 and 2, respectively.

Then, for $t \geq t_c$ the coupled system obeys the combined equations of motion

$$\begin{aligned} M_1(q_1)q_1'' &= f_1(q_1, q_1', u_1) - g_{q_1}'(q_1, q_2)^\top \lambda, \\ M_2(q_2)q_2'' &= f_2(q_2, q_2', u_2) - g_{q_2}'(q_2, q_2)^\top \lambda, \\ 0 &= g(q_1, q_2). \end{aligned}$$

Herein, the generalized forces $g_{q_j}'(q_1, q_2)^\top \lambda$, $j = 1, 2$, can be viewed as contact forces that apply at the contact point.

At $t = t_c$ we assume the consistency conditions

$$0 = g(q_1(t_c), q_2(t_c))$$

and

$$0 = g_{q_1}'[t_c]q_1'(t_c) + g_{q_2}'[t_c]q_2'(t_c).$$

Differentiating the algebraic constraint twice and suppressing arguments in g yields

$$0 \equiv \sum_{j=1}^{2} g_{q_j}' \cdot q_j' \tag{4.2}$$

and

$$0 \equiv \sum_{j=1}^{2} \left(g_{q_j}' \cdot q_j'' + \sum_{\ell=1}^{2} \left(g_{q_j}' \cdot q_j' \right)_{q_\ell}' \cdot q_\ell' \right). \tag{4.3}$$

In the separated case the mixed second derivatives vanish in (4.3). Introducing the differential equations in (4.3) yields

$$\begin{aligned} 0 &\equiv \sum_{j=1}^{2} \left(g_{q_j}' M_j^{-1} \left(f_j - (g_{q_j}')^\top \lambda \right) + \sum_{\ell=1}^{2} \left(g_{q_j}' \cdot q_j' \right)_{q_\ell}' \cdot q_\ell' \right) \\ &= \sum_{j=1}^{2} \left(g_{q_j}' M_j^{-1} f_j + \sum_{\ell=1}^{2} \left(g_{q_j}' \cdot q_j' \right)_{q_\ell}' \cdot q_\ell' \right) - \left(\sum_{j=1}^{2} g_{q_j}' M_j^{-1} (g_{q_j}')^\top \right) \lambda. \end{aligned} \tag{4.4}$$

Since M_j, $j = 1, 2$, are supposed to be symmetric and positive definite and since at least one of the Jacobians g'_{q_j}, $j = 1, 2$, is supposed to be of full rank, the equation can be solved for λ. Hence, the index of the coupled system is three.

A natural application of this type occurs in satellite docking maneuvers, see, e.g., [38], for instance if a service satellite approaches a target object and connects physically to it at a defined docking point.

A physical docking will cause internal coupling forces, which essentially have to be compensated by the mechanical parts, i.e., the material. One may also be interested in non-physical couplings. In this case, internal forces vanish as there is no real connection of the mechanical parts, that is, $\lambda = 0$. Then, the algebraic constraints have to be obeyed by the controls and we are in a similar situation as in Examples 3.1 and 3.2. The following example considers a physical connection between bodies but aims to minimize the internal forces by choosing the controls appropriately.

Example 4.1 We now consider two mobile robotic platforms of type (4.1) with generalized coordinates $q_j = (x_j, y_j, \psi_j)^\top$, controls $u_j = (f_{x,j}, f_{y,j}, u_{\omega,j})^\top$, mass matrices $M_j = diag(m_j, m_j, J_{z,j})$, and generalized force vectors

$$f_j(q_j, q_j', u_j) = \begin{pmatrix} f_{x,j} \cos \psi_j - f_{y,j} \sin \psi_j \\ f_{x,j} \sin \psi_j + f_{y,j} \cos \psi_j \\ u_{\omega,j} \end{pmatrix}$$

for $j = 1, 2$. Herein, (x_j, y_j) is the position of robot j, ψ_j its yaw angle, m_j its mass, and $J_{z,j}$ its moment of inertia. The control u_j consists of the forces $f_{x,j}$ and $f_{y,j}$ and the yaw rate $u_{\omega,j}$.

Let the robots be coupled by the algebraic constraint

$$g(q_1, q_2) = \begin{pmatrix} x_1 - \ell \sin \psi_1 \\ y_1 + \ell \cos \psi_1 \\ \psi_1 \end{pmatrix} - \begin{pmatrix} x_2 \\ y_2 \\ \psi_2 \end{pmatrix}.$$

where $\ell > 0$ defines a docking point on the y-axis of the robot's local coordinate system. The Jacobian of g reads

$$g'_{(q_1,q_2)}(q_1, q_2) = \left(\begin{array}{ccc|ccc} 1 & 0 & -\ell \cos \psi_1 & -1 & 0 & 0 \\ 0 & 1 & -\ell \sin \psi_1 & 0 & -1 & 0 \\ 0 & 0 & 1 & 0 & 0 & -1 \end{array}\right).$$

The coupled motion of the two robots is described by the DAE

$$m_1 x_1'' = f_{x,1} \cos \psi_1 - f_{y,1} \sin \psi_1 - \lambda_1, \tag{4.5}$$

$$m_1 y_1'' = f_{x,1} \sin \psi_1 + f_{y,1} \cos \psi_1 - \lambda_2, \tag{4.6}$$

$$J_{z,1} \psi_1'' = u_{\omega,1} - (\lambda_3 - \ell\lambda_1 \cos \psi_1 - \ell\lambda_2 \sin \psi_1), \tag{4.7}$$

$$m_2 x_2'' = f_{x,2} \cos \psi_2 - f_{y,2} \sin \psi_2 + \lambda_1, \tag{4.8}$$

$$m_2 y_2'' = f_{x,2} \sin \psi_2 + f_{y,2} \cos \psi_2 + \lambda_2, \tag{4.9}$$

$$J_{z,2} \psi_2'' = u_{\omega,2} + \lambda_3, \tag{4.10}$$

$$0 = x_1 - \ell \sin \psi_1 - x_2, \tag{4.11}$$

$$0 = y_1 + \ell \cos \psi_1 - y_2, \tag{4.12}$$

$$0 = \psi_1 - \psi_2. \tag{4.13}$$

The constraints on velocity level in (4.2) read

$$\begin{aligned} 0 &= x_1' - \ell \psi_1' \cos \psi_1 - x_2', \\ 0 &= y_1' - \ell \psi_1' \sin \psi_1 - y_2', \\ 0 &= \psi_1' - \psi_2', \end{aligned}$$

and those on acceleration level in (4.3) are given by

$$\begin{aligned} 0 &= x_1'' - \ell \psi_1'' \cos \psi_1 + \ell (\psi_1')^2 \sin \psi_1 - x_2'', \\ 0 &= y_1'' - \ell \psi_1'' \sin \psi_1 - \ell (\psi_1')^2 \cos \psi_1 - y_2'', \\ 0 &= \psi_1'' - \psi_2''. \end{aligned}$$

The non-singular matrix $\left(\sum_{j=1}^{2} g'_{q_j} M_j^{-1} (g'_{q_j})^\top \right)$ in (4.4) reads

$$\begin{pmatrix} \frac{1}{m_1} + \frac{\ell^2}{J_{z,1}} \cos^2 \psi_1 + \frac{1}{m_2} & \frac{\ell^2}{J_{z,1}} \sin \psi_1 \cos \psi_1 & -\frac{\ell}{J_{z,1}} \cos \psi_1 \\ \frac{\ell^2}{J_{z,1}} \sin \psi_1 \cos \psi_1 & \frac{1}{m_1} + \frac{\ell^2}{J_{z,1}} \sin^2 \psi_1 + \frac{1}{m_2} & -\frac{\ell}{J_{z,1}} \sin \psi_1 \\ -\frac{\ell}{J_{z,1}} \cos \psi_1 & -\frac{\ell}{J_{z,1}} \sin \psi_1 & \frac{1}{J_{z,1}} + \frac{1}{J_{z,2}} \end{pmatrix}.$$

Please note that the choice of the controls will influence the magnitude of the internal coupling forces in the above model (the coupling will persist owing to the algebraic constraints). Hence one may search for controls that perform a desired task while minimizing the internal forces. Figure 19 shows the result of a two-phase optimal control problem with the data $m = m_1 = m_2 = 10\,[\text{kg}]$, $J_{z,1} = J_{z,2} = m(a^2 + b^2)/12$, $a = 0.58\,[\text{m}]$, $b = 0.376\,[\text{m}]$, $\ell = 0.476\,[\text{m}]$. In the first phase, robot 2 moves from a given initial position to a target position next to robot 1, which is fixed at a given position during the first phase. The path of the second robot is colored in blue. The second phase starts at $t_c \approx 7.19495\,[\text{s}]$ (which is the end of phase one) and aims to move the coupled robots to a terminal position within $10\,[\text{s}]$. The path of the first robot is colored in red, the one of the second robot in blue. The following two optimal control problems are used to realize both phases:

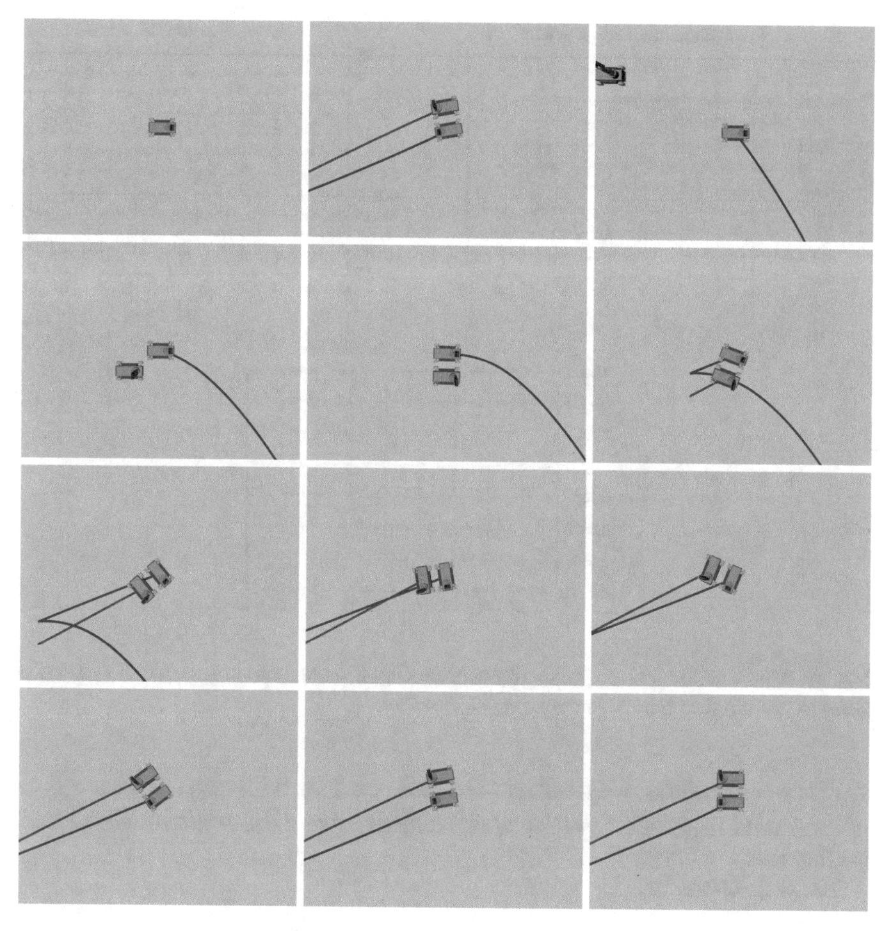

Fig. 19 Snapshots of the motion of the two mobile robot for phase one (rows 1 and 2) and phase two (rows 3 and 4). The lines indicate the history of the paths of the two robots

Phase 1: Minimize

$$t_c + \int_0^{t_c} \|u_1(t)\|^2 + \|u_2(t)\|^2 \, dt$$

subject to (4.5)–(4.10) with $\lambda_1 = \lambda_2 = \lambda_3 = 0$, $u_1 = 0$, $u_2 \in [-10, 10] \times [-10, 10] \times [-\pi/2, \pi/2]$, *initial and terminal conditions* $q_1(0) = v_1(0) = (0, 0, 0)^\top$, $q_2(0) = (5, -5, 0)^\top$, $v_2(0) = v_2(t_c) = (0, 0, 0)^\top$, $q_2(t_c) = (0, 0.476, 0)^\top$, *and the state constraint*

$$\left(\frac{x_2(t)}{3}\right)^2 + y_2(t) \geq r^2 \qquad \text{withr} = 0.45.$$

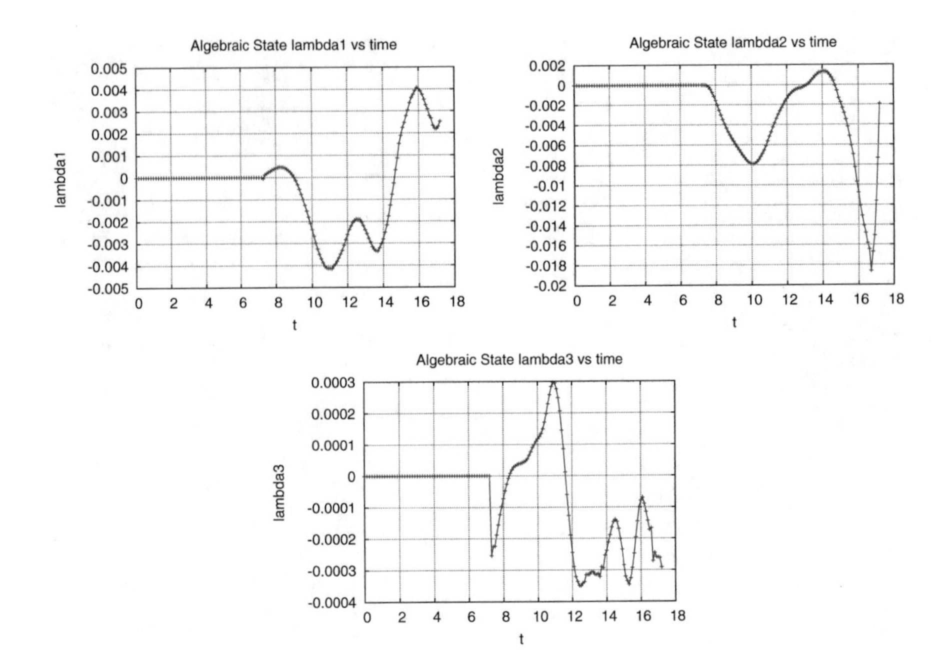

Fig. 20 Algebraic states λ_1, λ_2, and λ_3 for both phases with contact point $t_c \approx 7.19495$ [s]. These states are the physical contact forces acting on the robots

The state constraint serves as an anti-collision constraint as it restricts the positions of the second robot to the outside of an ellipsoid around the origin which contains the first robot.

Phase 2: Minimize

$$100 \int_{t_c}^{t_c+10} \lambda_1(t)^2 + \lambda_2(t)^2 + \lambda_3(t)^2 \, dt + 0.1 \int_{t_c}^{t_c+10} \|u_1(t)\|^2 + \|u_2(t)\|^2 \, dt$$

subject to (4.5)–(4.13), $u_1, u_2 \in [-10, 10] \times [-10, 10] \times [-\pi/2, \pi/2]$, *initial and terminal conditions* $q_1(t_c) = v_1(t_c) = v_2(t_c) = (0, 0, 0)^\top$, $q_2(t_c) = (0, 0.476, 0)^\top$, $q_1(t_c + 10) = (10, 5, -\pi)^\top$, $v_1(t_c + 10) = (0, 0, 0)^\top$.

Figures 20 and 21 show the algebraic states λ_1, λ_2, λ_3, i.e. the physical contact forces acting on the robots, and the controls u_1, u_2 for both phases. The numerical results have been obtained with the software OCPID-DAE1 [28].

Remark 4.1 Piecewise defined DAEs with varying index occur as well in the indirect solution approach for optimal control problems in the presence of control and state constraints. Herein, a multipoint boundary value problem has to be solved. Depending on the sequence of active and inactive constraints and their order the index of the DAE may change.

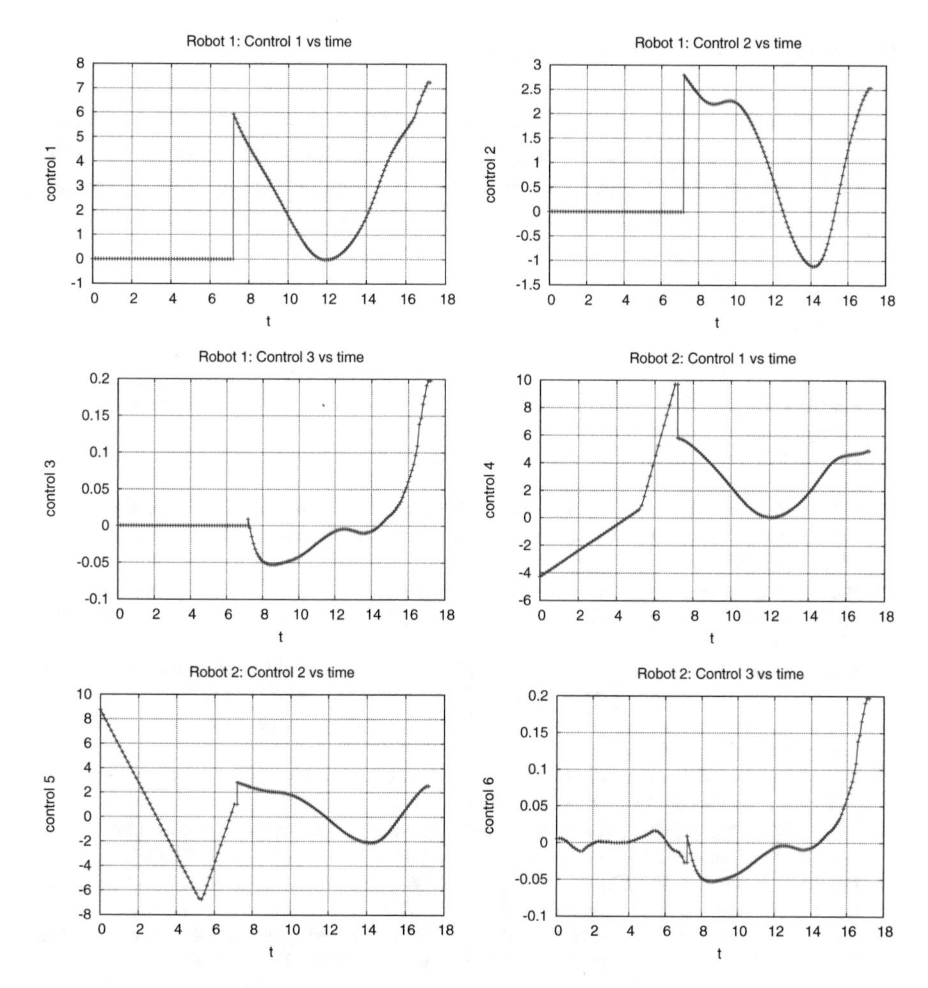

Fig. 21 Controls u_1 of robot 1 (top row and second row left) and u_2 of robot 2 (second row right and bottom row) for phases one and two with contact point $t_c \approx 7.19495$ [s]

5 Conclusions

The paper discusses selected applications in mobile robotics and vehicle dynamics. To this end DAEs arise by imposing path constraints for the robots. Herein, suitable control components need to be identified which serve as algebraic states. Moreover, higher index DAEs arise very naturally for such type of problems, sometimes with a mixed index. The remaining controls (if any) can be chosen such that an optimal performance is achieved which leads to a DAE optimal control problem. Although the paper is restricted to typical applications in mobile robotics, similar problems arise in flight path optimization, docking maneuvers for satellites etc.

6 Source Codes for Examples

Source codes for the numerical examples are attached to this paper. The additional libraries `OCPID-DAE1` and `sqpfiltertoolbox` are required to run the examples. These libraries are freely available for download on http://www.optimal-control.de after registration on the website.

References

1. Altmann, R., Heiland, J.: Simulation of multibody systems with servo constraints through optimal control. Multibody Syst. Dyn. **40**, 75–98 (2016)
2. Arnold, M.: DAE aspects of multibody system dynamics. In: Surveys in Differential-Algebraic Equations IV, pp. 41–106. Springer, Cham (2017)
3. Arnold, M., Burgermeister, B., Führer, C., Hippmann, G., Rill, G.: Numerical methods in vehicle system dynamics: state of the art and current developments. Veh. Syst. Dyn. **49**(7), 1159–1207 (2011)
4. Bastos, G., Seifried, R., Brüls, O.: Inverse dynamics of serial and parallel underactuated multibody systems using a DAE optimal control approach. Multibody Syst. Dyn. **30**(3), 359–376 (2013)
5. Bastos, G., Seifried, R., Brüls, O.: Analysis of stable model inversion methods for constrained underactuated mechanical systems. Mech. Mach. Theory **111**, 99–117 (2017)
6. Blajer, W.: Dynamics and control of mechanical systems in partly specified motion. J. Franklin Inst. **334B**(3), 407–426 (1997)
7. Blajer, W., Kołodziejczyk, K.: Control of underactuated mechanical systems with servo-constraints. Nonlinear Dyn. **50**(4), 781–791 (2007)
8. Blajer, W., Kołodziejczyk, K.: A geometric approach to solving problems of control constraints: theory and a DAE framework. Multibody Syst. Dyn. **11**(4), 343–364 (2004)
9. Blajer, W., Seifried, R., Kołodziejczyk, K.: Servo-constraint realization for underactuated mechanical systems. Arch. Appl. Mech. **85**, 1191–1207 (2015)
10. Brenan, K. E.: Numerical simulation of trajectory prescribed path control problems by the backward differentiation formulas. IEEE Trans. Autom. Control **31**, 266–269 (1986)
11. Brenan, K.E., Campbell, S.L., Petzold, L.R.: Numerical Solution of Initial-Value Problems in Differential-Algebraic Equations. Classics in Applied Mathematics, vol. 14. SIAM, Philadelphia (1996)
12. Burger, M.: Optimal control of dynamical systems: calculating input data for multibody system simulation. Ph.D. thesis, Fachbereich Mathematik, Technische Universität Kaiserslautern (2011)
13. Burger, M.: Calculating road input data for vehicle simulation. Multibody Syst. Dyn. **31**(1), 93–110 (2014)
14. Burger, M., Gerdts, M.: A survey on numerical methods for the simulation of initial value problems with DAEs. In: Surveys in Differential-Algebraic Equations IV, pp. 221–300. Springer, Cham (2017)
15. Campbell, S.L., Kunkel, P.: Completions of nonlinear DAE flows based on index reduction techniques and their stabilization. J. Comput. Appl. Math. **233**, 1021–1034 (2009)
16. Campbell, S.L., Kunkel, P.: General Nonlinear Differential Algebraic Equations and Tracking Problems. DAE Forum (2018, to appear)
17. Daoutidis, P.: DAEs in model reduction of chemical processes: an overview. In: Surveys in Differential-Algebraic Equations II, pp. 69–102. Springer, Cham (2015)

18. Diepolder, J., Bittner, M., Pipreck, P., Grüter, B., Holzapfel, F.: Facilitating aircraft optimal control based on numerical nonlinear dynamic inversion. In: 25th Mediterranean Conference on Control and Automation (MED), Valletta, Malta, July 3–6 2017
19. Duff, I.S., Gear, C.W.: Computing the structural index. SIAM J. Algebraic Discret. Methods **7**(4), 594–603 (1986)
20. Eich-Soellner, E., Führer, C.: Numerical Methods in Multibody Dynamics. B.G. Teubner, Stuttgart (1998)
21. Esterl, B.: Modulare echtzeitfähige Simulation des Fahrzeug-Mehrkörpersystems. Dissertation, TU Graz (2011)
22. Esterl, B., Keßler, T.: Kinematic coupling methods for the modular real-time simulation of the vehicle multi-body system. In: Proceedings of the Multibody Dynamics 2011, Eccomas Thematic Conference (2011)
23. Esterl, B., Butz, T., Simeon, B., Burgermeister, B.: Real-time capable vehicle-trailer coupling by algorithms for differential-algebraic equations. Veh. Syst. Dyn. **45**, 819–834 (2007)
24. Fliess, M., Levine, J.L., Martin, P., Rouchon, P.: Flatness and defect of non-linear systems: introductory theory and examples. Int. J. Control. **61**(6), 1327–1361 (1995)
25. Frasch, J.: Parallel algorithms for optimization of dynamic systems in real-time. Ph.D. thesis, Arenberg Doctoral School, Faculty of Engineering Science, KU Leuven (2014)
26. Gear, C.W.: Differential-algebraic equation index transformations. SIAM J. Sci. Stat. Comput. **9**, 39–47 (1988)
27. Gerdts, M.: Optimal control of ODEs and DAEs. Walter de Gruyter, Berlin/Boston (2012)
28. Gerdts, M.: OCPID-DAE1 – optimal control and parameter identification with differential-algebraic equations of index 1. Technical Report, User's Guide, Engineering Mathematics, Department of Aerospace Engineering, University of the Federal Armed Forces at Munich (2013). http://www.optimal-control.de
29. Gerdts, M.: A survey on optimal control problems with differential-algebraic equations. In: Surveys in Differential-Algebraic Equations II, pp. 103–161. Springer, Cham (2015)
30. Hairer, E., Wanner, G.: Solving Ordinary Differential Equations II: Stiff and Differential-Algebraic Problems, vol. 14, 2nd edn. Springer Series in Computational Mathematics, Berlin (1996)
31. Hairer, E., Lubich, Ch., Roche, M.: The numerical solution of differential-algebraic systems by Runge-Kutta methods. Lecture Notes in Mathematics, vol. 1409. Springer, Berlin (1989)
32. Kunkel, P., Mehrmann, V.: Regular solutions of nonlinear differential-algebraic equations and their numerical determination. Numer. Math. **79**, 581–600 (1998)
33. Kunkel, P., Mehrmann, V.: Differential-Algebraic Equations. Analysis and Numerical Solution. European Mathematical Society Publishing House, Zürich. viii, 377, pp. EUR 58.00 (2006)
34. Kunkel, P., Mehrmann, V., Seufer, I.: GENDA: A Software Package for the Numerical Solution of GEneral Nonlinear Differential-Algebraic Equations. Preprint 730-02, Institut für Mathematik, TU Berlin (2002)
35. Lamour, R., März, R., Tischendorf, C.: Differential-Algebraic Equations: A Projector Based Analysis. Differential-Algebraic Equations Forum. Springer, Berlin (2013)
36. Lot, R., Biral, F.: A curvilinear abscissa approach for the lap time optimization of racing vehicles. In: IFAC Proceedings, vol. 19, pp. 7559–7565 (2014)
37. Martin, P., Murray, R.M., Rouchon, P.: Flat systems, equivalence and trajectory generation. Technical report, CDS Technical Report (2003). http://www.cds.caltech.edu/~murray/papers/2003d_mmr03-cds.html
38. Michael, J., Chudej, K., Gerdts, M., Pannek, J.: Optimal rendezvous path planning to an uncontrolled tumbling target. In: Proceedings of the 19th IFAC Symposium on Automatic Control in Aerospace, September 2–6, 2013, University of Würzburg, Würzburg, Germany
39. Michael, J., Gerdts, M.: Pro-active optimal control for semi-active vehicle suspension based on sensitivity updates. Veh. Syst. Dyn. **53**(12), 1721–1741 (2015)
40. Otto, S., Seifried, R.: Real-time trajectory control of an overhead crane using servo-constraints. Multibody Syst. Dyn. **42**, 1–17 (2017)
41. Popp, K., Schiehlen, W.: Ground Vehicle Dynamics. Springer, Berlin (2010)

42. Rill, G.: Road Vehicle Dynamics: Fundamentals and Modeling. Taylor & Francis Group, Abingdon (2011)
43. Rill, G., Chucholowski, C.: Real time simulation of large vehicle systems. In: Proceedings of the Multibody Dynamics 2007, Eccomas Thematic Conference (2007)
44. Roberson, R.E., Schwertassek, R.: Dynamics of Multibody Systems. Springer, Berlin (1988)
45. Rotella, F., Carrillo, F.J., Ayadi, M.: Polynomial controller design based on flatness. Kybernetika **38**(5), 571–584 (2002)
46. Shabana, A.: Computational Dynamics, 2nd edn. Wiley, New York (2001)
47. Simeon, B., Führer, C. , Rentrop, P.: Differential-algebraic equations in vehicle system dynamics. Surv. Math. Ind. **1**, 1–37 (1991)
48. Simeon, B., Grupp, F., Führer, C., Rentrop, P.: A nonlinear truck model and its treatment as a multibody system. J. Comput. Appl. Math. **50**(1–3), 523–532 (1994)
49. Schwerin, R.v.: MultiBody System SIMulation. Lecture Notes in Computational Science and Engineering, vol. 7 (Springer, Berlin, 1999)
50. Trenn, S.: Solution concepts for linear DAEs: a survey. In: Ilchmann, A., Reis, T. (eds.) Surveys in Differential-Algebraic Equations I. Differential-Algebraic Equations Forum, pp. 137–172. Springer, Berlin (2013)
51. Woernle, C.: Mehrkörpersysteme. Springer, Berlin (2011)
52. Wojtyra, M., Fraczek, J.: Comparison of selected methods of handling redundant constraints in multibody systems simulations. J. Comput. Nonlinear Dyn. **8**(2), 021007 (1–9) (2012)
53. Wojtyra, M., Fraczek, J.: Joint reactions in rigid or flexible body mechanisms with redundant constraints. Bull. Pol. Acad. Sci. Tech. Sci. **60**(3), 617–626 (2012)
54. Wojtyra, M., Fraczek, J.: Solvability of reactions in rigid multibody systems with redundant nonholonomic constraints. Multibody Syst. Dyn. **30**(2), 153–171 (2013)

Open-Loop Control of Underactuated Mechanical Systems Using Servo-Constraints: Analysis and Some Examples

Svenja Otto and Robert Seifried

Abstract A classical trajectory tracking control approach combines feedforward control with a feedback loop. Since both parts can be designed independently, this is called a two degree of freedom control structure. Feedforward control is ideally an inverse model of the system. In case of underactuated mechanical systems the inverse model often cannot be derived analytically, or the derivation cannot follow a systematic approach. Then, the numerical approach based on servo-constraints has shown to be effective. In this approach, the equations of motion are appended by algebraic equations constraining the output to follow a specified output trajectory, representing the servo-constraints. The arising differential-algebraic equations (DAEs) are solved for the desired open-loop control input. An additional feedback loop stabilizes the system around the specified trajectories. This contribution reviews the use of servo-constraints in mechanical open-loop control problems. Since the arising set of DAEs is usually of higher index, index reduction and analysis methods are reviewed for flat as well as non-flat systems. Some typical examples are given and numerical results are presented.

Keywords Feedforward Control · Inverse model · Servo-constraints · Underactuated System

Mathematics Subject Classification (2010) 70E55, 70E60, 70Q05

S. Otto (✉) · R. Seifried
Institute of Mechanics and Ocean Engineering, Hamburg University of Technology, Hamburg, Germany
e-mail: svenja.otto@tuhh.de; robert.seifried@tuhh.de

S. Campbell et al. (eds.), *Applications of Differential-Algebraic Equations: Examples and Benchmarks*, Differential-Algebraic Equations Forum,
https://doi.org/10.1007/11221_2018_4

1 Introduction

A popular controller design for mechanical systems is a two degree of freedom control structure. In this approach, a feedforward and a feedback controller are designed independently for tracking control and disturbance rejection [30]. The feedforward part is responsible for tracking, while the feedback part stabilizes the system around the specified output trajectory. A feedforward controller can be designed in terms of an inverse model of the system. This is straightforward in case of fully actuated systems. However, for underactuated systems, this task becomes more difficult. Underactuated systems have less control inputs than degrees of freedom. For some special cases, called differentially flat systems, an algebraic solution of the inverse model exists. However, there is no systematic approach to find such an algebraic solution. In addition, for general underactuated multibody systems, the inverse model might be a dynamical system itself. The analytical derivation of such an inverse model might be possible using the Byrnes/Isidori input-output normal form [22] for multibody systems formulated in generalized coordinates. Recently, the Byrnes/Isidori form has been generalized for multibody systems formulated in redundant coordinates in [5]. However, this is often burdensome as the equations tend to become complex even for simple systems.

The method of servo-constraints was proposed to avoid such problems and gives a straightforward numerical representation of the inverse model. It is a relatively new approach introduced in [8]. It is also called control, path or program constraints, see e.g. [13]. A general framework for the use of servo-constraints has been introduced in [9] and is extended in [11]. The equations of motion of a multibody system are extended by constraints enforcing the system output to stay on a specified output trajectory. The inverse model is then the numerical solution to the set of arising differential-algebraic equations (DAEs). As part of the numerical solution, the open-loop control input is obtained. Moreover, the desired state trajectories are part of the solution. A state feedback controller can then extend the feedforward control loop to stabilize the system around the trajectory. Thereby, the desired state trajectories are helpful in calculating the current tracking error for state feedback control.

So far, the servo-constraints approach is mainly applied to differentially flat systems. A system is differentially flat if its system input can be expressed as a function of the system output and a finite number of its derivatives [17]. Typical examples of flat systems controlled by servo-constraints are rotary cranes [6], overhead gantry cranes [9, 24] and 2-mass-spring systems or their generalization to infinitely many masses [1, 18]. In case of differentially flat systems, the inverse model is purely algebraic and can be used as a reference solution for the numerical solution obtained by servo-constraints.

In case of non-flat systems, the inverse model itself is a dynamical system. This is closely related to the internal dynamics in nonlinear control theory [22, 26]. Thus, stability of the arising internal dynamics has to be considered. Its stability can be analyzed in terms of zero dynamics [22, 26, 31]. Zero dynamics is the nonlinear extension of transmission zeros in linear systems, as for example shown in

[32]. Zero dynamics allows a simplified stability analysis of the internal dynamics. Systems with asymptotically stable zero dynamics are called minimum phase systems, systems with unstable zero dynamics are non-minimum phase systems. First results of applying servo-constraints to non-flat systems are presented in [23] and [27]. In [27], results for minimum phase systems are presented, whereby the arising DAEs are solved by forward time integration. In case of unstable internal dynamics with a hyperbolic equilibrium point, stable inversion proposed in [15] can be applied [27]. This requires the solution of a boundary value problem, which is solved by a multiple shooting algorithm for a planar serial manipulator in [12]. Instead of solving for the internal dynamics and solving the boundary value problem, a reformulation to an optimization problem was proposed for a planar serial manipulator in [4]. Alternatively to solving the non-minimum phase system, the output can be redefined to obtain a minimum phase system [31].

Besides fully actuated or underactuated systems, servo-constraints have been applied to overactuated multibody systems with more independent controls than degrees of freedom. In this case, there is no unique solution to the inverse model problem and other requirements can be enforced. For example, minimizing the acceleration energy of a multibody system compared to its free motion is proposed in [3] to obtain a new set of servo-constraints.

Adding servo-constraints to the system dynamics usually results in a set of higher index DAEs [13]. Since higher index DAEs are more difficult to solve numerically, index reduction methods are used to reduce the differentiation index and therefore numerical complexity [19]. Index reduction by means of a projection onto the constrained and unconstrained directions is proposed in [9]. Index reduction by minimal extension is proposed in [2, 7].

Common DAE solving algorithms applicable for higher index DAEs are presented in [19]. Due to its simplicity, a fixed step implicit Euler scheme is proposed for servo-constraints problems in [9]. To ensure real-time applicability, a linear implicit Euler scheme is compared to an implicit Euler scheme in [24]. First experimental results and real-time capability of the servo-constraints approach are also shown in [24] for an overhead crane. In order to avoid the numerical issues arising with high-index DAEs, a reformulation of the problem as an optimization problem is proposed in [1]. The constraints are considered in the cost function. Thus, small errors in tracking are allowed in favor for smaller control inputs and relaxed smoothness conditions on the desired trajectory.

This contribution reviews the concept of servo-constraints for use in open-loop control of multibody systems. In Sect. 2, trajectory control and the modeling of mechanical systems are briefly reviewed. The servo-constraints approach, its solution approaches and index reduction methods are shown in Sect. 3. Various examples and simulation results are presented and analyzed in detail in Sect. 4. These include the mass-on-car system originally introduced in [28], an extended mass-on-car system and a mass-spring-damper chain with a finite number of masses. Moreover, the influence of different actuator models on the resulting DAE system is demonstrated. The results are summarized in Sect. 5.

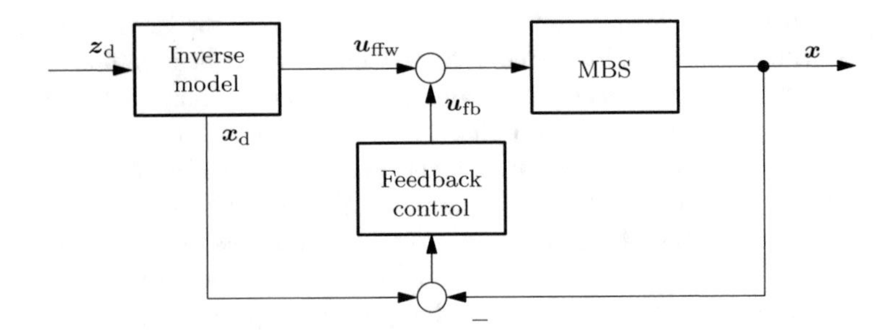

Fig. 1 Two degree of freedom control approach for multibody systems (MBS)

2 Trajectory Control and Modeling of Multibody Systems

For trajectory tracking control of multibody systems, a two degree of freedom control structure is an efficient control method [30]. In this case, a feedforward control input $\boldsymbol{u}_{\text{ffw}}$ and a feedback control input $\boldsymbol{u}_{\text{fb}}$ are designed to track a desired trajectory z_{d}, as visualized in Fig. 1. The feedforward control part is responsible for large motion tracking and is here based on an inverse model of the system. If there were no modeling errors or disturbances, the inverse model would cancel the model dynamics exactly, and there would be perfect tracking. However, there are usually some modeling errors and disturbances and a feedback control loop is necessary on top of the feedforward part. In order to reduce noise effects and to reduce corrections of the feedback path, an accurate inverse model is fundamental. For state feedback it is helpful that the inverse model not only provides the feedforward control input $\boldsymbol{u}_{\text{ffw}}$, but also the reference state trajectory $\boldsymbol{x}_{\text{d}}$. With the reference trajectory $\boldsymbol{x}_{\text{d}}$, the state tracking error is computed and used for state feedback control. Many different state feedback controllers are applicable. Simple feedback control strategies such as linear quadratic regulators might be sufficient due to small open-loop tracking errors with an accurate inverse model [24]. In case a prescribed tracking error performance is desired, a funnel controller might be applied, see e.g. [21].

In order to design the inverse model, a model of the multibody system is introduced first. Systems with n degrees of freedom, m inputs and m outputs are considered. Underactuated systems are included for which hold $n > m$. Two formulations, based on minimal coordinates as well as redundant coordinates are introduced. The following presentations are restricted to holonomic multibody systems, which represent the most important ones in engineering applications.

2.1 *Generalized Coordinates*

For multibody systems without kinematic loops, it is possible to select a set of generalized coordinates $\boldsymbol{y} \in \mathbb{R}^n$ and generalized velocities $\boldsymbol{v} \in \mathbb{R}^n$ and to project the Newton and Euler equations of each body onto the direction of free motion. Due to Jourdain's principle, this projection eliminates the reaction forces and yields a set of $2\,n$ ordinary differential equations (ODEs)

$$\dot{\boldsymbol{y}} = \boldsymbol{Z}(\boldsymbol{y})\,\boldsymbol{v} \tag{2.1}$$

$$\boldsymbol{M}(\boldsymbol{y},t)\,\dot{\boldsymbol{v}} = \boldsymbol{q}(\boldsymbol{y},\boldsymbol{v},t) + \boldsymbol{B}\boldsymbol{u} \tag{2.2}$$

with the kinematics matrix $\boldsymbol{Z} : \mathbb{R}^n \to \mathbb{R}^{n\times n}$, the invertible generalized mass matrix $\boldsymbol{M} : \mathbb{R}^n \times \mathbb{R} \to \mathbb{R}^{n\times n}$, the vector of generalized forces $\boldsymbol{q} : \mathbb{R}^n \times \mathbb{R}^n \times \mathbb{R} \to \mathbb{R}^n$ and the input distribution matrix $\boldsymbol{B} \in \mathbb{R}^{n\times m}$. Note that the functions $\boldsymbol{Z}$, $\boldsymbol{M}$ and $\boldsymbol{q}$ are assumed to be continuously differentiable. In most cases the generalized velocities are simply chosen as $\boldsymbol{v} = \dot{\boldsymbol{y}}$, thus $\boldsymbol{Z}$ is the identity matrix. However, the general case is used here such that generalized coordinates and redundant coordinates can be treated in a unified manner. For underactuated systems, the equations of motion are sometimes separated to

$$\begin{bmatrix} \dot{\boldsymbol{y}}_\mathrm{a} \\ \dot{\boldsymbol{y}}_\mathrm{u} \end{bmatrix} = \begin{bmatrix} \boldsymbol{Z}_\mathrm{aa} & \boldsymbol{Z}_\mathrm{au} \\ \boldsymbol{Z}_\mathrm{ua} & \boldsymbol{Z}_\mathrm{uu} \end{bmatrix} \begin{bmatrix} \boldsymbol{v}_\mathrm{a} \\ \boldsymbol{v}_\mathrm{u} \end{bmatrix} \tag{2.3}$$

$$\begin{bmatrix} \boldsymbol{M}_\mathrm{aa} & \boldsymbol{M}_\mathrm{au} \\ \boldsymbol{M}_\mathrm{ua} & \boldsymbol{M}_\mathrm{uu} \end{bmatrix} \begin{bmatrix} \dot{\boldsymbol{v}}_\mathrm{a} \\ \dot{\boldsymbol{v}}_\mathrm{u} \end{bmatrix} = \begin{bmatrix} \boldsymbol{q}_\mathrm{a} \\ \boldsymbol{q}_\mathrm{u} \end{bmatrix} + \begin{bmatrix} \boldsymbol{B}_\mathrm{a} \\ \boldsymbol{0} \end{bmatrix} \boldsymbol{u}\,, \tag{2.4}$$

where the indices a and u refer to actuated and unactuated parts respectively. These equations have $2\,n$ unknowns, namely the generalized coordinates $\boldsymbol{y}$ and velocities $\boldsymbol{v}$. Introducing the state vector

$$\boldsymbol{x} = \begin{bmatrix} \boldsymbol{x}_1 \\ \boldsymbol{x}_2 \end{bmatrix} = \begin{bmatrix} \boldsymbol{y} \\ \boldsymbol{v} \end{bmatrix} \in \mathbb{R}^{2n}\,, \tag{2.5}$$

the ODE formulation in state space form is

$$\dot{\boldsymbol{x}} = \begin{bmatrix} \boldsymbol{Z}\,\boldsymbol{x}_2 \\ \boldsymbol{M}^{-1}\,\boldsymbol{q} \end{bmatrix} + \begin{bmatrix} \boldsymbol{0} \\ \boldsymbol{M}^{-1}\boldsymbol{B} \end{bmatrix} \boldsymbol{u}\,. \tag{2.6}$$

The system output $\boldsymbol{z} : \mathbb{R}^n \to \mathbb{R}^m$ of the multibody system is mostly chosen on position level, providing

$$\boldsymbol{z} = \boldsymbol{z}(\boldsymbol{y})\,. \tag{2.7}$$

In case the multibody system includes kinematic loops, it is usually not possible to select a set of minimal coordinates $\boldsymbol{y}$. Then, the kinematic loop can be cut. It is then possible to select minimal coordinates for the resulting open chain, providing the dynamic equations (2.1)–(2.2). In order to close the loop again, loop closing constraints have to be added, yielding a DAE similar to the use of redundant coordinates.

2.2 Redundant Coordinates

Sometimes, it might be advantageous to select redundant coordinates to simplify the modeling process. For a system with n degrees of freedom and n_c geometric constraints $\boldsymbol{c}_\mathrm{g}(\boldsymbol{y}) : \mathbb{R}^{n+n_\mathrm{c}} \rightarrow \mathbb{R}^{n_\mathrm{c}}$ let now $\boldsymbol{y} \in \mathbb{R}^{n+n_\mathrm{c}}$ be the vector of redundant coordinates and $\boldsymbol{v} \in \mathbb{R}^{n+n_\mathrm{c}}$ the vector of redundant velocities. Then, the equations of motion in DAE form are

$$\dot{\boldsymbol{y}} = \boldsymbol{Z}(\boldsymbol{y})\, \boldsymbol{v} \tag{2.8}$$

$$\boldsymbol{M}(\boldsymbol{y}, t)\, \dot{\boldsymbol{v}} = \boldsymbol{q}(\boldsymbol{y}, \boldsymbol{v}, t) + \boldsymbol{C}_\mathrm{g}^\mathrm{T}(\boldsymbol{y})\, \boldsymbol{\lambda} + \boldsymbol{B}\boldsymbol{u} \tag{2.9}$$

$$\boldsymbol{c}_\mathrm{g}(\boldsymbol{y}) = \boldsymbol{0}\,, \tag{2.10}$$

where the geometric constraints $\boldsymbol{c}_\mathrm{g}(\boldsymbol{y})$ are enforced by the Lagrange multipliers $\boldsymbol{\lambda} \in \mathbb{R}^{n_\mathrm{c}}$ which represent the generalized reaction forces of the multibody system. Equations (2.8)–(2.10) are a set of DAEs and have $2n + 3n_\mathrm{c}$ unknowns, namely $\boldsymbol{y}$, $\boldsymbol{v}$ and $\boldsymbol{\lambda}$ and the same number of equations. Due to the invertibility of the mass matrix $\boldsymbol{M}$, one can show that the set of DAEs (2.8)–(2.10) is of differentiation index 3 for general multibody systems. The differentiation index of a DAE system is defined as the maximum number of differentiations of the algebraic constraint necessary to transform the set of equations to a set of ordinary differential equations [14, 19]. Besides the most often used differentiation index, there are other index concepts. See [14] for a discussion and comparison of the index concepts for general nonlinear DAEs. Both the ODE formulation of Eqs. (2.1)–(2.2) and the DAE formulation of Eqs. (2.8)–(2.10) can be used in the modeling process of multibody systems. For the following model inversion using servo-constraints, the underlying modeling approach for multibody systems is secondary. For the inverse model, a DAE will always be obtained, irrespectively of the modeling approach. Therefore, for a unified treatment of servo-constraints, there is no distinction made in notation for redundant coordinates or generalized coordinates.

3 Servo-Constraints Approach

For trajectory tracking of the output z, the inverse model of the multibody system is sought. The input to the inverse model is the desired output trajectory $z_\mathrm{d}(t)$ of the actual multibody system, while the inverse model provides the necessary input $\boldsymbol{u}_\mathrm{d} = \boldsymbol{u}_\mathrm{ffw}$ for the multibody system to follow the desired trajectory, see Fig. 1.

For fully actuated systems where each degree of freedom can be controlled with an independent control input, the inverse model can be split into inverse kinematics and inverse dynamics [29]. In the inverse kinematics, the desired state trajectory $\boldsymbol{x}_\mathrm{d}$ is determined from the desired output trajectory z_d, while the inverse dynamics provides the system input $\boldsymbol{u}_\mathrm{d}$ which is necessary to generate the desired state motion $\boldsymbol{x}_{\mathrm{d}}$. For underactuated systems with more degrees of freedom than independent control inputs, this separation is not possible and both problems must be solved simultaneously. The servo-constraints approach is a method to solve for the inverse model numerically.

Motivated from the DAE representation of the multibody dynamics in Eqs. (2.8)–(2.10), the equations of motion are appended to include constraints which enforce the output z to be equal to the desired output trajectory z_{d}. This extension is possible for either the ODE representation of Eqs. (2.1)–(2.2) in minimal coordinates or the DAE representation of Eqs. (2.8)–(2.10) in redundant coordinates. This yields a new set of equations

$$\dot{\boldsymbol{y}} = \boldsymbol{Z}(\boldsymbol{y})\,\boldsymbol{v} \tag{3.1}$$

$$\boldsymbol{M}(\boldsymbol{y},t)\,\dot{\boldsymbol{v}} = \boldsymbol{q}(\boldsymbol{y},\boldsymbol{v},t) + \boldsymbol{C}_{\mathrm{g}}^{\mathrm{T}}(\boldsymbol{y})\,\boldsymbol{\lambda} + \boldsymbol{B}\boldsymbol{u} \tag{3.2}$$

$$\boldsymbol{c}_{\mathrm{g}}(\boldsymbol{y}) = \boldsymbol{0} \tag{3.3}$$

$$\boldsymbol{c}(\boldsymbol{y},t) = \boldsymbol{z}(\boldsymbol{y}) - \boldsymbol{z}_{\mathrm{d}}(t) = \boldsymbol{0} \tag{3.4}$$

with servo-constraints $\boldsymbol{c}(\boldsymbol{y},t) : \mathbb{R}^n \times \mathbb{R} \rightarrow \mathbb{R}^m$. The set of equations has $2n + 3n_{\mathrm{c}} + m$ unknowns, namely the coordinates $\boldsymbol{y}$ and $\boldsymbol{v}$, the Lagrange multipliers $\boldsymbol{\lambda}$ and the inputs $\boldsymbol{u}$. In the following, it is assumed that the servo-constraints are compatible with the geometric constraints. Thus, motion that is already constrained by the geometric constraints cannot be forced on a desired trajectory that is not compatible with the geometric constraints. In case of minimal coordinates, the constraints $\boldsymbol{c}_{\mathrm{g}}$ and the Lagrange multipliers $\boldsymbol{\lambda}$ vanish. Thus, the servo-constraints are automatically compatible with the geometric constraints. The set of DAEs must be solved to yield the feedforward control input $\boldsymbol{u}_{\mathrm{d}}$. Preferably, this should be done in real-time to allow trajectories to be altered in real-time.

The set of equations including the additional constraints $\boldsymbol{c}(\boldsymbol{y})$ remains structurally similar to the forward dynamics of Eqs. (2.8)–(2.10). While for the forward dynamics, the geometric constraints $\boldsymbol{c}_{\mathrm{g}}(\boldsymbol{y})$ are enforced by the Lagrange multipliers $\boldsymbol{\lambda}$, in the inverse model case the servo-constraints $\boldsymbol{c}(\boldsymbol{y})$ are enforced by the system inputs $\boldsymbol{u}$. Thus, there are some similarities between the terms $\boldsymbol{B}\boldsymbol{u}$ and $\boldsymbol{C}_{\mathrm{g}}^{\mathrm{T}}\boldsymbol{\lambda}$. The generalized reaction forces act orthogonal to the tangent of the constraint manifold and the system is in a so-called ideal orthogonal realization. However, in contrast to the generalized reaction forces $\boldsymbol{C}_{\mathrm{g}}^{\mathrm{T}}\boldsymbol{\lambda}$, the system inputs $\boldsymbol{B}\boldsymbol{u}$ are not necessarily perpendicular to the tangent space of the constraint manifold and different configurations can be distinguished [9]. An inverse model can be in ideal and non-ideal orthogonal and tangential configuration, which is visualized in Fig. 2. In case of multi-input-multi-output (MIMO) systems, there can be orthogonal as well as tangential configurations for individual inputs, yielding a mixed tangential-orthogonal configuration.

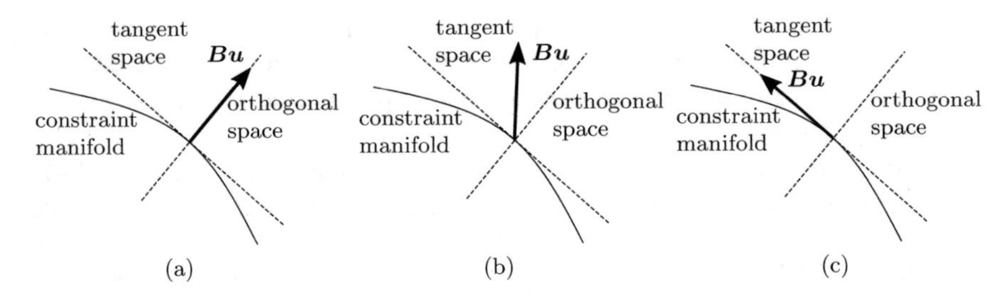

Fig. 2 Possible realizations of the system input $\boldsymbol{Bu}$ with respect to the tangent of the constraint manifold defined by servo-constraints $\boldsymbol{c}(\boldsymbol{y})$. (**a**) Orthogonal configuration. (**b**) Non-ideal orthogonal configuration. (**c**) Tangential configuration

Following nonlinear control theory methods [22] and differentiating the output z from Eq. (2.7) twice, yields

$$\dot{z} = \widetilde{\boldsymbol{H}}(\boldsymbol{y})\dot{\boldsymbol{y}} = \underbrace{\widetilde{\boldsymbol{H}}(\boldsymbol{y})\,\boldsymbol{Z}(\boldsymbol{y})}_{\boldsymbol{H}}\,\boldsymbol{v} \tag{3.5}$$

$$\ddot{z} = \boldsymbol{H}(\boldsymbol{y})\dot{\boldsymbol{v}} + \bar{z}(\boldsymbol{y}, \boldsymbol{v}), \tag{3.6}$$

where the matrix $\widetilde{\boldsymbol{H}} = \frac{\partial z(\boldsymbol{y})}{\partial \boldsymbol{y}} : \mathbb{R}^n \to \mathbb{R}^{m\times(n+n_c)}$ describes the Jacobian of the output and all other derivatives are collected in $\bar{z} = \dot{\boldsymbol{H}}\boldsymbol{v}$. Due to invertibility of the mass matrix $\boldsymbol{M}$, the accelerations $\dot{\boldsymbol{v}}$ are substituted by the dynamic equation (2.9) yielding

$$\ddot{z} = \boldsymbol{H}\boldsymbol{M}^{-1}\left(\boldsymbol{q} + \boldsymbol{B}\,\boldsymbol{u}\right) + \bar{z}. \tag{3.7}$$

From this follows the relation between input $\boldsymbol{u}$ and output z as

$$\boldsymbol{H}\boldsymbol{M}^{-1}\boldsymbol{B}\,\boldsymbol{u} = -\boldsymbol{H}\boldsymbol{M}^{-1}\boldsymbol{q} + \ddot{z} - \bar{z}. \tag{3.8}$$

The system input $\boldsymbol{u}$ is therefore connected to the output acceleration $\ddot{z}$ through the matrix

$$\boldsymbol{P} = \boldsymbol{H}\boldsymbol{M}^{-1}\boldsymbol{B}. \tag{3.9}$$

The rank p of the matrix $\boldsymbol{P} \in \mathbb{R}^{m\times m}$ is used as a measure of system configuration [9, 28]. Note that the matrix $\boldsymbol{P}$ corresponds to the matrix $\boldsymbol{C}_{\mathrm{g}}\boldsymbol{Z}\boldsymbol{M}^{-1}\boldsymbol{C}_{\mathrm{g}}^{\mathrm{T}}$ for geometric constraints which is of full rank for nonredundant geometric constraints.

In case of full rank $p = m$, the system input $\boldsymbol{u}$ can directly reach all components of the output and the inverse model is of ideal or non-ideal orthogonal realization. For orthogonal realizations, the arising DAEs are of index 3, similar to classical geometrically constrained systems. This can be seen by differentiating the output equation one more time and solving for $\dot{\boldsymbol{u}}$. In case of rank deficiency $0 < p < m$, only p components of the output are directly actuated by the input and there is a mixed tangential-orthogonal realization. In case of $p = 0$, no system output can be directly actuated and the inverse model is in tangential realization. Note that this case is characteristic for differentially flat systems but not restricted to them [28]. For the configurations with singular matrix $\boldsymbol{P}$, the differentiation index of the inverse model DAE is of index larger than 3.

3.1 Relationship Between Relative Degree and Differentiation Index of Inverse Model DAEs

The differentiation index as defined in Sect. 2.2 of the servo-constraints DAEs (3.1)–(3.4) is closely related to the relative degree from nonlinear control theory. For nonlinear systems, the relative degree is presented e.g. in [22, 31]. For single-input-single-output (SISO) systems, the relative degree is the number of differentiations of the output until the system input appears for the first time. For MIMO systems, this concept can be extended to a vector relative degree, see [31] for details. If the relative degree r of a system equals the number of states in state space ($2n$ for common multibody systems), the system is input-state linearizable. This is typical for the case of fully actuated systems, e.g. industrial robots. In case that the relative degree r is smaller than the number of states, the system is input-output linearizable. Thus, a linear relationship between input and output is possible by coordinate transformation and state feedback. However, there remains dynamics which is not observable from the new input-output relationship and is therefore called internal dynamics [31].

The internal dynamics is usually nonlinear and difficult to analyze. Therefore, the concept of zero dynamics is applied, see e.g. [31]. Holding the system output identically to zero at all time, $z(t) = \boldsymbol{0} \; \forall t$, provides the zero dynamics. Local stability of the zero dynamics can be analyzed by using for example Lyapunov's indirect method. For tracking problems, the local exponential stability of the zero dynamics guarantees stability of the internal dynamics if the desired output trajectory and its first $r-1$ derivatives are small in magnitude [31]. For stabilization problems with $z_{\mathrm{d}}(t) = 0 \; \forall t$, the result can be relaxed, so that local asymptotic stability of the zero dynamics guarantees local asymptotic stability of the internal dynamics [31].

The relative degree r and zero dynamics are nonlinear extensions of the pole excess and transmission zeros of linear systems, see e.g. [32]. For linear controllable and observable SISO systems in state space form

$$\dot{\boldsymbol{x}} = \boldsymbol{A}\,\boldsymbol{x} + \boldsymbol{b}\,u \tag{3.10}$$

$$z = \boldsymbol{C}\,\boldsymbol{x} + d\,u\,, \tag{3.11}$$

the relative degree equals the difference between the number of poles n_p and number of zeros n_z [32]. Thus, it holds

$$r = n_\mathrm{p} - n_\mathrm{z}\,. \tag{3.12}$$

The poles of a linear system in form of Eqs. (3.10)–(3.11) are the eigenvalues of the state space matrix $\boldsymbol{A}$. Thus, for mechanical systems without integral type force laws the number of poles is $n_\mathrm{p} = 2n$. The transmission zeros are extracted from the transfer function

$$G(s) = \boldsymbol{C}\,(s\boldsymbol{I} - \boldsymbol{A})^{-1}\,\boldsymbol{b} + d = \frac{p_\mathrm{d}(s)}{p_\mathrm{n}(s)} \tag{3.13}$$

of system (3.10)–(3.11), where s is the variable in frequency domain and $\boldsymbol{I}$ is the identity matrix of respective size. The transmission zeros are the zeros of the denominator polynomial $p_\mathrm{d}(s)$. For linear systems, the eigenvalues of the zero dynamics correspond exactly to the position of the transmission zeros. Linear analysis can be applied for linear systems to verify the zero dynamics and to determine stability of the internal dynamics. Note that the inverse model is also represented by the inverse transfer function $G(s)^{-1}$. Thus, with no transmission zeros in the transfer function $G(s)$, the inverse will not have any dynamics and is just an algebraic inverse model. If there are zeros, the inverse model features dynamics itself.

The relationship between the relative degree and the differentiation index for servo-constraints problems in form of Eqs. (3.1)–(3.4) has been analyzed in [13] for various mechanical examples. It was shown that the differentiation index is the relative degree plus one for the case that the system dynamics or at least the internal dynamics is modeled as an ODE. Otherwise, the differentiation index can be larger than $r + 1$, since the relative degree only concerns the input-output behavior, but the differentiation index concerns the complete DAE system (3.1)–(3.4). In the complete DAE system, it is possible for the internal dynamics to be modeled by a higher index DAE.

3.2 *Index Reduction and Analysis Methods*

The inverse model represented by the differential algebraic equations (3.1)–(3.4) is generally nonlinear and its differentiation index might be larger than 3. This complicates numerical treatment and analysis. Index reduction approaches have been proposed to reduce the DAE index, see e.g. [19]. Methods that are used so far in context of servo-constraints are the projection approach [9] and minimal extension [2]. Besides the index reduction methods, further analysis methods are available which are helpful for non-flat systems. In that case, the internal dynamics needs to be extracted from the DAEs and analyzed with respect to its stability. For that reason, a coordinate projection approach was proposed in [28]. Moreover, a reformulation of the model dynamics using redundant coordinates might be helpful for numerical treatment and analysis [11, 24]. In the following, the projection approach, coordinate transformation approach and a reformulation of the problem are briefly reviewed and applied to the examples in Sect. 4.

3.2.1 Projection Approach

For multibody systems in ODE form, the projection approach is presented in [9] and is expanded for systems in DAE form in [11]. Here, the projection for ODE systems given by Eqs. (2.1)–(2.2) is summarized. By defining suitable projection matrices, the equations are projected onto a constrained and an unconstrained subspace which are complementary subspaces. The constrained subspace is orthogonal to the tangent space of the constraint manifold and its projection matrix follows from the Jacobian of the servo-constraints $\boldsymbol{H} : \mathbb{R}^n \to \mathbb{R}^{m\times n}$. The unconstrained subspace is tangential to the constraint manifold and spanned by the matrix $\boldsymbol{G} : \mathbb{R}^n \to \mathbb{R}^{n\times(n-m)}$ obtained from

$$\boldsymbol{H}\boldsymbol{G} = \boldsymbol{0}\,, \quad \boldsymbol{G}^{\mathrm{T}}\boldsymbol{H}^{\mathrm{T}} = \boldsymbol{0}\,. \tag{3.14}$$

Projecting the dynamic equations onto the respective subspaces yields

$$\begin{bmatrix} \boldsymbol{G}^{\mathrm{T}} \\ \boldsymbol{H}\boldsymbol{M}^{-1} \end{bmatrix} \big(\boldsymbol{M}(\boldsymbol{y},t)\,\dot{\boldsymbol{v}} = \boldsymbol{q}(\boldsymbol{y},\boldsymbol{v},t) + \boldsymbol{B}\,\boldsymbol{u}\big)\,. \tag{3.15}$$

The term $\boldsymbol{H}\dot{\boldsymbol{v}}$ is replaced by the output equation (3.6), where also the servo-constraint is substituted to yield $\boldsymbol{H}\dot{\boldsymbol{v}} = \ddot{z}_{\mathrm{d}} - \bar{z}$. By this substitution, the number of differential equations is reduced. Reordering the projected dynamic equations and adding the servo-constraint Eq. (3.4) yields a new set of DAEs

$$\dot{\boldsymbol{y}} = \boldsymbol{Z}\boldsymbol{v} \tag{3.16}$$

$$\boldsymbol{G}^{\mathrm{T}}\boldsymbol{M}\dot{\boldsymbol{v}} = \boldsymbol{G}^{\mathrm{T}}\boldsymbol{q} + \boldsymbol{G}^{\mathrm{T}}\boldsymbol{B}\,\boldsymbol{u} \tag{3.17}$$

$$\boldsymbol{0} = \boldsymbol{H}\boldsymbol{M}^{-1}\boldsymbol{q} + \boldsymbol{H}\boldsymbol{M}^{-1}\boldsymbol{B}\boldsymbol{u} - \ddot{z}_{\mathrm{d}} + \dot{\boldsymbol{H}}\boldsymbol{v} \tag{3.18}$$

$$\boldsymbol{0} = z - z_{\mathrm{d}}\,. \tag{3.19}$$

Note that the dimension of differential equations is reduced. The differential equations (3.16) and (3.17) are of dimension n and $n-m$ respectively, while each of the algebraic equations (3.18) and (3.19) is of dimension m. For the overhead crane example which is often used for demonstration purposes, this projection was sufficient to reduce the DAE index from 5 to 3 and simplify the numerical solution [9, 24].

3.2.2 Coordinate Transformation

For analysis purposes of the system behavior, it might be convenient to apply a coordinate transformation to the equations of motion. With the transformation, ODEs describing the internal dynamics can be extracted and stability in terms of zero dynamics can be analyzed. While the coordinate transformation is here shown for dynamic systems in ODE form, a generalization is possible for systems in DAE form [5]. Motivated by nonlinear control theory, it is useful to rewrite the equations of motion given by Eqs. (2.1)–(2.2) in a new set of coordinates

$$\boldsymbol{y}' = \begin{bmatrix} \boldsymbol{z} \\ \boldsymbol{y}_{\mathrm{u}} \end{bmatrix} = \begin{bmatrix} \boldsymbol{z}(\boldsymbol{y}) \\ \boldsymbol{y}_{\mathrm{u}} \end{bmatrix}, \tag{3.20}$$

which include the system output $\boldsymbol{z}$. For the new velocities, it follows

$$\boldsymbol{v}' = \begin{bmatrix} \dot{\boldsymbol{z}} \\ \boldsymbol{v}_{\mathrm{u}} \end{bmatrix} = \begin{bmatrix} \boldsymbol{H}_{\mathrm{a}} & \boldsymbol{H}_{\mathrm{u}} \\ \boldsymbol{0} & \boldsymbol{I} \end{bmatrix} \begin{bmatrix} \boldsymbol{v}_{\mathrm{a}} \\ \boldsymbol{v}_{\mathrm{u}} \end{bmatrix}. \tag{3.21}$$

Note that the output Jacobian $\boldsymbol{H}$ is split into two parts such that $\boldsymbol{H} = \begin{bmatrix} \boldsymbol{H}_{\mathrm{a}} & \boldsymbol{H}_{\mathrm{u}} \end{bmatrix}$, according to actuated and unactuated generalized coordinates. Also note that the submatrix $\boldsymbol{H}_{\mathrm{a}}$ must be invertible for an admissible coordinate transformation [28]. This holds if the output equation $\boldsymbol{z}(\boldsymbol{y})$ depends on all actuated coordinates $\boldsymbol{y}_{\mathrm{a}}$. For the acceleration level holds accordingly

$$\dot{\boldsymbol{v}}' = \begin{bmatrix} \ddot{\boldsymbol{z}} \\ \dot{\boldsymbol{v}}_{\mathrm{u}} \end{bmatrix} = \begin{bmatrix} \boldsymbol{H}_{\mathrm{a}} & \boldsymbol{H}_{\mathrm{u}} \\ \boldsymbol{0} & \boldsymbol{I} \end{bmatrix} \begin{bmatrix} \dot{\boldsymbol{v}}_{\mathrm{a}} \\ \dot{\boldsymbol{v}}_{\mathrm{u}} \end{bmatrix} + \begin{bmatrix} \bar{\boldsymbol{z}} \\ \boldsymbol{0} \end{bmatrix}, \tag{3.22}$$

where the acceleration $\dot{\boldsymbol{v}} = \begin{bmatrix} \dot{\boldsymbol{v}}_{\mathrm{a}} & \dot{\boldsymbol{v}}_{\mathrm{u}} \end{bmatrix}^{\mathrm{T}}$ is substituted from Eq. (2.2) to yield

$$\dot{\boldsymbol{v}}' = \begin{bmatrix} \boldsymbol{P}\boldsymbol{u} + \boldsymbol{H}\boldsymbol{M}^{-1}\boldsymbol{q} + \bar{\boldsymbol{z}} \\ [\boldsymbol{0} \vdots \boldsymbol{I}]\boldsymbol{M}^{-1}(\boldsymbol{B}\boldsymbol{u} + \boldsymbol{q}) \end{bmatrix}. \tag{3.23}$$

In these equations, the original coordinates $\boldsymbol{y}_{\mathrm{a}}$, $\boldsymbol{v}_{\mathrm{a}}$ must be substituted by the new coordinates $\boldsymbol{z}$, $\dot{\boldsymbol{z}}$. This involves solving the generally nonlinear equation (3.20) for $\boldsymbol{y}_{\mathrm{a}}$. In Sect. 4.2 it will be shown how this coordinate transformation helps extracting and evaluating the internal dynamics.

3.2.3 Using Redundant Coordinates for Servo-Constraints Approach

A method slightly similar to the coordinate transformation is presented in the following. As was proposed in [11], it might be convenient to derive the system dynamics in DAE form since a DAE system has to be solved anyways for the inverse model. This simplifies the modeling process. In the special case, where the system output z is directly part of the redundant coordinate vector $\boldsymbol{y}$, this also simplifies the solution process. Then, the redundant coordinate vector can be reordered so that the output is represented as

$$z = \begin{bmatrix} \boldsymbol{I} & \boldsymbol{0} \end{bmatrix} \boldsymbol{y}\,, \tag{3.24}$$

where $\boldsymbol{I}$ is an $m \times m$ identity matrix and here $\boldsymbol{0} \in \mathbb{R}^{m \times n-m}$. The servo-constraints $\boldsymbol{c}(\boldsymbol{y})$ then reduce to

$$\boldsymbol{c}(\boldsymbol{y}) = \begin{bmatrix} \boldsymbol{I} & \boldsymbol{0} \end{bmatrix} \boldsymbol{y} - z_{\mathrm{d}}(t) = \boldsymbol{0}\,. \tag{3.25}$$

Then, the servo-constraints and its derivatives

$$\boldsymbol{0} = \begin{bmatrix} \boldsymbol{I} & \boldsymbol{0} \end{bmatrix} \dot{\boldsymbol{y}} - \dot{z}_{\mathrm{d}}\,, \qquad \boldsymbol{0} = \begin{bmatrix} \boldsymbol{I} & \boldsymbol{0} \end{bmatrix} (\boldsymbol{Z}\dot{\boldsymbol{v}} + \dot{\boldsymbol{Z}}\boldsymbol{v}) - \ddot{z}_{\mathrm{d}}\,. \tag{3.26}$$

can be substituted in the set of DAEs (3.1)–(3.4). Thus, the first m equations of the differential equation (3.2) reduce to algebraic instead of differential equations, since $\ddot{z}_{\mathrm{d}}$ is specified by the trajectory. It has been shown in [24], that this substitution reduces the DAE index from 5 to 3 for the overhead crane system.

This analysis method is related to the coordinate transformation because the output is also represented as part of the coordinate vector and is useful for analysis and index reduction purposes.

3.3 DAE Solver

The set of original differential algebraic equations (3.1)–(3.4) or projected equations (3.16)–(3.19) is solved for the desired control input $\boldsymbol{u}_{\mathrm{d}}$ and the desired trajectories of all coordinates. Note that the existence of a solution and especially a unique solution is not guaranteed for general nonlinear DAEs. Also, in order to calculate a solution, the set of initial conditions $\boldsymbol{y}_0$, $\boldsymbol{v}_0$, $\boldsymbol{u}_0$ must be consistent with the constraints (3.3), (3.4) or (3.18), (3.19) respectively. Consistent initial conditions may be determined by solving the static multibody system in the starting equilibrium position. Note that in case the solution is not unique or not existent, the solver might not even notify the user [16]. Therefore, the calculated solution must always be carefully monitored, see e.g. the proposed method in [16]. A solution might not exist in case the desired trajectory cannot be generated by the system states because of e.g. conflicting geometric constraints and servo-constraints. Multiple solutions arise when the desired output trajectory can be reached by different state trajectories. This is for example already the case for a 2-arm manipulator. Redundant multibody systems with more system inputs than degrees of freedom also show this property. In that case, any solution is fine for open-loop control as long as the constraints on the output trajectory are fulfilled. The remaining free parameters might then be utilized to optimize the motion e.g. by constraining or optimizing the energy input to the system [3]. The calculated solution can always be verified by applying the determined control input $\boldsymbol{u}_{\mathrm{d}}$ in a forward-time integration. In the following, we assume the solution to be existent and unique.

Applicable DAE solvers for higher index DAEs must be chosen carefully. For detailed presentation of applicable integration schemes refer to [19]. DAE integrators either use a fixed or variable step size. The choice of a suitable method depends on the desired application.

The servo-constraints method is applicable for evaluating desired trajectories and estimating maximum control inputs. In this offline usage, a variable-step size solver with potentially high convergence order is an appropriate choice to yield accurate solutions. For this usage, a reformulation of the DAE problem into an optimization problem can be helpful as well, as proposed in [1].

For online usage as a feedforward control with possibly time-varying trajectories, a fixed-step size solver with real-time applicability is necessary. In order to ensure real-time capability, one or few iterations in an underlying nonlinear equations solver are necessary. An implicit Euler scheme has the necessary stability properties and is applied in [10, 24, 28]. In the Euler scheme, nonlinear equations have to be solved using an iterative scheme, for example Newton's method. In order to ensure real-time capability, the scheme can be restricted to one iteration, resulting in the linear implicit Euler scheme. For the overhead crane example, it was shown in [24] that both the linear implicit Euler scheme and the implicit Euler scheme with a maximum number of 10 iterations show comparable results and are both suitable for real-time implementation. A drawback of such a simple scheme is for example its numerical damping [20]. Moreover, ill-conditioning issues might arise when applying Newton's method. Such problems might be prevented by pre-scaling the algebraic equations [25]. Also, higher order schemes with fixed-step size such as implicit Runge-Kutta methods or backwards differencing formulas also seem applicable and yield higher convergence orders. The use of more advanced integration schemes for real-time application of servo-constraints is ongoing research. This paper concentrates on the application of the implicit Euler.

For the implicit Euler scheme, a discretization with time step size Δt is performed by finite differences

$$\dot{\boldsymbol{y}}_k \approx \frac{\boldsymbol{y}_{k+1} - \boldsymbol{y}_k}{\Delta t}, \tag{3.27}$$

where the indices $k+1$ and k describe the value of $\boldsymbol{y}$ at the time instants $k+1$ and k respectively and the division is element-wise. Applying the implicit Euler approximation then results in the nonlinear set of equations

$$\frac{\boldsymbol{y}_{k+1} - \boldsymbol{y}_k}{\Delta t} = \boldsymbol{Z}(\boldsymbol{y}_{k+1})\, \boldsymbol{v}_{k+1} \tag{3.28}$$

$$\boldsymbol{M}\left(\boldsymbol{y}_{k+1}, t_{k+1}\right) \frac{\boldsymbol{v}_{k+1} - \boldsymbol{v}_k}{\Delta t} = \boldsymbol{q}\left(\boldsymbol{y}_{k+1}, \boldsymbol{v}_{k+1}, t_{k+1}\right) + \boldsymbol{C}_{\mathrm{g}}^{\mathrm{T}}(\boldsymbol{y}_{k+1})\, \boldsymbol{\lambda}_{k+1} + \boldsymbol{B}\, \boldsymbol{u}_{k+1} \tag{3.29}$$

$$\mathbf{0} = \boldsymbol{c}_{\mathrm{g}}(\boldsymbol{y}_{k+1}) \tag{3.30}$$

$$\mathbf{0} = \boldsymbol{c}(\boldsymbol{y}_{k+1}) \tag{3.31}$$

which is collected in a single vector $\boldsymbol{F}$ for readability

$$\mathbf{0} = \boldsymbol{F}(\boldsymbol{y}_{k+1}, \boldsymbol{v}_{k+1}, \boldsymbol{u}_{k+1}, \boldsymbol{\lambda}_{k+1}) \, . \tag{3.32}$$

The set of nonlinear equations $\boldsymbol{F}$ is solved for the coordinates $\boldsymbol{y}$ and $\boldsymbol{v}$, generalized forces $\boldsymbol{\lambda}$ and inputs $\boldsymbol{u}$ at time instant t_{k+1} using Newton's method. Note that the implicit Euler scheme is also applicable to the projected DAE system of Eqs. (3.16)–(3.19) and results in an analogous nonlinear set of equations.

4 Illustrative Examples

Examples of some multibody systems are presented in the following. They include a mass-on-car system derived in minimal and redundant coordinates. With this system it is shown how small changes in the model properties can change the inverse model configuration from orthogonal to tangential realization. Also, the substitution from Sect. 3.2.3 is applied to enhance the DAE solution. Then, an extended mass-on-car system demonstrates the existence of first order as well as second order internal dynamics. The internal dynamics is extracted in ODE form based on the coordinate transformation from Sect. 3.2.2. Moreover, a mass-spring-damper chain is presented to show the influence of an increasing DAE index and how the projection method from Sect. 3.2.1 improves the DAE solution. Finally, it is illustrated how the choice of actuator models influences the resulting differentiation index.

4.1 Mass-on-Car System

The mass-on-car system is analyzed in detail in [28] to illustrate the different phenomena arising in inverse model problems. The results using minimal coordinates from [28] are briefly reviewed and compared to the use of redundant coordinates. The SISO system consists of two masses connected by a linear spring-damper combination. The first mass moves horizontally and is driven by the force F. The second mass is inclined by an angle α and is connected by a spring-damper combination. The two degree of freedom system is underactuated and shown in Fig. 3.

The minimal coordinates are chosen as

$$\boldsymbol{y} = \begin{bmatrix} \boldsymbol{y}_{\mathrm{a}} \\ \hdashline \boldsymbol{y}_{\mathrm{u}} \end{bmatrix} = \begin{bmatrix} x_1 \\ \hdashline s_1 \end{bmatrix} , \tag{4.1}$$

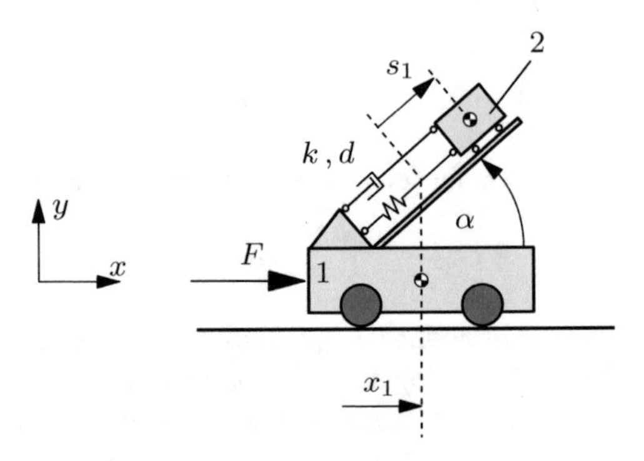

Fig. 3 SISO mass-on-car system

where x_1 describes the position of mass 1 and s_1 denotes the relative motion of mass 2 with respect to mass 1. For equilibrium holds $s_1 = 0$. The system input is the force F applied to mass 1 in horizontal direction. The equations of motion in ODE form arise as

$$\dot{\boldsymbol{y}} = \boldsymbol{v} \tag{4.2}$$

$$\begin{bmatrix} m_1 + m_2 & m_2 \cos\alpha \\ m_2 \cos\alpha & m_2 \end{bmatrix} \dot{\boldsymbol{v}} = - \begin{bmatrix} 0 \\ k\, s_1 + d\, \dot{s}_1 \end{bmatrix} + \begin{bmatrix} 1 \\ 0 \end{bmatrix} F\,, \tag{4.3}$$

where the masses of bodies 1 and 2 are m_1 and m_2 and the spring and damper coefficients are k and d respectively. The system output is the horizontal position of mass 2 such that

$$z = x_1 + s_1 \cos\alpha\,. \tag{4.4}$$

Accordingly, the servo-constraint is

$$c(\boldsymbol{y}, t) = x_1 + s_1 \cos\alpha - z_\mathrm{d}(t) = 0\,. \tag{4.5}$$

Thus, Eqs. (4.2), (4.3) and (4.5) form the inverse model. The various configurations of the inverse model are derived in [28] for different angles α. For analysis, the matrix $\boldsymbol{P}$ from Eq. (3.9) is evaluated as

$$P = \frac{\sin^2\alpha}{m_1 + m_2 \sin^2\alpha}\,. \tag{4.6}$$

A configuration with angle $\alpha = 90°$ decouples the motion between mass 1 and mass 2. It can be shown that the relative degree is $r = 2$ and the inverse model has an ideal orthogonal realization, due to regularity of P. The motion of mass 2 remains as internal dynamics of second order and does not contribute to the output z. In case the angle is $0° < \alpha < 90°$, the motion between mass 1 and 2 is coupled. However, the system has still a relative degree $r = 2$ and is in non-ideal orthogonal realization.

Table 1 Overview of different mass-on-car cases depending on angle α

Case	Angle	Damping ($\mathrm{Nsm^{-1}}$)	Relative degree r	DAE index	Dimension internal dynamics
1	$\alpha = 90°$	$d > 0$	2	3	2
		$d = 0$	2	3	2
2	$0° < \alpha < 90°$	$d > 0$	2	3	2
		$d = 0$	2	3	2
3	$\alpha = 0°$	$d > 0$	3	4	1
		$d = 0$	4	5	Diff. flat

For an angle $\alpha = 0°$, the system has relative degree $r = 3$ and represents tangential realization due to $P = 0$. The internal dynamics is first order dynamics in this case. In case of $\alpha = 0°$ and no damping $d = 0\,\mathrm{Nsm^{-1}}$, the system becomes differentially flat and has a relative degree $r = 4$. There is no internal dynamics and the analytical solution is derived in [28] as

$$F_{\mathrm{exact}} = (m_1 + m_2)\, \ddot{z}_{\mathrm{d}} + \frac{m_1\, m_2}{k}\, z_{\mathrm{d}}^{(4)} \,. \tag{4.7}$$

These various cases are summarized in Table 1. So far, the system was analyzed using the minimal coordinates of Eq. (4.1). As described in Sect. 3.2.3, it might be advantageous to rewrite the system dynamics using redundant coordinates. Here, the redundant coordinates are chosen as

$$\boldsymbol{y} = \begin{bmatrix} x_1 \\ x_2 \\ y_2 \end{bmatrix}, \tag{4.8}$$

where x_i and y_i denote the horizontal and vertical position of mass 1 and 2 respectively. Then, the model dynamics arises in DAE form as

$$\dot{\boldsymbol{y}} = \boldsymbol{v} \tag{4.9}$$

$$\begin{bmatrix} m_1 & 0 & 0 \\ 0 & m_2 & 0 \\ 0 & 0 & m_2 \end{bmatrix} \dot{\boldsymbol{v}} = \begin{bmatrix} F_{\mathrm{SD}} \cos\alpha \\ -F_{\mathrm{SD}} \cos\alpha \\ -F_{\mathrm{SD}} \sin\alpha - m_2\, g \end{bmatrix} + \boldsymbol{C}_{\mathrm{g}}^{\mathrm{T}} \lambda + \begin{bmatrix} 1 \\ 0 \\ 0 \end{bmatrix} F \tag{4.10}$$

$$c_{\mathrm{g}}(\boldsymbol{y}) = (x_2 - x_1)\, \tan\alpha - (y_2 - h) = 0 \,. \tag{4.11}$$

Here, the generalized reaction force between the inclined plane and mass 2 is denoted by λ, the vertical position of mass 2 in equilibrium is h and the applied force F_{SD} from the spring-damper combination is

$$F_{\mathrm{SD}} = k\, s_1 + d\, \dot{s}_1 - m_2\, g \sin\alpha \,. \tag{4.12}$$

Using redundant coordinates, the relative position s_1 is

$$s_1 = \begin{cases} \sqrt{(x_2 - x_1)^2 + (y_2 - h)^2} & x_2 > x_1 \\ 0 & x_1 = x_2 \\ -\sqrt{(x_2 - x_1)^2 + (y_2 - h)^2} & x_2 < x_1 \end{cases} \tag{4.13}$$

where the constraint $c_g(\boldsymbol{y})$ from Eq. (4.11) is substituted in Eq. (4.13) yielding

$$s_1 = (x_2 - x_1)\,\sqrt{1 + \tan^2\alpha}\,. \tag{4.14}$$

Thus, singularities in $\dot{s}$ are avoided. Note that the geometric constraint $\boldsymbol{c}_g(\boldsymbol{y})$ of Eq. (4.11) forces mass 2 to stay on the inclined plane and its Jacobian is

$$\boldsymbol{C}_g = \begin{bmatrix} -\tan\alpha & \tan\alpha & -1 \end{bmatrix}. \tag{4.15}$$

The system output is still the horizontal position of mass 2, $z = x_2$. The servo-constraint is then given by

$$c(\boldsymbol{y}, t) = x_2 - z_d(t) = 0\,. \tag{4.16}$$

Together with Eqs. (4.9)–(4.11), this forms the inverse model DAE. Note that with the redundant coordinates of Eq. (4.8), the system output is part of the coordinate vector with

$$z = \begin{bmatrix} 0 & 1 & 0 \end{bmatrix} \boldsymbol{y} = x_2\,. \tag{4.17}$$

As proposed in Sect. 3.2.3, the servo-constraint Eq. (4.16) is directly substituted into the model dynamics in order to reduce numeric complexity. This yields the inverse model

$$\begin{bmatrix} m_1 & 0 & 0 \\ 0 & m_2 & 0 \\ 0 & 0 & m_2 \end{bmatrix} \begin{bmatrix} \ddot{x}_1 \\ \ddot{z}_d \\ \ddot{y}_2 \end{bmatrix} = \begin{bmatrix} F_{SD}\cos\alpha \\ -F_{SD}\cos\alpha \\ -F_{SD}\sin\alpha - m_2\,g \end{bmatrix} + \boldsymbol{C}_g^{\mathrm{T}}\lambda + \begin{bmatrix} 1 \\ 0 \\ 0 \end{bmatrix} F \tag{4.18}$$

$$c_g(\boldsymbol{y}) = (z_d - x_1)\tan\alpha - (y_2 - h) = 0\,, \tag{4.19}$$

where the explicit servo-constraint $c(\boldsymbol{y}, t)$ is dropped due to its substitution into Eqs. (4.18)–(4.19).

For comparison, the inverse model is solved by an implicit Euler scheme in form of minimal coordinates given by Eqs. (4.2), (4.3) and (4.5), redundant coordinates given by Eqs. (4.9)–(4.11) and (4.16) and the substituted redundant

Table 2 Overview of parameters for mass-on-car system

Parameter	m_1	m_2	k	d
Value	1 kg	1 kg	$5\,\mathrm{Nm}^{-1}$	$0\,\mathrm{Nsm}^{-1}$

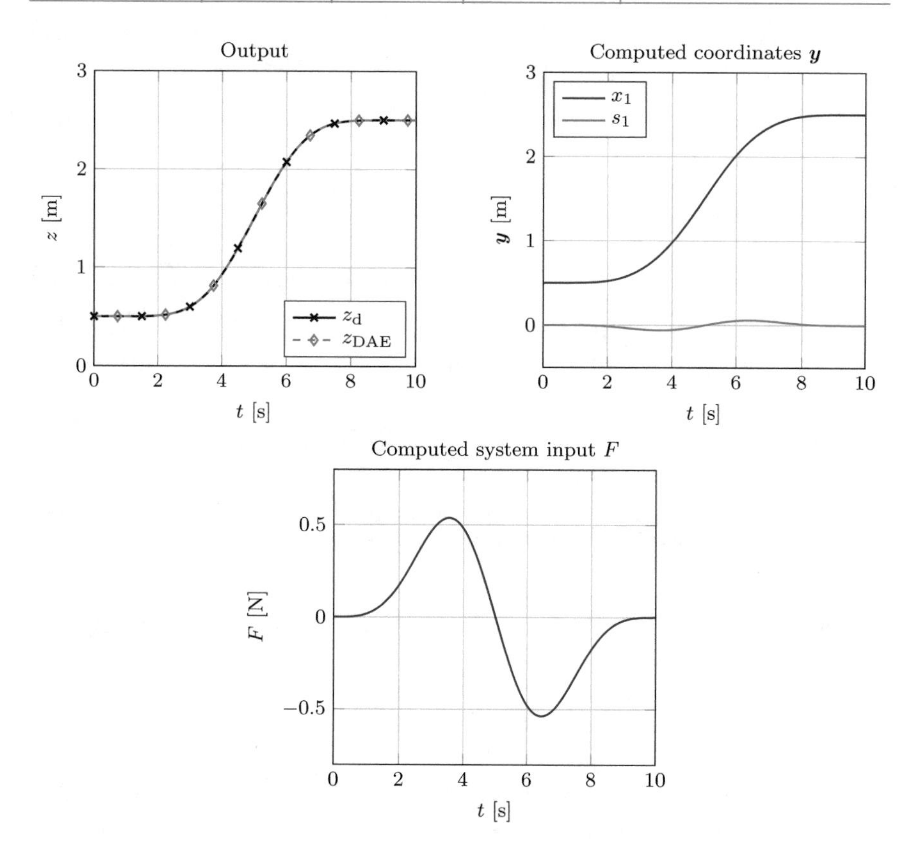

Fig. 4 Desired trajectory z_d and numerical results for the mass-on-car system with time step size $\Delta t = 0.01$ s

coordinates given by Eqs. (4.18) and (4.19). Moreover, index reduction by projection as described in Sect. 3.2.1 is applied to the minimal coordinates formulation of Eqs. (4.2)–(4.3) and the solution is compared to the other formulations. The model parameters are summarized in Table 2 and the differentially flat case with $\alpha = 0\,°$ and $d = 0\,\mathrm{Nsm}^{-1}$ is chosen. The desired trajectory $z_\mathrm{d}(t)$ from $z_\mathrm{d}(0) = 0.5\,\mathrm{m}$ to $z_\mathrm{d}(t_\mathrm{f}) = 2.5\,\mathrm{m}$ as well as the computed minimal coordinates x_1 and s_1 and input F are shown in Fig. 4 for step size $\Delta t = 0.01$ s and transition time $t_\mathrm{f} = 10$ s. Irrespectively of the formulation, no obvious difference is seen from these plots. Therefore, only one representative solution is shown here.

Convergence of the implicit Euler scheme is shown in Fig. 5, where the maximum error of the computed system input is plotted over different time steps Δt. The

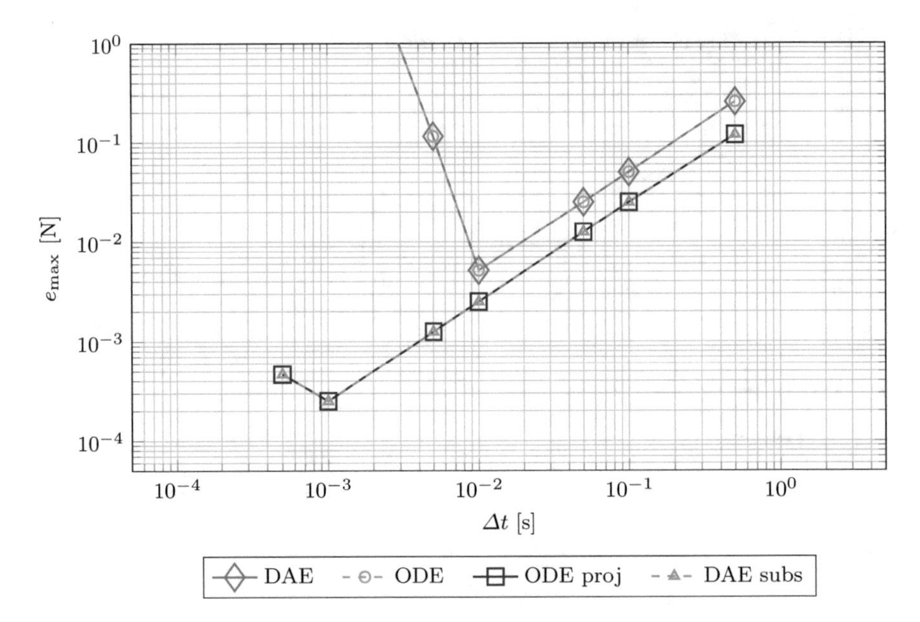

Fig. 5 Maximum error e_{max} between numerical and analytical solution of the mass-on-car system for different time step sizes Δt and various inverse model formulations

maximum error e_{max} between the analytic system input F_{exact} given by Eq. (4.7) and the computed input F is

$$e_{max} = \max_t \{|F_{exact}(t) - F(t)|\} . \tag{4.20}$$

All results show first order convergence behavior. Both the formulation in redundant coordinates given by Eqs. (4.9)–(4.11) and (4.16) and the minimal coordinates formulation of Eqs. (4.2)–(4.3) and (4.5), referenced by *DAE* and *ODE* respectively, run into numerical rounding errors for step sizes smaller than $\Delta t = 0.01$ s. In contrast, the substituted Eqs. (4.18)–(4.19) and the projected equations, referenced by *DAE subs* and *ODE proj* respectively, show stable numerical results up to a step size $\Delta t = 0.001$ s due to index reduction. Thus, a substitution or projection should be favored over the other two formulations for solving the inverse model problem. Plotting the computed system input F over time also shows the numerical instability for small step sizes. In Fig. 6, the computed system input is shown for a step size $\Delta t = 0.005$ s for the redundant coordinate formulation as well as the substituted formulation. Numerical rounding errors result in numerical noise that amplifies over time for the DAE, while the substituted formulation is smooth.

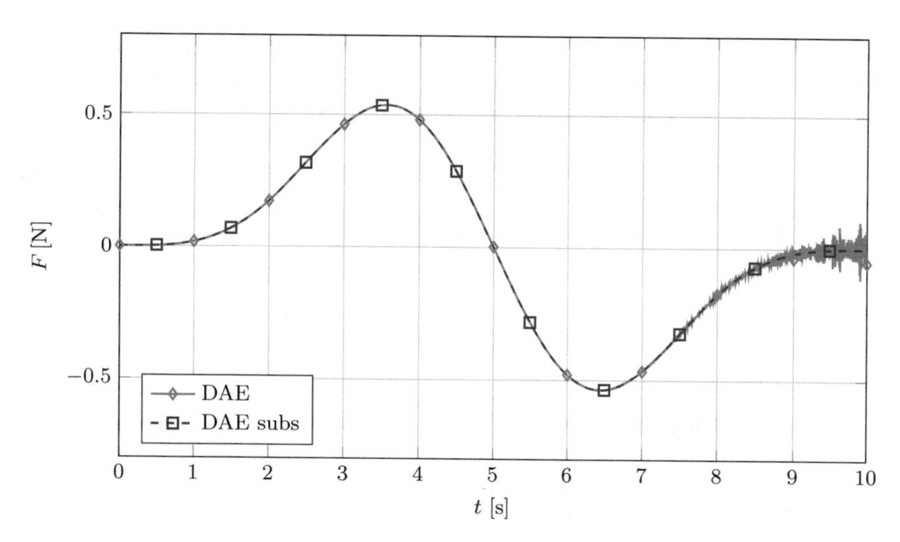

Fig. 6 Computed system input F for the mass-on-car system for time step size $\Delta t = 0.005$ s

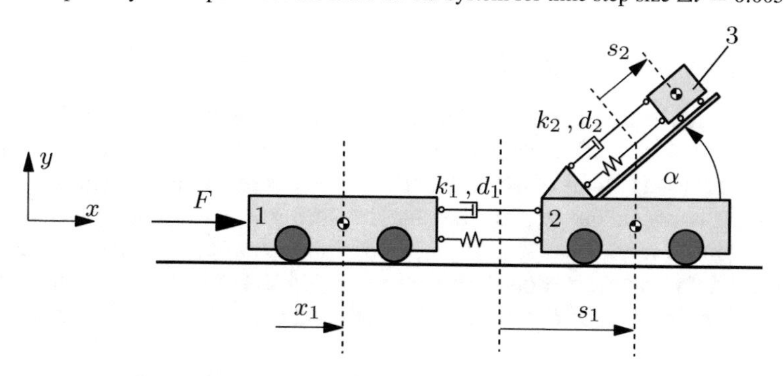

Fig. 7 Extended mass-on-car system

4.2 Extension of Mass-on-Car System

The previously described mass-on-car system is extended by an additional mass and is visualized in Fig. 7. The additional body 1 with mass m_1 is driven by the force F and is connected to body 2 by a linear spring-damper combination with coefficients k_1 and d_1 respectively. Body 3 moves on a plane inclined by an angle α and is connected to body 2 by a linear spring-damper combination with coefficients k_2 and d_2. The SISO system has $n = 3$ of freedom and is underactuated.

The dynamics of the extended mass-on-car system is described by minimal coordinates

$$\boldsymbol{y} = \begin{bmatrix} \boldsymbol{y}_\mathrm{a} \\ \hdashline \boldsymbol{y}_\mathrm{u} \end{bmatrix} = \begin{bmatrix} x_1 \\ \hdashline s_1 \\ s_2 \end{bmatrix}, \tag{4.21}$$

where the position of body 1 is described by x_1 and the relative positions of bodies 2 and 3 are described by s_1 and s_2 respectively. Applying the Newton-Euler formalism, the equations of motion in ODE form arise as

$$\dot{\boldsymbol{y}} = \boldsymbol{v} \tag{4.22}$$

$$\begin{bmatrix} m_1 + m_2 + m_3 & m_2 + m_3 & m_3 \cos\alpha \\ m_2 + m_3 & m_2 + m_3 & m_3 \cos\alpha \\ m_3 \cos\alpha & m_3 \cos\alpha & m_3 \end{bmatrix} \dot{\boldsymbol{v}} = - \begin{bmatrix} 0 \\ d_1 \dot{s}_1 + k_1 s_1 \\ d_2 \dot{s}_2 + k_2 s_2 \end{bmatrix} + \begin{bmatrix} 1 \\ 0 \\ 0 \end{bmatrix} F \, . \tag{4.23}$$

The system output is defined as the horizontal position of mass 3, such that

$$z = x_1 + s_1 + s_2 \cos\alpha \, . \tag{4.24}$$

Thus, the servo-constraints follow as

$$c(\boldsymbol{y}, t) = x_1 + s_1 + s_2 \cos\alpha - z_{\mathrm{d}} = 0 \, . \tag{4.25}$$

In order to determine the constraint realization, the matrix $\boldsymbol{P}$ from Eq. (3.9) is evaluated for this SISO system as $P = 0$ for all configuration angles α. This is because the system input F never directly influences the output z. There is always a coupling between the two masses by the spring k_1. Thus, the inverse model is in tangential realization for all angles α. Further analysis is performed using the coordinate transformation described in Sect. 3.2.2. Accordingly, the actuated coordinate x_1 is replaced by the output z, such that the new set of coordinates is

$$\boldsymbol{y}' = \begin{bmatrix} z \\ s_1 \\ s_2 \end{bmatrix} = \begin{bmatrix} x_1 + s_1 + s_2 \cos\alpha \\ s_1 \\ s_2 \end{bmatrix} = \begin{bmatrix} 1 & 1 & \cos\alpha \\ 0 & 1 & 0 \\ 0 & 0 & 1 \end{bmatrix} \boldsymbol{y} = \begin{bmatrix} \boldsymbol{H}_{\mathrm{a}} & \boldsymbol{H}_{\mathrm{u}} \\ \boldsymbol{0} & \boldsymbol{I} \end{bmatrix} \boldsymbol{y} \, . \tag{4.26}$$

The transformation matrix is invertible for all angles α, since $\boldsymbol{H}_{\mathrm{a}}$ is invertible for all angles α. Following the procedure described in Sect. 3.2.2, the dynamics of the new set of generalized coordinates $\boldsymbol{y}'$ arises as

$$\dot{\boldsymbol{y}}' = \begin{bmatrix} \dot{z} \\ \dot{s}_1 \\ \dot{s}_2 \end{bmatrix} = \boldsymbol{v}' \tag{4.27}$$

$$\dot{\boldsymbol{v}}' = \begin{bmatrix} 0 \\ -\dfrac{1}{m_1} \\ 0 \end{bmatrix} F + \begin{bmatrix} -\dfrac{m_3\, F_{\mathrm{SD},1}\, \sin^2\alpha + m_2\, F_{\mathrm{SD},2}\, \cos\alpha}{m_3\big(m_2 + m_3\, \sin^2\alpha\big)} \\ \dfrac{F_{\mathrm{SD},2}\, \cos\alpha}{m_2 + m_3 \sin^2\alpha} - \dfrac{F_{\mathrm{SD},1}\, (m_1 + m_2 + m_3 \sin^2\alpha)}{m_1\, \big(m_2 + m_3 \sin^2\alpha\big)} \\ \dfrac{F_{\mathrm{SD},1}\, \cos\alpha}{m_2 + m_3 \sin^2\alpha} - \dfrac{(m_2 + m_3)\, F_{\mathrm{SD},2}}{m_3\big(m_2 + m_3 \sin^2\alpha\big)} \end{bmatrix}, \tag{4.28}$$

with the abbreviations

$$F_{\mathrm{SD},1} = d_1\, \dot{s}_1 + k_1\, s_1 \tag{4.29}$$

$$F_{\mathrm{SD},2} = d_2\, \dot{s}_2 + k_2\, s_2\,. \tag{4.30}$$

From these equations, the system properties such as internal dynamics and the relative degree are analyzed in the following for three cases of angles α. Case 1 describes the configuration in which $\alpha = 90°$, while case 2 covers $0° < \alpha < 90°$ and case 3 stands for $\alpha = 0°$.

Case 1 The angle of the inclined plane is set to $\alpha = 90°$. In this case, the dynamics of body 3 is entirely decoupled from the motion of bodies 1 and 2 and the output reduces to $z = x_1 + s_1$. The dynamics of body 3 cannot be seen from the output and is thus internal dynamics. The motion of bodies 1 and 2 is equivalent to the motion of a 2-mass-spring-damper chain. From analysis of Eq. (4.28), a relationship between input F and output z is sought. Analyzing the second part of Eq. (4.28) shows that the input F is a function of $\ddot{s}_1$

$$F = -m_1\ddot{s}_1 - \frac{m_1 + m_2 + m_3}{m_2 + m_3}\, F_{\mathrm{SD},1}\,, \tag{4.31}$$

where $\alpha = 90°$ is substituted. From the first part of Eq. (4.28) follows that $\dot{s}_1$ is a function of $\ddot{z}$, which can be differentiated once. Thus, the third derivative of the output $\dddot{z}$ influences the input F and the system is of relative degree $r = 3$.

According to Eq. (3.12), the dimension of the internal dynamics is therefore 3. The states of the internal dynamics are the state s_1 of the first order dynamics of body 2 and the states $[s_2\ \ \dot{s}_2]^{\mathrm{T}}$ of the second order dynamics for body 3, collected in $\boldsymbol{\eta} = [s_1\ \ s_2\ \ \dot{s}_2]^{\mathrm{T}}$. Reordering the first and third equation of Eq. (4.28), yields the internal dynamics

$$\dot{\boldsymbol{\eta}} = \begin{bmatrix} -\dfrac{k_1}{d_1} s_1 - \dfrac{m_2 + m_3}{d_1}\ddot{z} \\ \dot{s}_2 \\ -\dfrac{d_2\dot{s}_2 + k_2 s_2}{m_3} \end{bmatrix} = \underbrace{\begin{bmatrix} -\dfrac{k_1}{d_1} & 0 & 0 \\ 0 & 0 & 1 \\ 0 & -\dfrac{k_2}{m_3} & -\dfrac{d_2}{m_3} \end{bmatrix}}_{\boldsymbol{A}_1} \boldsymbol{\eta} + \boldsymbol{f}(\ddot{z})\,. \tag{4.32}$$

Note that the dependence on the second derivative of the output z is collected in the function $\boldsymbol{f}(\ddot{z})$. Stability of the free internal dynamics is analyzed as zero dynamics with $\ddot{z} = \dot{z} = z = 0 \; \forall \; t$, such that $\boldsymbol{f}(\ddot{z}) = \mathbf{0}$. Thus, the zero dynamics is linear and stability is read off the eigenvalues $\boldsymbol{\lambda}_{\mathrm{A}_1}$ of the matrix $\boldsymbol{A}_1$. The eigenvalues

$$\boldsymbol{\lambda}_{\mathrm{A}_1} = \begin{bmatrix} -\dfrac{k_1}{d_1} \\ -\dfrac{d_2 + \sqrt{d_2^2 - 4k_2 m_3}}{2\,m_3} \\ -\dfrac{d_2 - \sqrt{d_2^2 - 4k_2 m_3}}{2\,m_3} \end{bmatrix} \tag{4.33}$$

have negative real parts for all $d_i > 0\,\mathrm{Nsm}^{-1}$ and thus the zero dynamics is asymptotically stable. The first eigenvalue is also reflected in the transfer function of the model dynamics

$$G(s) = \frac{k_1 + d_1 s}{a_4\, s^4 + a_3\, s^3 + a_2\, s^2}\,, \tag{4.34}$$

with the polynomial coefficients

$$\begin{aligned} a_4 &= m_1 m_2 + m_1 m_3 \\ a_3 &= d_1 m_1 + d_1 m_2 + d_1 m_3 \\ a_2 &= k_1 m_1 + k_1 m_2 + k_1 m_3\,. \end{aligned}$$

Note that due to complete decoupling the dynamics of mass 3 cannot be seen from the transfer function $G(s)$. The relative degree $r = 3$ is represented as the pole excess of 4 poles versus 1 transmission zero of the transfer function.

For neglectable damping, $d_i = 0\,\mathrm{Nsm}^{-1}$, the undamped dynamics of body 3 is still decoupled and of second order. However, the first part of Eq. (4.28) reduces to the algebraic relationship

$$s_1 = -\frac{m_2 + m_3}{k_1}\ddot{z} \tag{4.35}$$

between the variable s_1 and the second derivative of the output $\ddot{z}$, while Eq. (4.31) still holds. Therefore, the dynamics of bodies 1 and 2 have now relative degree $r = 4$. The internal dynamics is only the second order dynamics of body 3.

Case 2 For the general case of $0° < \alpha < 90°$, the motion of body 3 and bodies 1 and 2 is coupled. Again, the relationship between input and output is analyzed with Eq. (4.28). The second part shows again the relationship $F = F(\ddot{s}_1)$ and from the first part follows that the coordinate $\dot{s}_1$ is a function of the second derivative $\ddot{z}$. However, this time the dynamics of s_1 and s_2 are coupled, so that $\dot{s}_1 = \dot{s}_1(\ddot{z}, \dot{s}_2, s_2)$. Once more, the input F is a function of $\dddot{z}$ and the relative degree is $r = 3$. The internal dynamics must therefore have 3 states. Similar to case 1, the states of the internal dynamics are $\boldsymbol{\eta} = [s_1 \;\; s_2 \;\; \dot{s}_2]^{\mathrm{T}}$. Once more reordering the first and third part of Eq. (4.28) highlights the internal dynamics

$$\dot{\boldsymbol{\eta}} = \begin{bmatrix} -\dfrac{m_2 F_{\mathrm{SD},2} \cos\alpha}{d_1\, m_3 \sin^2\alpha} - \dfrac{k_1}{d_1} s_1 + \left(-\dfrac{m_3}{d_1} - \dfrac{m_2}{d_1 \sin^2\alpha}\right)\ddot{z} \\ \dot{s}_2 \\ -\dfrac{F_{\mathrm{SD},2}}{m_3 \sin^2\alpha} - \dfrac{\cos\alpha}{\sin^2\alpha}\ddot{z} \end{bmatrix} \tag{4.36}$$

$$= \underbrace{\begin{bmatrix} -\dfrac{k_1}{d_1} & -\dfrac{k_2\, m_2 \cos\alpha}{d_1\, m_3 \sin^2\alpha} & -\dfrac{d_2\, m_2 \cos\alpha}{d_1\, m_3 \sin^2\alpha} \\ 0 & 0 & 1 \\ 0 & -\dfrac{k_2}{m_3 \sin^2\alpha} & -\dfrac{d_2}{m_3 \sin^2\alpha} \end{bmatrix}}_{\boldsymbol{A}_2} \boldsymbol{\eta} + \boldsymbol{f}(\ddot{z})\,, \tag{4.37}$$

which are again driven by the second order derivative $\ddot{z}$. For the zero dynamics, it is $\ddot{z} = \dot{z} = z = 0$ and thus $\boldsymbol{f}(\ddot{z}) = \boldsymbol{0}$. The eigenvalues $\boldsymbol{\lambda}_{\mathrm{A}_2}$ of the state space matrix $\boldsymbol{A}_2$ are again analyzed for stability and arise as

$$\boldsymbol{\lambda}_{\mathrm{A}_2} = \begin{bmatrix} -\dfrac{k_1}{d_1} \\ -\dfrac{d_2 + \sqrt{d_2^2 - 4k_2 m_3 \sin^2\alpha}}{2\, m_3 \sin^2\alpha} \\ -\dfrac{d_2 - \sqrt{d_2^2 - 4k_2 m_3 \sin^2\alpha}}{2\, m_3 \sin^2\alpha} \end{bmatrix}. \tag{4.38}$$

They have negative real parts for all $d_i > 0\,\mathrm{Nsm}^{-1}$ and all angles $\alpha > 0°$ and are visualized in the root locus plot in Fig. 8. The system is analyzed with the parameters shown in Table 3. While the first eigenvalue is constant at $\lambda_{\mathrm{A}_2,1} = -\frac{k_1}{d_1}$, the other two eigenvalues $\lambda_{\mathrm{A}_2,2}$ and $\lambda_{\mathrm{A}_2,3}$ can be interpreted of describing the dynamics of a 2-mass oscillator. Its dynamics changes depending on the damping parameter d_2. The system is lightly damped for $d_2 < 2\sqrt{k_2\, m_3}\sin\alpha = \hat{d}_2$ and the eigenvalues are complex conjugates. The value $d_2 = \hat{d}_2$ describes the limit case for light damping

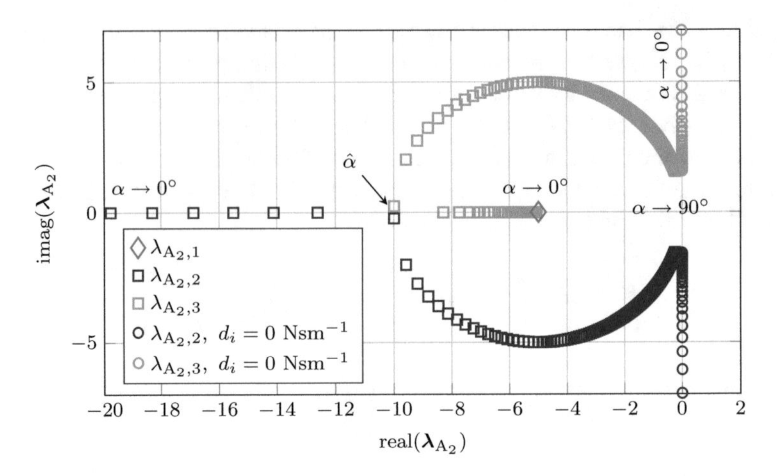

Fig. 8 Eigenvalues $\boldsymbol{\lambda}_{A_2}$ of the extended mass-on-car system in configuration case 2 where $0° < \alpha < 90°$

Table 3 Overview of parameters for extended mass-on-car system

Parameter	Value
m_1	1 kg
m_2	1 kg
m_3	2 kg
k_1	5 Nm^{-1}
k_2	5 Nm^{-1}
d_1	0 Nsm^{-1} and 1 Nsm^{-1}
d_2	0 Nsm^{-1} and 1 Nsm^{-1}

and for larger damping parameters d_2, the 2-mass oscillator is strongly damped as the eigenvalues $\lambda_{A_2,2}$ and $\lambda_{A_2,3}$ become real. This change of behavior is reflected in Fig. 8 at the value $\hat{\alpha} = \arcsin\left(\frac{d_2}{2\sqrt{k_2 m_3}}\right)$. For the limits $\alpha \to 0°$ and $\alpha \to 90°$, the eigenvalues tend to

$$\lim_{\alpha \to 0} \lambda_{A_2,2} = -\infty \tag{4.39}$$

$$\lim_{\alpha \to 0} \lambda_{A_2,3} = -\frac{k_2}{d_2} \tag{4.40}$$

$$\lim_{\alpha \to 90°} \lambda_{A_2,2} = -\frac{d_2 + \sqrt{d_2^2 - 4k_2 m_3}}{2m_3} \tag{4.41}$$

$$\lim_{\alpha \to 90°} \lambda_{A_2,3} = -\frac{d_2 - \sqrt{d_2^2 - 4k_2 m_3}}{2m_3} \tag{4.42}$$

respectively. Note that the limit $\lim_{\alpha \to 0} \lambda_{A_{2,3}}$ is obtained from Eq. (4.38) and applying the rule of L'Hospital twice. Case 1 is obtained in the limit $\alpha \to 90°$. This shows that, depending on the angle α, the characteristic behavior of the internal dynamics might be fundamentally different.

For the system in the same configuration without damping, $d_i = 0\,\mathrm{Nsm}^{-1}$, the differential equation for $\dot{s}_1$ from the first part of Eq. (4.28) reduces to the algebraic relation

$$s_1 = -\frac{m_2 + m_3 \sin^2 \alpha}{k_1 \sin^2 \alpha} \ddot{z} - \frac{k_2 m_2 \cos \alpha}{k_1 m_3 \sin^2 \alpha} s_2 \,. \tag{4.43}$$

Thus, s_1 is a function of $\ddot{z}$ and the system input then depends on $z^{(4)}$. Therefore, the system with $0° < \alpha < 90°$ and no damping $d_i = 0\,\mathrm{Nsm}^{-1}$ has relative degree $r = 4$. This is also reflected in the eigenvalues. Due to the algebraic relation for s_1, the dimension of the internal dynamics is only 2 and thus the eigenvalues associated with the coordinate s_1 vanish. The two remaining eigenvalues stay on the imaginary axis, see Fig. 8.

The numerical solution of the extended mass-on-car system for case 2 with damping $d_i > 0\,\mathrm{Nsm}^{-1}$ and an angle $\alpha = 45°$ is presented in the following. The desired output trajectory $z_\mathrm{d}(t)$ is shown in Fig. 9 from $z_\mathrm{d}(0) = 1$ m to $z_\mathrm{d}(t_\mathrm{f}) = 4$ m with $t_\mathrm{f} = 15$ s. Solving the servo-constraints problem given by Eqs. (4.22), (4.23) and (4.25) yields the desired control inputs $\boldsymbol{u}_\mathrm{d}(t)$ which are shown in Fig. 9. The solution was calculated using an implicit Euler scheme with time step size $\Delta t = 0.01$ s and a maximum number of 10 iterations in the Newton scheme to ensure real-time capability. Note that during the calculations, the convergence criterion was fulfilled before the maximum amount of iterations was reached.

In order to verify the accuracy of the solution, the system is afterwards simulated with the desired control inputs $\boldsymbol{u}_\mathrm{d}$ and the simulated system output is also shown in Fig. 9. As there are no disturbances and no initial errors in the simulation, the simulated system output is identical to the desired one. The system features internal dynamics, which are integrated over time by the DAE solver. Due to the coordinate transformation, it was possible to obtain an ODE formulation of the internal dynamics, given by Eq. (4.37). In order to evaluate the solution, the internal dynamics computed by the inverse model given by Eqs. (4.22), (4.23) and (4.25) is compared to the respective ODE solution which is integrated with a Runge-Kutta 45 scheme in Matlab as a reference. Both lead to similar results as shown in Fig. 10 for a step size of $\Delta t = 0.01$ s. In order to demonstrate significant integration errors of the simple Euler scheme with a larger step size, Fig. 11 shows the same comparison of the internal dynamics. In this case the inverse model solution was obtained with a step size $\Delta t = 0.1$ s. This leads to notable integration errors.

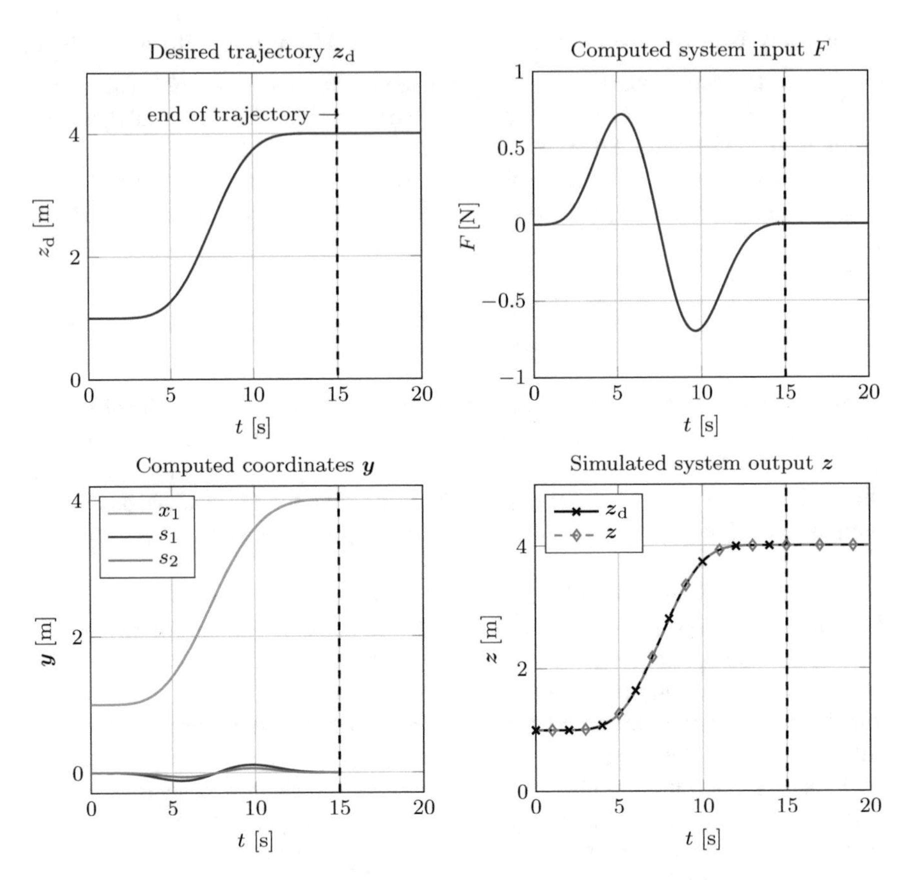

Fig. 9 Numerical results for the extended mass-on-car system in case 2 with angle $\alpha = 45°$ and time step size $\Delta t = 0.01$ s

Case 3 For angle $\alpha = 0°$, the system reduces to a free 3-mass-spring-damper chain. In this case, Eq. (4.28) reduces to

$$\dot{\boldsymbol{v}}' = \begin{bmatrix} 0 \\ -\dfrac{1}{m_1} \\ 0 \end{bmatrix} F + \begin{bmatrix} -\dfrac{F_{\mathrm{SD},2}}{m_3} \\ \dfrac{F_{\mathrm{SD},2}}{m_2} - \dfrac{m_1 + m_2}{m_1 m_2} F_{\mathrm{SD},1} \\ \dfrac{F_{\mathrm{SD},1}}{m_2} - \dfrac{m_2 + m_3}{m_2 m_3} F_{\mathrm{SD},2} \end{bmatrix} . \tag{4.44}$$

These equations are again analyzed to find the relation between input F and output z. From the second part follows that input F depends on $\ddot{s}_1$ with

$$F = -m_1 \ddot{s}_1 + \frac{m_1}{m_2} F_{\mathrm{SD},2} - \frac{m_1 + m_2}{m_2} F_{\mathrm{SD},1} . \tag{4.45}$$

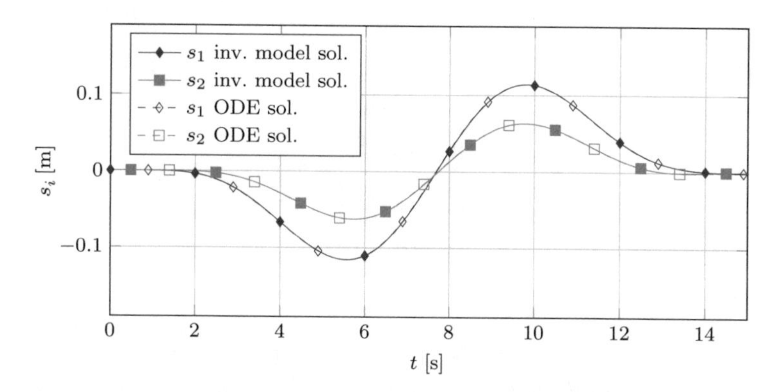

Fig. 10 Internal states s_1 and s_2 for case 2 with angle $\alpha = 45°$ and step size $\Delta t = 0.01$ s for the inverse model solution as well as reference ODE solution

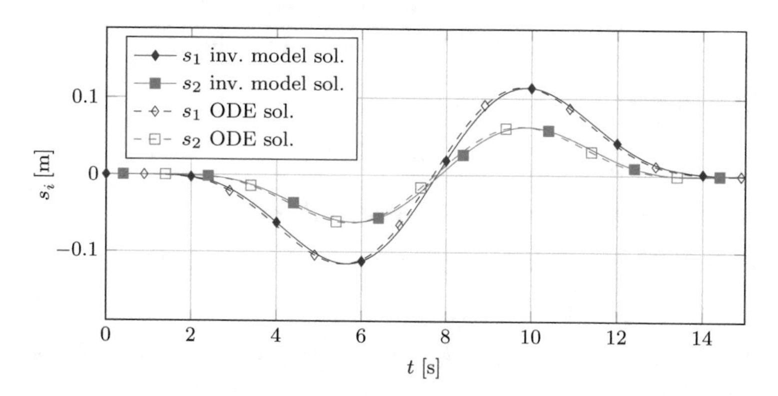

Fig. 11 Internal states s_1 and s_2 for case 2 with angle $\alpha = 45°$ and step size $\Delta t = 0.1$ s for the inverse model solution as well as reference ODE solution

From the third part follows the coordinate $\dot{s}_1$ as a function of $\dddot{s}_2$. With the relationship $\dot{s}_2 = \dot{s}_2(\ddot{z})$ from the first equation follows that the input F is a function of the fourth derivative of the output and hence the system has relative degree $r = 4$. Thus, there remain two internal dynamics states. Reordering and transforming the first and third part of Eq. (4.44) yields

$$\dot{s}_2 = -\frac{1}{d_2}\,(m_3\,\ddot{z} + k_2\,s_2) \tag{4.46}$$

$$\dddot{s}_2 = -F_{\mathrm{SD},2}\left(\frac{1}{m_2} + \frac{1}{m_3}\right) + \frac{1}{m_2}F_{\mathrm{SD},1}\,. \tag{4.47}$$

From Eq. (4.47) follows a differential equation for $\dot{s}_1$, where $\dddot{s}_2$ and $\dot{s}_2$ are substituted from Eq. (4.46) and its derivative. Then, the internal dynamics in the states

$\eta = [s_1 \;\; s_2]^{\mathrm{T}}$ is

$$\dot{\eta} = \underbrace{\begin{bmatrix} -\dfrac{k_1}{d_1} & \dfrac{k_2^2\, m_2}{d_1\, d_2^2} \\ 0 & -\dfrac{k_2}{d_2} \end{bmatrix}}_{A_3} \eta + f(\dddot{z}), \tag{4.48}$$

which is driven by the third derivative $\dddot{z}$. The zero dynamics is obtained by setting $\dddot{z} = \ddot{z} = \dot{z} = z = 0$ and therefore $f(\dddot{z}) = \mathbf{0}$. The eigenvalues $\boldsymbol{\lambda}_{A_3}$ of the zero dynamics are read off the main diagonal to have negative real parts for all $d_i > 0\,\mathrm{Nsm}^{-1}$. The equivalent transfer function of the system dynamics reads

$$G(s) = \frac{(k_1 + d_1\, s)\,(k_2 + d_2\, s)}{a_6\, s^6 + a_5\, s^5 + a_4\, s^4 + a_3\, s^3 + a_2\, s^2} \tag{4.49}$$

with the coefficients

$$\begin{aligned} a_6 &= m_1\, m_2\, m_3 \\ a_5 &= d_1\, m_1\, m_3 + d_2\, m_1\, m_2 + d_1\, m_2\, m_3 + d_2\, m_1\, m_3 \\ a_4 &= d_1\, d_2\, m_1 + d_1\, d_2\, m_2 + d_1\, d_2\, m_3 + k_1\, m_1\, m_3 + k_2\, m_1\, m_2 \\ &\quad + k_1\, m_2\, m_3 + k_2\, m_1\, m_3 \\ a_3 &= d_1\, k_2\, m_1 + d_2\, k_1\, m_1 + d_1\, k_2\, m_2 + d_2\, k_1\, m_2 + d_1\, k_2\, m_3 + d_2\, k_1\, m_3 \\ a_2 &= k_1\, k_2\, m_1 + k_1\, k_2\, m_2 + k_1\, k_2\, m_3\,. \end{aligned}$$

Stability of the zero dynamics is reflected by the transmission zeros of the transfer function $G(s)$ which are in the left half plane. Moreover, the transfer function has a pole excess of 4, which corresponds to relative degree $r = 4$.

Without damping, the internal dynamics vanishes completely and thus the relative degree is $r = 6$. The system is then differentially flat and the system input depends on $z^{(6)}$. The analytic inverse model can be extracted by reordering Eq. (4.44). The system coordinates $\boldsymbol{y}$ are functions of the output z and a finite number of derivatives, such that

$$x_1 = z - s_1 - s_2 \tag{4.50}$$

$$s_1 = \frac{m_2}{k_1}\left(-\frac{m_3}{k_2}\, z^{(4)} + \frac{m_2 + m_3}{m_2\, m_3}\, k_2\, s_2\right) \tag{4.51}$$

$$s_2 = -\frac{m_3}{k_2}\,\ddot{z} \tag{4.52}$$

Table 4 Overview of different extended mass-on-car configurations

Case	Angle	Damping (Nsm^{-1})	Relative degree r	DAE index	Dimension internal dynamics
1	$\alpha = 90°$	$d_i > 0$	3	4	3
		$d_i = 0$	4	5	2
2	$0° < \alpha < 90°$	$d_i > 0$	3	4	3
		$d_i = 0$	4	5	2
3	$\alpha = 0°$	$d_i > 0$	4	5	2
		$d_i = 0$	6	7	Diff. flat

and the input F is a function of the coordinates and the inputs

$$F = \frac{m_1\, k_2}{m_2} s_2 - \frac{m_1 + m_2}{m_2}\, k_1\, s_1 - m_1\, \ddot{s}_1\,. \tag{4.53}$$

Substituting all information in Eq. (4.53) shows the relation $F = F\left(z^{(6)}\right)$.

The different phenomena of the extended mass-on-car system are summarized in Table 4. It is again noted that the differentiation index of the corresponding servo-constraint problem is larger by one than the relative degree of the system.

4.3 *Mass-Spring-Damper Chain*

A mass-spring-damper chain is a generalization of the mass-on-car system with $\alpha = 0°$ and an additional spring-damper pair that connects the first body to a wall. It is a popular example and also analyzed in the context of servo-constraints in e.g. [1, 18] and [28]. The mass-spring-damper chain has several properties, which turn it into a suitable example to analyze the servo-constraints approach. First of all, the system is linear in case of linear spring-damper combinations so that a linear reference analysis is possible. For the case of neglectable damping, the model is differentially flat. Thus, there is an analytic reference solution suitable for comparison and convergence studies. Moreover, its relative degree changes with the number of masses.

Let the chain have f masses and the position coordinates $x_1, \ldots x_f$, as shown in Fig. 12. The masses are connected by linear spring-damper combinations with coefficients k_i and d_i respectively. It is a SISO system, where the output is the position x_1 of the first mass and the input is the force F applied on the last mass. It has f degrees of freedom and is underactuated for $f > 1$. The chain with translational degrees of freedom is equivalent to systems with torsional degrees of freedom such as a drive train.

The dynamics of a general damped system with f masses and linear spring-damper combinations is of the linear ODE form

$$\dot{\boldsymbol{y}} = \boldsymbol{v} \tag{4.54}$$

$$\boldsymbol{M}\dot{\boldsymbol{v}} + \boldsymbol{D}\boldsymbol{v} + \boldsymbol{K}\boldsymbol{y} = \boldsymbol{B}F\,, \tag{4.55}$$

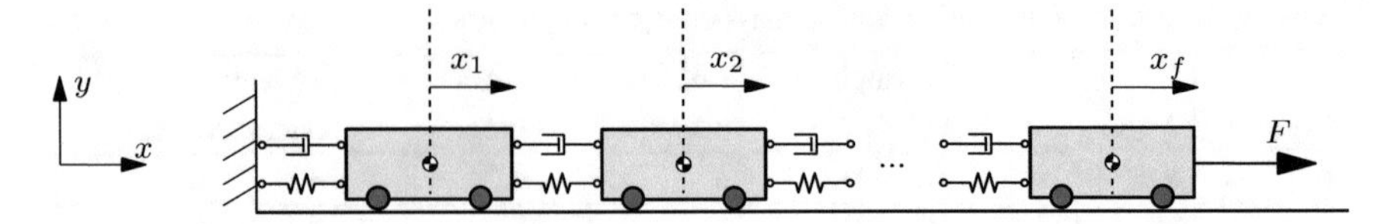

Fig. 12 Mass-spring-damper chain with f masses

with the mass matrix $\boldsymbol{M}$, the damping matrix $\boldsymbol{D}$, the stiffness matrix $\boldsymbol{K}$, the input distribution matrix $\boldsymbol{B}$ and the generalized coordinate vector

$$\boldsymbol{y} = \begin{bmatrix} \boldsymbol{y}_\mathrm{u} \\ \hdashline y_\mathrm{a} \end{bmatrix} = \begin{bmatrix} x_1 \\ x_2 \\ \vdots \\ x_{f-1} \\ \hdashline x_f \end{bmatrix}. \tag{4.56}$$

The dynamics matrices arise as

$$\boldsymbol{M} = \begin{bmatrix} m_1 & 0 & & & \\ 0 & m_2 & 0 & & \\ & \ddots & \ddots & \ddots & \\ & & & m_{f-1} & 0 \\ & & & 0 & m_f \end{bmatrix} \tag{4.57}$$

$$\boldsymbol{D} = \begin{bmatrix} d_1+d_2 & -d_2 & 0 & & \\ -d_2 & d_2+d_3 & -d_3 & & \\ & \ddots & \ddots & \ddots & \\ & & -d_{f-1} & d_{f-1}+d_f & -d_f \\ & & & -d_f & d_f \end{bmatrix} \tag{4.58}$$

$$\boldsymbol{K} = \begin{bmatrix} k_1+k_2 & -k_2 & 0 & & \\ -k_2 & k_2+k_3 & -k_3 & & \\ & \ddots & \ddots & \ddots & \\ & & -k_{f-1} & k_{f-1}+k_f & -k_f \\ & & & -k_f & k_f \end{bmatrix} \tag{4.59}$$

$$\boldsymbol{B} = \begin{bmatrix} 0 \\ 0 \\ \vdots \\ 0 \\ 1 \end{bmatrix}, \tag{4.60}$$

with the respective linear spring and damper coefficients k_i and d_i and masses m_i. The scalar system output z is the position of mass 1, such that

$$z = \begin{bmatrix} 1 \; 0 \ldots 0 \; 0 \end{bmatrix} \boldsymbol{y} = x_1 \,. \tag{4.61}$$

Together with the servo-constraint

$$c(\boldsymbol{y}, t) = x_1 - z_\mathrm{d} = 0 \,, \tag{4.62}$$

Eqs. (4.54) and (4.55) form the inverse model. Note that in contrast to the formulation of the mass-on-car system, the positions of all masses are here defined by absolute coordinates. Thus, the system output z does not depend on the actuated coordinate x_f and the coordinate transformation as proposed in Sect. 3.2.2 is not applicable directly. Beforehand, a coordinate transformation to the relative coordinates s_i used in the previous example would be necessary. Then, the internal dynamics can again be derived in ODE form. On the other hand, absolute positions are useful in deriving the analytical solution to the inverse model and are hence pursued here. Defining the state vector

$$\boldsymbol{x} = \begin{bmatrix} \boldsymbol{y} \\ \boldsymbol{v} \end{bmatrix} , \tag{4.63}$$

the linear state space equations are

$$\dot{\boldsymbol{x}} = \underbrace{\begin{bmatrix} \boldsymbol{0} & \boldsymbol{I} \\ -\boldsymbol{M}^{-1}\boldsymbol{K} & -\boldsymbol{M}^{-1}\boldsymbol{D} \end{bmatrix}}_{\boldsymbol{A}} \boldsymbol{x} + \underbrace{\begin{bmatrix} \boldsymbol{0} \\ \boldsymbol{M}^{-1}\boldsymbol{B} \end{bmatrix}}_{\boldsymbol{b}} u \tag{4.64}$$

$$z = \underbrace{\begin{bmatrix} 1 \; 0 \ldots 0 \; 0 \end{bmatrix}}_{\boldsymbol{C}} \boldsymbol{x} + \underbrace{0}_{d} \, u \,. \tag{4.65}$$

The linear state space equations are transformed to a transfer function $G(s)$ by Eq. (3.13) and linear analysis is performed as a reference. In general, from the transfer function analysis is inferred that for an undamped chain with f masses, the relative degree is $r = 2f$ and there is no internal dynamics. Thus, the differentiation index of the undamped inverse model DAE also changes with the number of masses and is $2f + 1$. For every additional damper with $d_i \neq 0$ for $i = 2, \ldots f$, the numerator polynomial includes the multiplier term $(k_i + d_i\, s)$. Therefore, for a system with a number of n_d dampers $d_i \neq 0$ for $i = 2, \ldots f$, the relative degree is $r = 2f - n_\mathrm{d}$. The existence of the damper d_1 does not influence the relative degree of the system, because this damper is not located between input F and output $z = x_1$. Accordingly, a fully damped system features internal dynamics and its relative degree is $r = 2f - (f - 1) = f + 1$. The internal dynamics is of first order and the total number of internal dynamics states is $f - 1$.

This is demonstrated for the case of $f = 2$ masses with the dynamic equations

$$\begin{bmatrix} m_1 & 0 \\ 0 & m_2 \end{bmatrix} \begin{bmatrix} \ddot{x}_1 \\ \ddot{x}_2 \end{bmatrix} + \begin{bmatrix} d_1 + d_2 & -d_2 \\ -d_2 & d_2 \end{bmatrix} \begin{bmatrix} \dot{x}_1 \\ \dot{x}_2 \end{bmatrix} + \begin{bmatrix} k_1 + k_2 & -k_2 \\ -k_2 & k_2 \end{bmatrix} \begin{bmatrix} x_1 \\ x_2 \end{bmatrix} = \begin{bmatrix} 0 \\ 1 \end{bmatrix} F \tag{4.66}$$

$$z = \begin{bmatrix} 1 & 0 \end{bmatrix} \begin{bmatrix} x_1 \\ x_2 \end{bmatrix}. \tag{4.67}$$

Applying Eq. (3.13), the transfer function reads for $f = 2$

$$G(s) = \frac{k_2 + d_2 s}{a_4\, s^4 + a_2\, s^3 + a_2\, s^2 + a_1\, s + a_0}, \tag{4.68}$$

with

$$\begin{aligned}
a_4 &= m_1\, m_2 \\
a_3 &= m_1\, d_2 + m_2\, d_1 + m_2\, d_2 \\
a_2 &= m_1\, k_2 + m_2\, k_1 + m_2\, k_2 + d_1\, d_2 \\
a_1 &= d_1\, k_2 + d_2\, k_1 \\
a_0 &= k_1\, k_2\,.
\end{aligned}$$

The transfer function $G(s)$ has one transmission zero and 4 poles, which is a pole excess of 3. This corresponds to a relative degree $r = 3$. For neglectable damping $d_i = 0\,\mathrm{Nsm}^{-1}$, the transfer function reduces to

$$G(s) = \frac{k_2}{m_1\, m_2\, s^4 + (m_1\, k_2 + m_2\, k_1 + m_2\, k_2)\, s^2 + k_1\, k_2}, \tag{4.69}$$

which shows a pole excess of 4, corresponding to a relative degree $r = 4$. There remains no internal dynamics. In fact, an analytical solution for the inverse model is derived in the following. This is shown for $f = 2$ masses and can be automated for any f masses. Differentiating the output equation (4.67) twice yields

$$\ddot{z} = \ddot{x}_1 = \frac{1}{m_1}\Big(- (k_1 + k_2)\, x_1 + k_2\, x_2 \Big), \tag{4.70}$$

which is reordered for an expression for the position of the last mass

$$x_2(x_1, \ddot{z}) = \frac{1}{k_2}\Big(m_1 \ddot{z} + (k_1 + k_2)\, x_1 \Big). \tag{4.71}$$

Table 5 Overview of parameters of mass-spring-damper chain

Parameter	m	k	d
Value	1 kg	$5\,\mathrm{Nm}^{-1}$	$0\,\mathrm{Nsm}^{-1}$

Differentiating equation (4.70) two more times shows the relation between input F and the fourth derivative of the output z

$$z^{(4)} = x_1^{(4)} = \frac{1}{m_1}\Big(- (k_1 + k_2)\, \ddot{x}_1 + k_2\, \ddot{x}_2\Big) \tag{4.72}$$

$$= \frac{1}{m_1}\left(-(k_1 + k_2)\, \ddot{x}_1 + \frac{k_2}{m_2}\, (F + k_2\, x_1 - k_2\, x_2)\right), \tag{4.73}$$

where x_2 can be substituted by Eq. (4.71). Solving for F gives the analytic solution

$$F_{\mathrm{exact}} = \frac{m_1 m_2}{k_2}\, z^{(4)} + \frac{(k_1 + k_2)\, m_2}{k_2}\, \ddot{x}_1 - k_2\, (x_1 - x_2) \tag{4.74}$$

for the inverse model and shows that the input F is a function of the output $z = x_1$ and four derivatives of z. Thus, the system is differentially flat and has relative degree $r = 4$. This procedure of differentiating the output z can be repeated for f masses until the input F appears.

The analytical solution for the inverse model, given by Eq. (4.74) for 2 masses and respective equations for additional masses, is used for convergence studies of the undamped servo-constraints inverse model. The simulation parameters are shown in Table 5. It is assumed that all masses, spring and damper coefficients are equal, $m_i = m$, $k_i = k$, $d_i = d$.

The desired path z_{d} from $z_{\mathrm{d}}(0) = 0\,\mathrm{m}$ to $z_{\mathrm{d}}(t_{\mathrm{f}}) = 2\,\mathrm{m}$ is shown in Fig. 13, where the transition time is $t_{\mathrm{f}} = 15\,\mathrm{s}$. The inverse model DAE is solved by the implicit Euler method with $\Delta t = 0.01\,\mathrm{s}$. The computed coordinates and input are shown in Fig. 13. Due to equal masses and springs, the mass m_2 travels twice the distance of mass m_1.

The first order convergence of the implicit Euler scheme is verified by the analytical solution of Eq. (4.74) and solving the inverse model with different times steps Δt. The maximum error is calculated according to Eq. (4.20). The inverse model is solved in the original formulation given by Eqs. (4.54), (4.55) and (4.62) and the projected formulation obtained by applying Eqs. (3.16)–(3.19). The original and projected formulations are referenced by indices *orig* and *proj* respectively. Note that similar to the substitution process presented in Sect. 3.2.3, the system output z is simply one of the coordinates $\boldsymbol{y}$. Thus, the servo-constraint Eq. (4.62) and its derivatives are substituted in the dynamic equations (4.54) and (4.55) as proposed in Sect. 3.2.3 for redundant coordinates. This reduces the differentiation index by 2. The substituted and reduced equations for $f = 2$ read

$$\begin{bmatrix} m_1 & 0 \\ 0 & m_2 \end{bmatrix}\begin{bmatrix} \ddot{z}_{\mathrm{d}} \\ \ddot{x}_2 \end{bmatrix} + \begin{bmatrix} d_1 + d_2 & -d_2 \\ -d_2 & d_2 \end{bmatrix}\begin{bmatrix} \dot{z}_{\mathrm{d}} \\ \dot{x}_2 \end{bmatrix} + \begin{bmatrix} k_1 + k_2 & -k_2 \\ -k_2 & k_2 \end{bmatrix}\begin{bmatrix} z_{\mathrm{d}} \\ x_2 \end{bmatrix} = \begin{bmatrix} 0 \\ 1 \end{bmatrix} F, \tag{4.75}$$

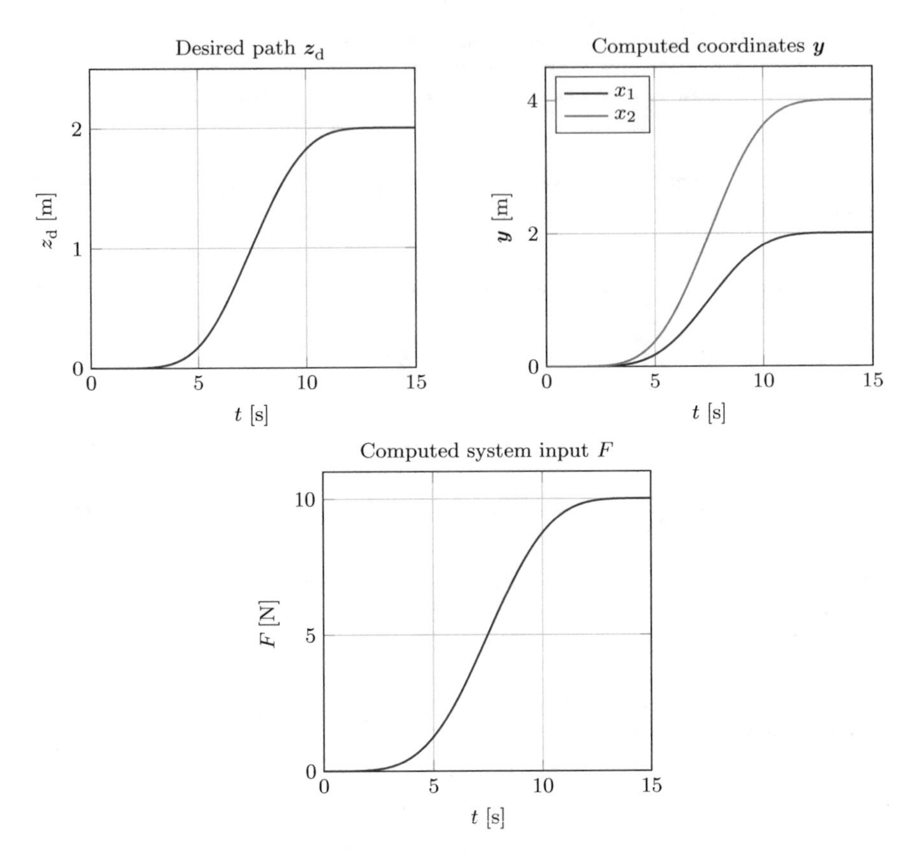

Fig. 13 Numerical results for a 2-mass-spring-damper combination without damping with $\Delta t = 0.01$ s

and respectively for more than 2 masses. Note that the servo-constraint Eq. (4.62) is dropped due to substitution. This approach is referenced by the index *subs* and compared to the projected and original solutions. The model is solved for a number of $f = 2$, 3 and 4 masses. Therefore, the original inverse model DAE has a differentiation index of 5, 7 and 9 respectively.

The convergence diagram in Fig. 14 shows the first order convergence behavior. Note that the substitution and projection approach yield similar results for $f = 2$, 3 and 4 masses. Both the projection and substitution approach yield numerically more stable results than the original models, respectively for each tested number of masses. This is due to index reduction. Considering more masses makes the numerical solution less stable, respectively for each inverse model formulation. This is due to a higher differentiation index of a mass-spring chain with more masses. The presented implementation with an implicit Euler scheme is able to solve a model with $f = 4$ masses with a step size of $\Delta t = 0.1$ s resulting in a maximum error of $e_{\max} \approx 0.06$ N. Better results with smaller step sizes are obtained by the projected and substituted versions respectively, when the original formulation is

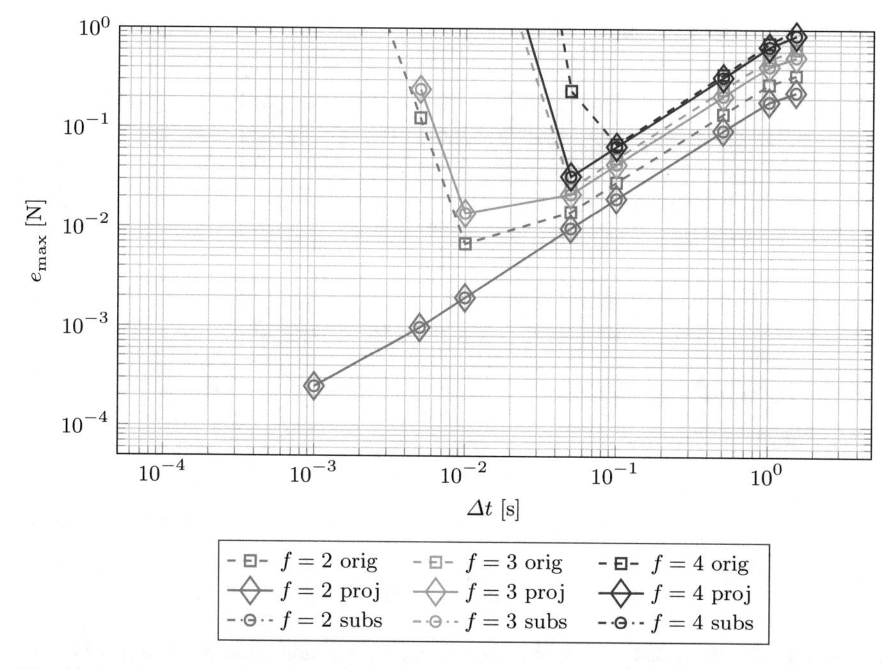

Fig. 14 Maximum error $e_{\max}$ between numerical and analytical solution for an f-mass-spring-damper chain for different time step sizes Δt and various inverse model formulations

already unstable. For more than $f = 4$ masses, the numerical solution becomes unstable too quickly using the implicit Euler scheme.

The numerical stability is also evaluated in terms of the resulting Jacobian matrix of the set of equations $\boldsymbol{F}$ from Eq. (3.32) that is solved with Newton's method in each time step. The Jacobian matrix is derived symbolically offline and then evaluated in each time step to improve calculation speed. The Jacobian matrix for the $f = 2$ undamped system is for the original unprojected equations

$$\boldsymbol{J}_{\text{orig}} = \begin{bmatrix} \frac{1}{\Delta t} & 0 & -1 & 0 & 0 \\ 0 & \frac{1}{\Delta t} & 0 & -1 & 0 \\ \frac{2k}{m} & -\frac{k}{m} & \frac{1}{\Delta t} & 0 & 0 \\ -\frac{k}{m} & \frac{k}{m} & 0 & \frac{1}{\Delta t} & -\frac{1}{m} \\ -1 & 0 & 0 & 0 & 0 \end{bmatrix}, \tag{4.76}$$

for the projected equations

$$\boldsymbol{J}_{\text{proj}} = \begin{bmatrix} -\frac{1}{\Delta t} & 0 & 1 & 0 & 0 \\ 0 & -\frac{1}{\Delta t} & 0 & 1 & 0 \\ k & -k & 0 & -\frac{m}{\Delta t} & 1 \\ -\frac{2k}{m} & \frac{k}{m} & 0 & 0 & 0 \\ -1 & 0 & 0 & 0 & 0 \end{bmatrix} \tag{4.77}$$

and for the substituted equations

$$\boldsymbol{J}_{\text{subs}} = \begin{bmatrix} -\frac{1}{\Delta t} & 1 & 0 \\ -k & 0 & 0 \\ k & \frac{m}{\Delta t} & -1 \end{bmatrix}. \tag{4.78}$$

Note that due to substitution, the size of equations and unknowns reduces to 3. The condition number κ, defined as the ratio between largest and smallest singular value of the Jacobian $\boldsymbol{J}$ is evaluated for each solution and shown in Fig. 15. Large condition numbers indicate bad conditioning for solving the set of equations by Newton's method. It can be read of Fig. 15 that the condition number increases for more masses and higher differentiation index and thus also supports the numerical instability due to rounding errors presented in Fig. 14. For a given step size, the condition numbers of the projected and substituted equations are smaller than the respective condition number of the original equations. Figure 15 also emphasizes that the condition number of the substituted system of equations behaves similar to the one of the projected equations with a small constant difference. The condition number can be used as an indicator for numerical solvability of the inverse model by means of servo-constraints.

4.4 Influence of Actuator Model on the DAE Index

Standard industrial actuators are usually velocity-controlled and not force-controlled as assumed in the previous examples. In a practical implementation, the system input is therefore usually a certain setpoint velocity instead of a force. This setpoint velocity is enforced by internal control loops of the actuators. The actuator dynamics is assumed to be much faster than the system dynamics. Thus, they are approximated by low order linear system dynamics, e.g. zero order, first order or second order system dynamics and the coefficients are fitted in system identification

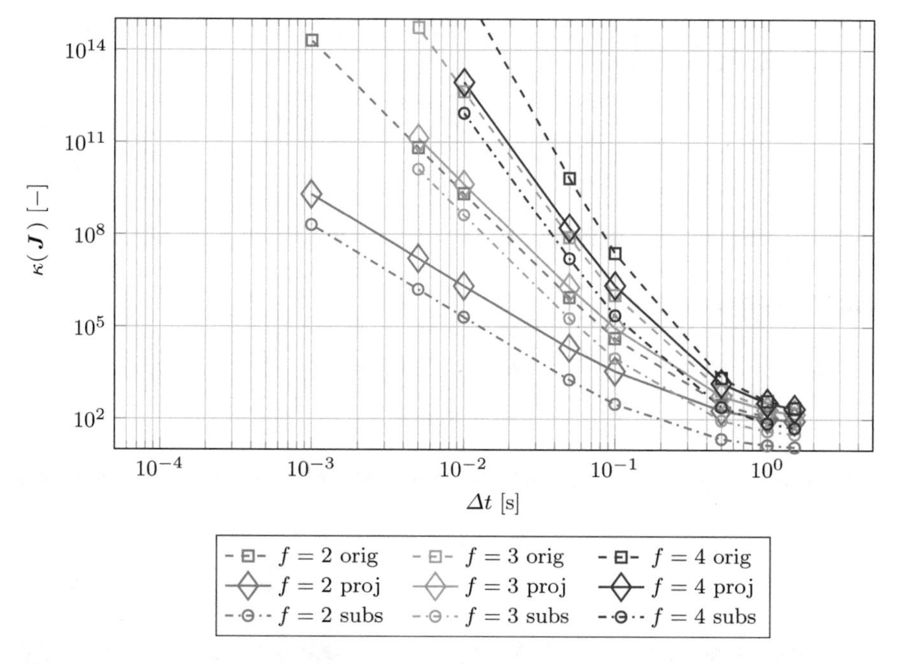

Fig. 15 Condition number κ of the respective Jacobians of f-mass-spring-damper chains for different time step sizes Δt and various inverse model formulations

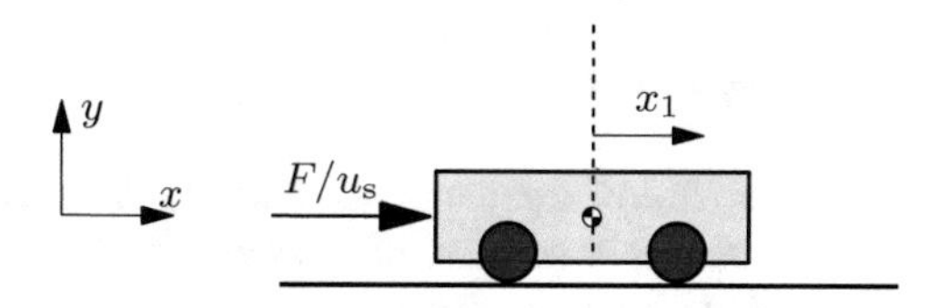

Fig. 16 Single mass with either force input F or setpoint velocity input u_s

tests. Depending on what type of system order is assumed, the differentiation index of the arising inverse model DAE changes. This is demonstrated with a simple single mass system shown in Fig. 16. The scalar system input F is applied to a single mass m and the system output is its position x_1.

For this system, the dynamic equations with a force-controlled actuator are

$$m\,\ddot{x}_1 = F \tag{4.79}$$

$$z = x_1\,, \tag{4.80}$$

where the system input is modeled as the force F. In this case, the relative degree of the system is $r = 2$ and there are no internal dynamics. The differentiation index of the servo-constraints DAE system is 3 respectively.

Assuming a velocity-controlled actuator, let the setpoint input velocity u_s be the system input instead of the input F. The velocity $\dot{x}_1$ is then a rheonomic constraint on the model. Modeling the actuator dynamics as zero order, which is direct feedthrough, yields

$$\dot{x}_1 = u_s \tag{4.81}$$

$$z = x_1 \,. \tag{4.82}$$

In this system, the first derivative of the output z is necessary for calculating the input velocity u_s. Therefore, the system has relative degree $r = 1$. For first order actuator dynamics holds

$$\tau \ddot{x}_1 = -\dot{x}_1 + u_s \tag{4.83}$$

$$z = x_1 \,, \tag{4.84}$$

with the time constant τ of the first order actuator model. Solving for the input yields

$$u_s = \tau \ddot{x}_1 + \dot{x}_1 \tag{4.85}$$

$$= \tau \ddot{z} + \dot{z} \,, \tag{4.86}$$

where the second derivative of the output z is necessary to calculate the control input. Therefore, the system has relative degree $r = 2$. The second order actuator dynamic equations are

$$K \omega_n^2 \, u_s = \dddot{x}_1 + 2\zeta \omega_n \, \ddot{x}_1 + \omega_n^2 \, \dot{x}_1 \,, \tag{4.87}$$

with the static gain K, the natural frequency ω_n and the damping ratio ζ. Again substituting the system output shows that the system in this configuration has relative degree $r = 3$. This example shows how modeling actuator dynamics influences the relative degree of a multibody system.

5 Conclusion

Servo-constraints are an efficient method in the context of open-loop control of underactuated multibody systems. They append the system dynamics and form a set of higher index DAEs, which can be solved numerically. The solution poses an inverse model for the underactuated multibody system, which is applicable for trajectory tracking controllers. A state feedback controller can be added for disturbance rejection which makes use of the desired state trajectory that is part of the inverse model solution.

Servo-constraints are closely related to classical geometric constraints. However, since the system input is not necessarily orthogonal to the tangent of the constraint manifold and might influence the output only indirectly, the differentiation index of the inverse model can be higher than three. The configuration is classified into orthogonal, mixed tangential-orthogonal and tangential realization, depending on how the input influences the system output.

In case the underactuated system is not differentially flat, there remains internal dynamics, which have to be accounted for. In that case, stability of the internal dynamics has to be investigated in terms of the stability of the zero dynamics.

The differentiation index of the arising set of DAEs can be reduced by index reduction methods such as the projection method. Moreover, a coordinate transformation and a reformulation of the original dynamics using redundant coordinates is useful for analysis of the servo-constraints problem. The inverse model DAEs are here solved using an implicit Euler scheme to ensure real-time capability.

Servo-constraints are applied on several examples. The example of a mass-on-car system is used to demonstrate the solution convergence of the inverse model solution, when using either minimal or redundant coordinates for modeling the system. Moreover, better convergence is possible when index reduction by projection is applied or the servo-constraints are directly substituted into the equations. This is possible when part of the coordinates directly forms the output. An extended version of the mass-on-car system illustrates how the coordinate transformation helps in extracting the internal dynamics in ODE form. Stability of the internal dynamics is analyzed based on the ODE formulation and the transfer function of the system. An f-mass-spring-damper chain shows that high index DAE problems arise even for simple multibody systems. Convergence of the original inverse model solution and index reduced formulations are compared. Finally, it is demonstrated how the choice of the actuator model can influence the differentiation index and relative degree of the inverse model.

References

1. Altmann, R., Heiland, J.: Simulation of multibody systems with servo constraints through optimal control. Multibody Syst. Dyn. **40**, 1–24 (2016)
2. Altmann, R., Betsch, P., Yang, Y.: Index reduction by minimal extension for the inverse dynamics simulation of cranes. Multibody Syst. Dyn. **36**(3), 295–321 (2016)
3. Bajodah, A.H., Hodges, D.H., Chen, Y.H.: Inverse dynamics of servo-constraints based on the generalized inverse. Nonlinear Dyn. **39**(1), 179–196 (2005)
4. Bastos, G., Seifried, R., Brüls, O.: Inverse dynamics of serial and parallel underactuated multibody systems using a DAE optimal control approach. Multibody Syst. Dyn. **30**(3), 359–376 (2013)
5. Berger, T.: The zero dynamics form for nonlinear differential-algebraic systems. IEEE Trans. Autom. Control **62**(8), 4131–4137 (2017). https://doi.org/10.1109/TAC.2016.2620561
6. Betsch, P., Quasem, M., Uhlar, S.: Numerical integration of discrete mechanical systems with mixed holonomic and control constraints. J. Mech. Sci. Technol. **23**(4), 1012–1018 (2009)
7. Betsch, P., Altmann, R., Yang, Y.: Numerical integration of underactuated mechanical systems subjected to mixed holonomic and servo constraints. Multibody Dyn. 1–18 (2016)
8. Blajer, W.: Index of differential-algebraic equations governing the dynamics of constrained mechanical systems. Appl. Math. Model. **16**(2), 70–77 (1992)

9. Blajer, W., Kolodziejczyk, K.: A geometric approach to solving problems of control constraints: theory and a DAE framework. Multibody Syst. Dyn. **11**(4), 343–364 (2004)
10. Blajer, W., Kolodziejczyk, K.: Motion planning and control of gantry cranes in cluttered work environment. IET Control Theory Appl. **1**(5), 1370–1379 (2007)
11. Blajer, W., Kołodziejczyk, K.: Improved DAE formulation for inverse dynamics simulation of cranes. Multibody Syst. Dyn. **25**(2), 131–143 (2011)
12. Brüls, O., Bastos, G.J., Seifried, R.: A stable inversion method for feedforward control of constrained flexible multibody systems. J. Comput. Nonlinear Dyn. **9**(1), 011014 (2013)
13. Campbell, S.L.: High-index differential algebraic equations. Mech. Struct. Mach. **23**(2), 199–222 (1995)
14. Campbell, S.L., Gear, C.W.: The index of general nonlinear DAEs. Numer. Math. **72**(2), 173–196 (1995). https://doi.org/10.1007/s002110050165
15. Devasia, S., Chen, D., Paden, B.: Nonlinear inversion-based output tracking. IEEE Trans. Autom. Control **41**(7), 930–942 (1996)
16. Estévez Schwarz, D., Lamour, R.: Diagnosis of singular points of properly stated daes using automatic differentiation. Numer. Algorithms **70**(4), 777–805 (2015). https://doi.org/10.1007/s11075-015-9973-x
17. Fliess, M., Lévine, J., Martin, P., Rouchon, P.: Flatness and defect of non-linear systems: introductory theory and examples. Int. J. Control **61**(6), 1327–1361 (1995)
18. Fumagalli, A., Masarati, P., Morandini, M., Mantegazza, P.: Control constraint realization for multibody systems. J. Comput. Nonlinear Dyn. **6**, 011002 (2010)
19. Hairer, E.: Stiff and Differential-Algebraic Problems. Springer, Berlin u.a. (2002)
20. Hairer, E.: Solving Ordinary Differential Equations: Nonstiff Problems, Second Revised edn., 3 printing edn. Springer, Berlin u.a. (2008)
21. Ilchmann, A., Ryan, E.P., Sangwin, C.J.: Tracking with prescribed transient behaviour. ESAIM Control Optim. Calc. Var. **7**, 471–493 (2002)
22. Isidori, A.: Nonlinear Control Systems, 3rd edn. Springer, Berlin [u.a.] (1996)
23. Masarati, P., Morandini, M., Fumagalli, A.: Control constraint of underactuated aerospace systems. J. Comput. Nonlinear Dyn. **9**(2), 021014 (2014)
24. Otto, S., Seifried, R.: Real-time trajectory control of an overhead crane using servo-constraints. Multibody Syst. Dyn. 1–17 (2017)
25. Petzold, L., Lötstedt, P.: Numerical solution of nonlinear differential equations with algebraic constraints ii: practical implications. SIAM J. Sci. Stat. Comput. **7**(3), 720–733 (1986). https://doi.org/10.1137/0907049
26. Sastry, S.: Nonlinear Systems Analysis, Stability, and Control. Springer, New York [u.a.] (1999)
27. Seifried, R.: Two approaches for feedforward control and optimal design of underactuated multibody systems. Multibody Syst. Dyn. **27**(1), 75–93 (2012)
28. Seifried, R., Blajer, W.: Analysis of servo-constraint problems for underactuated multibody systems. Mech. Sci. **4**(1), 113–129 (2013)
29. Siciliano, B., Oriolo, G., Sciavicco, L., Villani, L.: Robotics Modelling, Planning and Control. Springer, London (2009)
30. Skogestad, S.: Multivariable Feedback Control: Analysis and Design, reprinted edn. Wiley, Chichester u.a. (2004)
31. Slotine, J.J.E., Li, W.: Applied Nonlinear Control. Prentice Hall, Englewood Cliffs (1991)
32. Svaricek, F.: Nulldynamik linearer und nichtlinearer Systeme: Definition, Eigenschaften und Anwendungen. Automatisierungstechnik **54**(7), 310–322 (2006)

Systems of Differential Algebraic Equations in Computational Electromagnetics

Idoia Cortes Garcia, Sebastian Schöps, Herbert De Gersem, and Sascha Baumanns

Abstract Starting from space-discretisation of Maxwell's equations, various classical formulations are proposed for the simulation of electromagnetic fields. They differ in the phenomena considered as well as in the variables chosen for discretisation. This contribution presents a literature survey of the most common approximations and formulations with a focus on their structural properties. The differential-algebraic character is discussed and quantified by the differential index concept.

Keywords DAE index · Maxwell's equations · Quasistatic approximations

Mathematics Subject Classification (2010) 34A09, 35Q61, 78A25, 78M12, 65D30

1 Introduction

Electromagnetic theory has been established by Maxwell in 1864 and was reformulated into the language of vector calculus by Heavyside in 1891 [50, 64]. A historical overview can be found in the review article [77]. The theory is well understood and rigorously presented in many text books, e.g. [45, 49, 54]. More recently researchers have begun to formulate the equations in terms of exterior calculus and differential forms which expresses the relations more elegantly and metric-free, e.g. [51].

I. Cortes Garcia · S. Schöps (✉) · H. De Gersem
Technische Universität Darmstadt, Graduate School of Computational Engineering, Darmstadt, Germany
e-mail: cortes@gsc.tu-darmstadt.de; schoeps@gsc.tu-darmstadt.de; degersem@temf.tu-darmstadt.de

S. Baumanns
Universität zu Köln, Mathematisches Institut, Köln, Germany
e-mail: sbaumanns@math.uni-koeln.de

S. Campbell et al. (eds.), *Applications of Differential-Algebraic Equations: Examples and Benchmarks*, Differential-Algebraic Equations Forum,
https://doi.org/10.1007/11221_2018_8

The simulation of three-dimensional spatially distributed electromagnetic phenomena based on Maxwell's equations is roughly 50 years old. An early key contribution was the proposition of the finite difference time domain method (FDTD) by Yee to solve the high-frequency hyperbolic problem on equidistant grids in 1966 [100], and its subsequent generalisations and improvements, e.g. [90]. Among the most interesting generalisations are the Finite Integration Technique [97] and the Cell Method [4] because they can be considered as discrete differential forms [24]. Most finite-difference codes formulate the problem in terms of the electric and magnetic field strength and yield ordinary differential equations after space discretisation which are solved explicitly in time. FDTD is very robust and remarkable efficient [68] and is considered to be among the 'top rank of computational tools for engineers and scientists studying electrodynamic phenomena and systems' [91].

Around the same time at which FIT was proposed, circuit simulation programs became popular, e.g. [70, 96] and Albert Ruehli proposed the Partial Element Equivalent Circuit method (PEEC) [79, 80]. PEEC is based on an integral formulation of the equations and utilises Green's functions similarly to the Boundary Element Method (BEM) or the Method of Moments (MOM) as BEM is called in the electromagnetics community [48].

Historically, the Finite Element Method (FEM) was firstly employed to Maxwell's equations using nodal basis functions. For vectorial fields, this produces wrong results known as 'spurious modes' in the literature. Their violation of the underlying structure, or more specifically of the function spaces, is nowadays well understood. Nédélec proposed his edge elements in 1980 [71] which are also known as Whitney elements [20]. A rigorous mathematical discussion can be found in many text books, e.g. [3, 5, 67]. Albeit less wide spread, the application of nodal elements is still popular, for example in the context of discontinuous Galerkin FEM [44, 53]. Also equivalences among the methods have been shown, most prominently FIT can be interpreted on hexahedral meshes as lowest order Nédélec FEM with mass lumping [17, 24].

From an application point of view, electromagnetic devices may behave very differently, e.g. a transformer in a power plant and an antenna of a mobile phone are both described by the same set of Maxwell's equations but still feature different phenomena. Therefore, engineers often solve subsets (*simplifications*) of Maxwell's equation that are relevant for their problem, for example the well-known eddy-current problem [39, 49, 76] or the well-known wave equation [91]. For each, one or more *formulations* have been proposed. They are either distinguished by the use of different variables or gauging conditions [14, 15].

It follows from the variety of simplifications and formulations that discretisation methods have individual strengths and weaknesses for the different classes of applications. For example, the formulation used in FDTD relies on an explicit time integration method which is particularly efficient if the mass matrices are easily invertible, i.e., if they are diagonal or at least block diagonal [23]. This allows FDTD to solve problems with several billions of degrees of freedom. Classical FEM is less commonly applied in that case but one may also analyse high-frequency electromagnetic phenomena in the frequency domain, either to investigate resonance behaviour [98] or source problems with right-hand-sides that can be assumed to vary sinusoidally at a given frequency. In these cases, one solves smaller complex-valued linear systems but the reduced sparsity due to FEM is counteracted by the flexibility in the mesh generation [16]. Also coupling Maxwell's equations to other physics may require tailored formulations, see for example for applications in the field of semiconductors [81, 83].

In the low-frequency regime the situation is often more involved since one deals with degenerated versions of Maxwell's equations as certain contributions to the equation vanish in the static limit and the original system becomes unstable [42, 55, 73]. One often turns to approximations of Maxwell's equations as the well-known eddy-current problem. These approximate formulations are often more complicated as they may yield parabolic-semi-elliptic equations that become eventually systems of differential-algebraic equations (DAEs) after space discretisation. The resulting systems are commonly integrated in time domain by fully or linear-implicit methods, e.g. [31, 72]. Only recently, explicit method gained again interest [6, 40, 85].

Most circuit and electromagnetic field formulations yield DAE systems; the first mathematical treatment of such problems can be traced back to the 60s [60] but gained increased interest in the 80s, e.g. [75]. An important concept in the analysis of DAEs and their well-posedness are the various index concepts, which try to quantify the difficulty of the numerical time-domain solution, see e.g. [75]. This paper discusses the most important low and high-frequency formulations in computational electromagnetics with respect to their differential index. An detailed introduction of the index and its variants is not discussed here and the reader is referred to text books and survey articles [25, 47, 60, 65].

This paper summarises relevant discrete formulations stemming from Maxwell's equations. It collects the corresponding known DAE results from the literature, i.e., [7, 10, 72, 94], homogenises their notation and discusses a few missing cases. Each problem is concretised by a mathematical description and specification of an example. The corresponding source code is freely available such that these example can be used as *benchmarks*, e.g. for the development of time integrators or numerical tools to analyse differential equations.

The paper is organised as follows: Sect. 2 discusses Maxwell's equations, the relevant material relations and boundary conditions. The classical low-frequency approximations and electromagnetic potentials are introduced. Section 3 outlines the spatial discretisation in terms of the finite element method and the finite integration technique. After establishing the DAE index concept in Sect. 4, the various discrete formulations are derived. They are discussed separately for the high-frequency full-wave case in Sect. 5 and the quasistatic approximations in Sect. 6. Finally, conclusion are drawn in Sect. 7.

2 Maxwell's Equations

Electromagnetic phenomena are described on the macroscopic level by Maxwell's equations [45, 49, 50, 54, 64]. Those can be studied in a standstill frame of reference in integral form

$$\int_{\partial A} \mathbf{E} \cdot \mathrm{d}\mathbf{s} = -\int_{A} \frac{\partial \mathbf{B}}{\partial t} \cdot \mathrm{d}\mathbf{A} \,, \tag{2.1a}$$

$$\int_{\partial V} \mathbf{D} \cdot \mathrm{d}\mathbf{A} = \int_{V} \rho \mathrm{d}V \,, \tag{2.1b}$$

$$\int_{\partial A} \mathbf{H} \cdot \mathrm{d}\mathbf{s} = \int_{A} \left(\frac{\partial \mathbf{D}}{\partial t} + \mathbf{J} \right) \cdot \mathrm{d}\mathbf{A} \,, \tag{2.1c}$$

$$\int_{\partial V} \mathbf{B} \cdot \mathrm{d}\mathbf{A} = 0, \tag{2.1d}$$

for all areas A and volumes $V \subset \mathbb{R}^3$. Using Stokes and Gauß' theorems one derives a set of partial differential equations, see e.g. [5, Chapter 1.1.2] for a mathematical discussion on their equivalence,

$$\nabla \times \mathbf{E} = -\frac{\partial \mathbf{B}}{\partial t} \,, \tag{2.2a}$$

$$\nabla \times \mathbf{H} = \frac{\partial \mathbf{D}}{\partial t} + \mathbf{J} \,, \tag{2.2b}$$

$$\nabla \cdot \mathbf{D} = \rho \,, \tag{2.2c}$$

$$\nabla \cdot \mathbf{B} = 0, \tag{2.2d}$$

with $\mathbf{E}$ the electric field strength, $\mathbf{B}$ the magnetic flux density, $\mathbf{H}$ the magnetic field strength, $\mathbf{D}$ the electric flux density and $\mathbf{J}$ the electric current density composed of conductive and source currents, being vector fields $\mathcal{I} \times \Omega \to \mathbb{R}^3$ depending on space $\mathbf{r} \in \Omega$ and time $t \in \mathcal{I}$. The electric charge density $\rho : \mathcal{I} \times \Omega \to \mathbb{R}$ is the only scalar field. Finally A and V are all areas (respectively volumes) in Ω.

Assumption 2.1 (Domain) *The domain $\Omega \subset \mathbb{R}^3$ is open, bounded, Lipschitz and contractible (simply connected with connected boundary, see e.g., [20]).*

Maxwell's equations give raise to the so-called de Rham complex, see e.g. [20]. It describes abstractly the relation of the electromagnetic fields in terms of the images and kernels of the differential operators. A simple visualisation is given in Fig. 1. This diagram is sometimes called Tonti diagram [93], Deschamps diagram [38], or, in the special case of Maxwell's equations, Maxwell's house [19, 21].

Exploiting the fact that the divergence of a curl vanishes, one can derive from Ampère-Maxwell's law (2.2b)–(2.2c) the continuity equation

$$0 = \frac{\partial \rho}{\partial t} + \nabla \cdot \mathbf{J} \,, \tag{2.3}$$

which can be interpreted in the static case as Kirchhoff's current law.

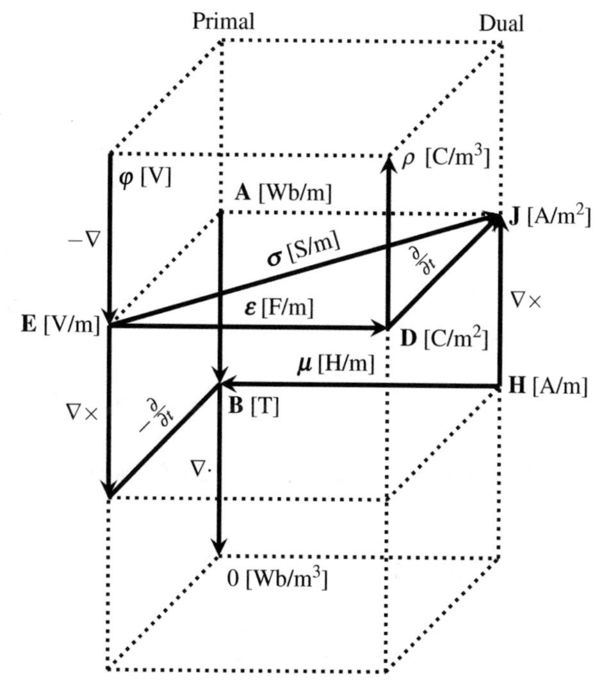

Fig. 1 Maxwell's house, based on similar diagrams in [19, 38, 93]. The concept of duality is for example discussed in the framework of differential forms in [51] and using traditional vector calculus in [54, Section 6.11]

2.1 Boundary Conditions and Material Relations

To mimic the behaviour of the electromagnetic field of an infinite domain on a truncated computational domain and to model field symmetries, boundary conditions are imposed on $\Gamma = \partial\Omega$. We restrict ourselves to homogeneous electric ('ebc') and magnetic boundary conditions ('mbc')

$$\begin{cases} \mathbf{n} \times \mathbf{E} = 0 & \text{in } \Gamma_{\text{ebc}} , \\ \mathbf{n} \times \mathbf{H} = 0 & \text{in } \Gamma_{\text{mbc}} , \end{cases} \tag{2.4}$$

where $\mathbf{n}$ is the outward normal to the boundary, $\Gamma_{\text{ebc}} \cup \Gamma_{\text{mbc}} = \Gamma$ and $\Gamma_{\text{ebc}} \cap \Gamma_{\text{mbc}} = \emptyset$.

Remark 1 Electrical engineers typically use the physical notation of electric ('ebc') or magnetic ('mbc') boundary conditions rather than the mathematical terminology of 'Dirichlet' or 'Neumann' conditions. The reason is that the mathematical distinction depends on the particular formulation, i.e. the variables chosen to describe the problem, while the physical point of view remains the same. For example in an E-based formulation, ebc and mbc correspond to Dirichlet and Neumann conditions, respectively, whereas in an H-based formulation, ebc and mbc correspond to Neumann and Dirichlet conditions, respectively.

Fig. 2 Sketch of domain

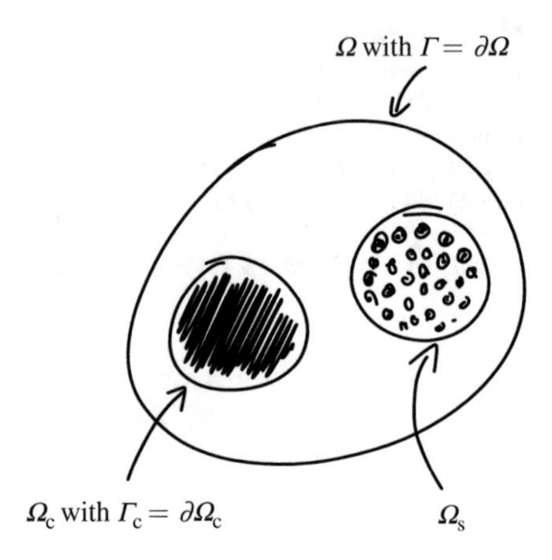

The fields in Maxwell's equations are further related to each other by the material relations

$$\mathbf{D} = \boldsymbol{\varepsilon}\mathbf{E}\,, \qquad \mathbf{J}_c = \boldsymbol{\sigma}\mathbf{E}\,, \qquad \mathbf{H} = \boldsymbol{\nu}\mathbf{B}\,, \tag{2.5}$$

where the permittivity $\boldsymbol{\varepsilon}$, conductivity $\boldsymbol{\sigma}$ and reluctivity (inverse permeability $\boldsymbol{\mu}$) $\boldsymbol{\nu}$ are rank-2 tensor fields, $\boldsymbol{\xi} : \Omega \to \mathbb{R}^{3\times3}$, $\boldsymbol{\xi} \in \{\boldsymbol{\varepsilon}, \boldsymbol{\sigma}, \boldsymbol{\mu}, \boldsymbol{\nu}\}$, whose possible polarisation or magnetisation and nonlinear or hysteretic dependencies on the fields are disregarded in the following for simplicity of notation and $\mathbf{J}_c : \mathcal{I} \times \Omega \to \mathbb{R}^3$ is the conduction current density. With these material relations one defines the total current density as

$$\mathbf{J} = \mathbf{J}_c + \mathbf{J}_s \tag{2.6}$$

where $\mathbf{J}_s$ is a given source current density that represents for example the current density impressed by a stranded conductor [86].

We assume the following material and excitation properties as shown in Fig. 2, see also [2, 86] for a more rigorous discussion.

Assumption 2.2 (Material) *The permittivity and permeability tensors, i.e., $\boldsymbol{\varepsilon}$ and $\boldsymbol{\mu}$, are positive definite on the whole domain Ω and only depend on space* $\mathbf{r}$. *The conductivity tensor is positive definite on a subdomain $\Omega_c \subset \Omega$ and vanishes elsewhere, i.e.,* $\text{supp}(\boldsymbol{\sigma}) = \Omega_c$. *The source current density is defined on the subdomain $\Omega_s \subset \Omega$ with $\Omega_c \cap \Omega_s = \emptyset$, such that* $\text{supp}(\mathbf{J}_s) = \Omega_s$.

This assumption describes the situation of an excitation given by one or several stranded conductors. The key assumption behind this model is a homogeneous current distribution which is justified in many situations, since the individual strands have diameters small than the skin depth and are therefore not affected by eddy currents, i.e. Ω_c and Ω_s are disjoint. Other models, e.g. solid and foil conductors, are not covered here. However, it can be shown that the various models can be transformed into each other and thus have similar properties [84].

2.2 Modelling of Excitations

The excitation has been given in (2.6) by the known source current density $\mathbf{J}_s$ which is typically either determined by given voltage drops $u_k : \mathscr{I} \to \mathbb{R}$ or lumped currents $i_k : \mathscr{I} \to \mathbb{R}$. The 3D-0D coupling is governed by so-called *conductor models*. Besides the *solid* and *stranded* models [12], also more elaborated conductors have been proposed, e.g., foil-conductor models [35].

The source current density $\mathbf{J}_s$ is not necessarily solenoidal, i.e.

$$\nabla \cdot \mathbf{J}_s \neq 0.$$

Divergence-freeness is only required for the total current density $\mathbf{J}$ in the absence of charge variations due to the continuity equation (2.3). This has been exploited e.g. in [86, Figure 3] to increase the sparsity of the coupling matrices. However, most conductor models enforce this property such that the source current can be given alternatively in terms of a source magnetic field strength

$$\mathbf{J}_s = \nabla \times \mathbf{H}_s.$$

In [86] the abstract framework of *winding density functions* was proposed. It unifies the individual stranded, solid and foil conductor models and denotes them abstractly by

$$\boldsymbol{\chi}_k : \Omega \to \mathbb{R}^3 \tag{2.7}$$

with an superscript if needed to distinguish among models, e.g. (i) for stranded and (u) solid conductors. In the simplest case they are characteristic functions with a given orientation.

Example 2 If $\Omega_s = \Omega_{s,1} \cup \Omega_{s,2} \cup \Omega_{s,3}$ consists of two parts of a winding oriented in z-direction, each with cross section A_k and made of N_k strands, and a massive bar with length ℓ_3 aligned with the z-direction, the source current is given by

$$\mathbf{J}_s = \sum_{k=1}^{2} \boldsymbol{\chi}_k^{(i)} i_k + \sigma \boldsymbol{\chi}_3^{(u)} u_3 \,. \tag{2.8}$$

The winding density functions for the stranded conductor model are

$$\boldsymbol{\chi}_k^{(i)}(\mathbf{r}) = \begin{cases} \frac{N_k}{A_k} \mathbf{n}_z & \mathbf{r} \in \Omega_{s,k} \\ 0 & \text{otherwise} \end{cases} \tag{2.9}$$

and the unit vector in z-direction is denoted by $\mathbf{n}_z$. The stranded conductor model distributes an applied current in a homogeneous way such that the individual strands are neither spatially resolved nor modelled as line currents which would cause a too high computational effort. There are many proposals in the literature on how to construct them, most often a Laplace-type problem is solved on the subdomain Ω_s, see e.g. [36, 41, 86]. The winding function for the solid conductor is

$$\boldsymbol{\chi}_3^{(\mathrm{u})}(\mathbf{r}) = \begin{cases} \frac{1}{\ell_3}\mathbf{n}_z & \mathbf{r} \in \Omega_{\mathrm{s},3} \\ 0 & \text{otherwise}\ . \end{cases} \tag{2.10}$$

The solid conductor model homogeneously distributes an applied voltage drop in the massive-conductor's volume.

The winding density functions allow to retrieve global quantities in a post-processing step, i.e., the current through a solid conductor model is calculated by

$$i_k = \int_\Omega \boldsymbol{\chi}_k^{(\mathrm{u})} \cdot \mathbf{J}\,\mathrm{d}V \tag{2.11}$$

and the voltage induced along a stranded conductor model follows from

$$u_k = -\int_\Omega \boldsymbol{\chi}_k^{(\mathrm{i})} \cdot \mathbf{E}\,\mathrm{d}V\ . \tag{2.12}$$

The expressions (2.8), (2.11) and (2.12) can also be used to set up a field-circuit coupled model [37].

An important property postulated in [86] is that winding functions should fulfil a partition of unity property. The integration of $\boldsymbol{\chi}_k(\mathbf{r})$ along a line ℓ_k between both electrodes of a solid conductor gives always 1 and analogously, $\boldsymbol{\chi}_k^{(\mathrm{i})}(\mathbf{r})$ integrated over any cross-sectional plane A_k of a stranded conductor should equal the number of turns N_k of the winding:

$$\int_{\ell_k} \boldsymbol{\chi}_k^{(\mathrm{u})} \cdot \mathrm{d}\mathbf{s} = 1\ , \quad \forall \ell_k \quad \text{and} \quad \int_{A_k} \boldsymbol{\chi}_k^{(\mathrm{i})} \cdot \mathrm{d}\mathbf{S} = N_k, \quad \forall A_k\ . \tag{2.13}$$

Furthermore, conductor models should not intersect, i.e., [7]

$$\boldsymbol{\chi}_i \cdot \boldsymbol{\chi}_j \equiv 0 \qquad \text{for} \quad i \neq j \tag{2.14}$$

where $\boldsymbol{\chi}_i$ and $\boldsymbol{\chi}_j$ are winding functions of any type.

For simplicity of notation, we will restrict us in the following to the case of non-intersecting stranded conductors models, i.e.

Assumption 2.4 (Excitation) *The source current density is given by n_{str} winding functions that fulfil* (2.13) *and* (2.14) *such that the excitation is given by*

$$\mathbf{J}_s = \sum_{k=1}^{n_{\mathrm{str}}} \boldsymbol{\chi}_k i_k \quad \textit{where} \quad \boldsymbol{\chi}_k \equiv \boldsymbol{\chi}_k^{(i)}.$$

2.3 Static and Quasistatic Fields

Following the common classification of slowly varying electromagnetic fields, [39], we introduce the following definition for quasistatic and static fields

Definition 3 (Simplifications) The fields in Eq. (2.2) are called

(a) static if the variation of the magnetic and electric flux densities is disregarded:

$$\frac{\partial}{\partial t}\mathbf{B} = 0 \quad \text{and} \quad \frac{\partial}{\partial t}\mathbf{D} = 0 \, ;$$

(b) electroquasistatic if the variation of the magnetic flux density is disregarded:

$$\frac{\partial}{\partial t}\mathbf{B} = 0 \, ;$$

(c) magnetoquasistatic if the variation of the electric flux density is disregarded:

$$\frac{\partial}{\partial t}\mathbf{D} = 0 \, ;$$

(d) full wave if no simplifications are made.

In contrast to the full Maxwell's equations, the classical quasistatic approximations above feature only first order derivatives w.r.t. to time. However, there is another model for slowly varying fields that does not fit into this categorisation, the so-called Darwin approximation, e.g. [61]. It considers the decomposition of the electric field strength $\mathbf{E} = \mathbf{E}_{\mathrm{irr}} + \mathbf{E}_{\mathrm{rem}}$ into an *irrotational part* $\mathbf{E}_{\mathrm{irr}}$ and a *remainder part* $\mathbf{E}_{\mathrm{rem}}$. In contrast to (a)–(c) the Darwin approximation only neglects the displacement currents related to $\mathbf{E}_{\mathrm{rem}}$ from the law of Ampère-Maxwell (2.2b). It still considers second order time derivatives.

The various approximations neglect the influence of several transient phenomena with respect to others, which implicitly categorises fields into primary and secondary ones. For example, let us consider a magnetoquasistatic situation, i.e., the displacement current density $\frac{\partial}{\partial t}\mathbf{D} = 0$ is disregarded. This still allows the electric field $\frac{\partial}{\partial t}\mathbf{E} \neq 0$ to vary. However, this variation implies that there is a secondary displacement current density $\frac{\partial}{\partial t}\mathbf{D} = \frac{\partial}{\partial t}\boldsymbol{\varepsilon}\mathbf{E} \neq 0$ which is in the formulation not further considered.

Remark 4 Depending on the application, an electrical engineer chooses the formulation that is best suited for the problem at hand. Typically the physical dimensions, the materials and the occurring frequency are used to estimate which simplification is acceptable, see e.g. [49, 82, 89].

2.4 Electromagnetic Potentials

Typically, one combines the relevant Maxwell equations into a *formulation* by defining appropriate *potentials*. One possibility is the $\mathbf{A}$–ϕ formulation [13, 20, 56], where a *magnetic vector potential* $\mathbf{A} : \mathscr{I} \times \Omega \to \mathbb{R}^3$ and an *electric scalar potential* $\phi : \mathscr{I} \times \Omega \to \mathbb{R}$ follow as integration constants from integrating the magnetic Gauss law and Faraday-Lenz' law in space, i.e.,

$$\mathbf{B} = \nabla \times \mathbf{A} \qquad \text{and} \qquad \mathbf{E} = -\frac{\partial \mathbf{A}}{\partial t} - \nabla \phi \,. \tag{2.15}$$

The magnetic flux density $\mathbf{B}$ defines the magnetic vector potential $\mathbf{A}$ only up to a gradient field. For a unique solution an additional gauging condition is required [13, 30, 62].

A different approach can be taken with the $\mathbf{T}$–Ω formulation in case of a magnetoquasistatic approximation (Definition 3(c)) [15, 26, 95]. Here, an *electric vector potential* $\mathbf{T} : \mathscr{I} \times \Omega \to \mathbb{R}^3$ and a *magnetic scalar potential* $\psi : \mathscr{I} \times \Omega \to \mathbb{R}$ describe the fields as

$$\mathbf{J}_\mathrm{c} = \nabla \times \mathbf{T} \qquad \text{and} \qquad \mathbf{H} = \mathbf{H}_\mathrm{s} + \mathbf{T} - \nabla \psi \,, \tag{2.16}$$

with $\nabla \times \mathbf{H}_\mathrm{s} = \mathbf{J}_\mathrm{s}$. Again, to ensure uniqueness of solution, an additional gauge condition is necessary for $\mathbf{T}$. In contrast to the $\mathbf{A}$–ϕ-formulation, the electric vector potential $\mathbf{T}$ is only non-zero on Ω_c.

Existence and uniqueness of the continuous solution will not be discussed in this contributions, see for example [2, 42] for several formulations in the frequency domain case with anisotropic materials and mixed boundary conditions.

The boundary conditions introduced in (2.4) can now be translated into expressions involving only the potentials. This yields for the A–ϕ-formulation

$$\begin{cases} \mathbf{n} \times \mathbf{A} = 0, \ \phi = 0 & \text{on } \Gamma_\mathrm{ebc} \,, \\ \mathbf{n} \times (\nu \nabla \times \mathbf{A}) = 0, & \text{on } \Gamma_\mathrm{mbc} \end{cases} \tag{2.17}$$

and the for the T–Ω one

$$\begin{cases} \mu \frac{\partial \psi}{\partial \mathbf{n}} = 0, & \text{on } \Gamma_\mathrm{ebc} \,, \\ \mathbf{n} \times \nabla \psi = 0, & \text{on } \Gamma_\mathrm{mbc} \,. \end{cases} \tag{2.18}$$

For the electric vector potential $\mathbf{T}$ in the $\mathbf{T}$–Ω formulation, boundary conditions have to be set on the corresponding subdomain where it is defined $\Gamma_\mathrm{c} = \partial \Omega_\mathrm{c}$. This leads to electric boundary conditions

$$\mathbf{n}_\mathrm{c} \times \mathbf{T} = 0 \qquad \text{on } \Gamma_\mathrm{c} \,,$$

with, analogous to the cases before, $\mathbf{n}_\mathrm{c}$ being the outward normal unit vector of Γ_c.

3 Spatial Discretisation

Starting from a differential formulation the Ritz-Galerkin the FE method can be applied using the appropriate Whitney basis functions [67]. Alternatively, FIT or similarly the Cell Method provide a spatial discretisation of Maxwell's equations based on the integral form [4, 97]. In the lowest order case FE and FIT only differ by quadrature, i.e., FIT uses the midpoint rule [17]. We derive in the following the discretisation of the partial differential operators in the terminology of FIT on an hexahedral grid since this allows a simple and explicit construction of divergence, curl and gradient matrices which will aid the following discussion.

3.1 Domain and Grid

The domain Ω is decomposed into an oriented simplicial complex that forms the computational grid. For the explanation, it is considered to be a brick and the grid is defined in cartesian coordinates as

$$G = \{V(i_x, i_y, i_z) \subset \mathbb{R}^3 | V(i_x, i_y, i_z) = [x_{i_x}, x_{i_x+1}] \times [y_{i_y}, y_{i_y+1}] \times [z_{i_z}, z_{i_z+1}],$$
$$\text{for } i_x = 1, \ldots, n_x - 1;\ i_y = 1, \ldots, n_y - 1;\ i_z = 1, \ldots, n_z - 1\}.$$

The elements $V(i_x, i_y, i_z) = V(n)$ are numbered consecutively with an index n:

$$n(i_x, i_y, i_z) = i_x k_x + (i_y - 1)k_y + (i_z - 1)k_z,$$

with $k_x = 1$, $k_y = n_x$ and $k_z = n_x n_y$. Our discrete field quantities can be defined on several geometrical objects such as points $P(n)$, edges $L_\omega(n)$ or facets $A_\omega(n)$. An edge $L_\omega(n)$ connects points $P(n)$ and $P(n+k_\omega)$ in $\omega = \{x, y, z\}$ direction. The facet $A_\omega(n)$ is defined by its smallest possible point $P(n)$ and directed such that its normal vector points towards ω. There are $N = n_x n_y n_z$ points and as each point defines three edges and facets, there are in total $N_{\text{dof}} = 3n_x n_y n_z$ edges and facets, ordered in x, y and finally z-direction.

Nowadays, inspired by the notation of differential forms, it is well understood that a consistent mimetic discretisation of Maxwell's equations requires a primal/dual mesh pair. Even the discretisation with Whitney FEs implicitly constructs a dual mesh [20]. This can be traced back to the inherent structure of Maxwell's equations which are formed with quantities being dual to each other (see [54, Section 6.11]) that are linked by material properties (hodge operators in the terminology of differential forms). This concept is for example rigorously introduced in [51].

In contrast to FEM, both FIT and the Cell Method define the second (dual) grid $\tilde{G}$ explicitly. It is obtained by taking the centre of the cells in G as dual grid points (see Fig. 3). Now the dual quantities can be defined on the dual points $\widetilde{P}(n)$, edges

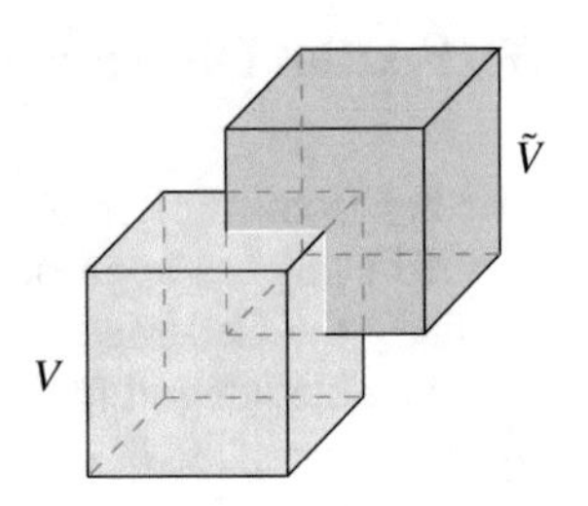

Fig. 3 Primal and dual grid cells

$\widetilde{L}_\omega(n)$, facets $\widetilde{A}_\omega(n)$ and volumes $\widetilde{V}_n$. Dual edges and facets are truncated at the boundary [99].

3.2 Maxwell's Grid Equations

To illustrate the construction of the operator matrices, Faraday-Lenz's law in integral form, i.e. Eq. (2.1a),

$$\int_{\partial A} \mathbf{E} \cdot \mathrm{d}\mathbf{s} = -\int_{A} \frac{\partial \mathbf{B}}{\partial t} \cdot \mathrm{d}\mathbf{A} \,,$$

is used as an example. The equality must be fulfilled for all areas A, in particular for each facet $A_\omega(i, j, k)$ of the computational grid G. For the case $\omega = z$,

$$\widehat{\mathbf{e}}_x(i, j, k) + \widehat{\mathbf{e}}_y(i+1, j, k) - \widehat{\mathbf{e}}_x(i, j+1, k) - \widehat{\mathbf{e}}_y(i, j, k) = -\frac{\mathrm{d}}{\mathrm{d}t}\widehat{\widehat{\mathbf{b}}}_z(i, j, k) \,,$$

with

$$\widehat{\mathbf{e}}_\omega(i, j, k) = \int_{L_\omega(i,j,k)} \mathbf{E} \cdot \mathrm{d}\mathbf{s} \quad \text{and} \quad \widehat{\widehat{\mathbf{b}}}_z(i, j, k) = -\int_{A_z(i,j,k)} \mathbf{B} \cdot \mathrm{d}\mathbf{A} \,.$$

This procedure is carried out for all the facets of G and the following matrix equation

$$\underbrace{\begin{bmatrix} & & \vdots & & \\ \cdots 1 & \cdots & -1 \; \cdots & -1 \; 1 & \cdots \\ & & \vdots & & \end{bmatrix}}_{\mathbf{C}} \widehat{\mathbf{e}} = -\frac{\mathrm{d}}{\mathrm{d}t}\widehat{\widehat{\mathbf{b}}}$$

is obtained, which describes Faraday's law in our grid. The matrix $\mathbf{C}$ applies the curl operator on quantities integrated along edges. Similarly, the divergence matrix $\mathbf{S}$, acting on surface integrated degrees of freedom and the gradient matrix $\mathbf{G}$ are built. The same strategy is followed to obtain the matrices for the dual grid $\widetilde{\mathbf{C}}$, $\widetilde{\mathbf{S}}$ and $\widetilde{\mathbf{G}}$. It can be shown that the matrices mimic all classical identities of vector field on the discrete level, e.g. [87] and [84, Appendix A]. With this, the semi-discrete Maxwell's Grid Equations

$$\mathbf{C}\widehat{\mathbf{e}} = -\frac{\mathrm{d}}{\mathrm{d}t}\widehat{\widehat{\mathbf{b}}}\,, \tag{3.1a}$$

$$\widetilde{\mathbf{C}}\widehat{\mathbf{h}} = \frac{\mathrm{d}}{\mathrm{d}t}\widehat{\widehat{\mathbf{d}}} + \widehat{\widehat{\mathbf{j}}}\,, \tag{3.1b}$$

$$\mathbf{S}\widehat{\widehat{\mathbf{b}}} = 0\,, \tag{3.1c}$$

$$\widetilde{\mathbf{S}}\widehat{\widehat{\mathbf{d}}} = \mathbf{q} \tag{3.1d}$$

are obtained which are closely resemble the system (2.2). The matrices $\mathbf{C}, \widetilde{\mathbf{C}} \in \{-1, 0, 1\}^{N_{\mathrm{dof}} \times N_{\mathrm{dof}}}$ are the discrete curl operators, $\mathbf{S}, \widetilde{\mathbf{S}} \in \{-1, 0, 1\}^{N \times N_{\mathrm{dof}}}$ the discrete divergence operators, which are all defined on the primal and dual grid, respectively. The fields are semi-discretely given by $\widehat{\mathbf{e}}, \widehat{\mathbf{h}}, \widehat{\widehat{\mathbf{d}}}, \widehat{\widehat{\mathbf{j}}}, \widehat{\widehat{\mathbf{b}}} : \mathscr{I} \to \mathbb{R}^{N_{\mathrm{dof}}}$ and $\mathbf{q} : \mathscr{I} \to \mathbb{R}^{N}$, and correspond to integrals of electric and magnetic voltages, electric fluxes, electric currents, magnetic fluxes and electric charges, respectively.

Lemma 1 *The operator matrices fulfil the following properties [99]*

- *divergence of the curl and curl of the gradient vanish on both grids*

$$\mathbf{SC} = 0\,, \quad \widetilde{\mathbf{S}}\widetilde{\mathbf{C}} = 0 \quad \textit{and} \quad \mathbf{CG} = 0\,, \quad \widetilde{\mathbf{C}}\widetilde{\mathbf{G}} = 0 \tag{3.2}$$

- *primal (dual) gradient and dual (primal) divergence fulfill*

$$\mathbf{G} = -\widetilde{\mathbf{S}}^{\top} \quad \textit{and} \quad \widetilde{\mathbf{G}} = -\mathbf{S}^{\top} \tag{3.3}$$

- *curl and dual curl are related by*

$$\widetilde{\mathbf{C}} = \mathbf{C}^{\top}. \tag{3.4}$$

Furthermore, potentials can be introduced on the primal grid, i.e.

$$\widehat{\mathbf{e}} = -\frac{\mathrm{d}}{\mathrm{d}t}\widehat{\mathbf{a}} - \mathbf{G}\boldsymbol{\Phi}\,, \tag{3.5}$$

where $\widehat{\mathbf{a}}$ is the line-integrated magnetic vector potential and $\boldsymbol{\Phi}$ the electric scalar potential located on primary nodes. This is similar to the definition of the potentials in the continuous case, i.e., (2.15). The properties stated in this Lemma have been proven in [8, 24, 84].

The numbering scheme explained in Sect. 3.1 yields matrices with a simple banded structure. The sparsity pattern is such that an efficient implementation may not construct those matrices explicitly but apply the corresponding operations as such to vectors. However, the numbering scheme introduces superfluous objects allocated outside of the domain Ω. For example in the case of the points located at the boundary where $i_x = n_x$, an edge in x direction $L_x(n_x, i_y, i_z) \notin \Omega$. Those objects are called *phantom objects*. However, the homogeneous Dirichlet boundary conditions explained in Sect. 2.1, as well as the deletion of the phantom objects can be incorporated either by removing them with (truncated) projection matrices or by setting the corresponding degrees of freedom to zero. For a more detailed description of the process, see [8] and [84, Appendix A].

Assumption 3.2 (Boundary Conditions) *The degrees of freedom and all the operators are projected to an appropriate subspace considering the homogeneous Dirichlet boundary ('ebc') conditions and disregarding any phantom objects in* $\widetilde{\mathbf{S}}$, $\widetilde{\mathbf{S}}^\top$, $\mathbf{C}$ *and* $\widetilde{\mathbf{C}}$. *Therefore* $\ker \widetilde{\mathbf{S}}^\top = 0$.

This assumption imposes boundary conditions directly on the system matrices and thus is a necessary condition to ensure uniqueness of solution. It is important to note that the reduced matrices keep the properties described in Lemma 1, see for example [8, Section 3.2.4].

Please note that identical operators (without phantom objects) are obtained when applying the FE method with lowest-order Whitney basis functions using the same primal grid [17, 24].

3.3 Material Matrices

The degrees of freedom have been introduced as integrals and thus the discretisation did not yet introduce any approximation error. This however happens when applying the matrices describing the material relations. In the FE case, the material matrices are given by the integrals

$$\left[\mathbf{M}_\xi\right]_{n,m} = \int_\Omega \mathbf{w}_n \cdot \xi \mathbf{w}_m \, \mathrm{d}\Omega \ ,$$

where $\xi \in \{\sigma, \boldsymbol{\nu}, \boldsymbol{\varepsilon}\}$ and $\mathbf{w}_\star$ are from an appropriate space, i.e., tangentially continuous Nédélec vectorial shape functions [18, 71] related to the nth edge of the grid for discretising $\boldsymbol{\varepsilon}$ and $\boldsymbol{\sigma}$ and normally continuous Raviart-Thomas vectorial shape functions [18, 78] for discretizing $\boldsymbol{\nu}$.

In FIT, the matrix construction is derived from the Taylor expansion of the material laws. In the following, only the construction of the conductivity matrix is explained. For simplicity of notation, the conductivity $\sigma(\mathbf{r})$ is assumed to be isotropic and conforming to the primal grid, i.e. $\sigma^{(n)} = \sigma(\mathbf{r}_n)$ is constant on each primal volume ($\mathbf{r}_n \in V(n)$). Consider a primal edge $L_z(i, j, k)$ and its associated dual facet $\widetilde{A}_z(i, j, k)$ (Fig. 4). The tangential component E_z of the electric field

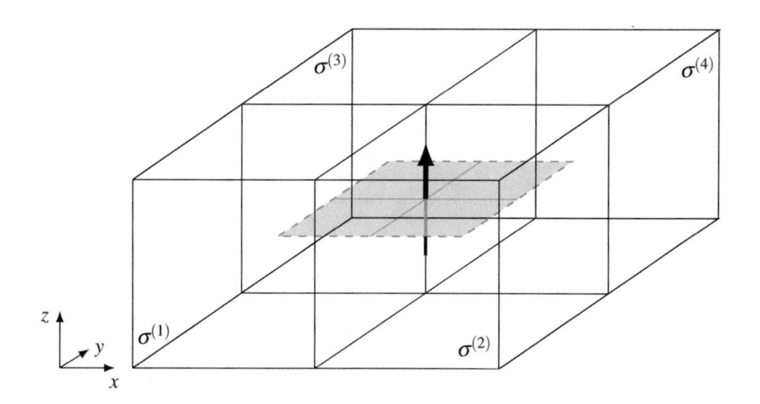

Fig. 4 Sketch of dual facet $\widetilde{A}_z(i, j, k)$ with its normal vector

strength is continuous along $L_z(i, j, k)$ and is found by approximation from

$$\widehat{\mathbf{e}}_z(i, j, k) = \int_{L_z(i,j,k)} \mathbf{E} \cdot \mathrm{d}\mathbf{s} \approx E_z \, |L_z(i, j, k)| \, ,$$

where $|\cdot|$ denotes the length, area or volume depending on the object. The current density integrated on the corresponding dual facet reads

$$\begin{aligned}\widehat{\widehat{\mathbf{j}}}_z(i, j, k) &= \int_{\widetilde{A}_z(i,j,k)} \mathbf{J} \cdot \mathrm{d}\mathbf{A} = \int_{\widetilde{A}_z(i,j,k)} J_z \, \mathrm{d}A = \sum_{q=1}^{4} \int_{\widetilde{A}_z^{(q)}(i,j,k)} \sigma^{(q)} E_z \, \mathrm{d}A \\ &\approx \sum_{q=1}^{4} \sigma^{(q)} E_z |\widetilde{A}_z^{(q)}(i, j, k)| = \mathbf{M}_{\sigma,i,j,k} \widehat{\mathbf{e}}_z(i, j, k) \, ,\end{aligned}$$

where the conductances $\mathbf{M}_{\sigma,i,j,k} = \bar{\sigma}(i, j, k) \frac{|\widetilde{A}_z(i,j,k)|}{|L_z(i,j,k)|}$ include the conductivities

$$\bar{\sigma}(i, j, k) = \sum_{q=1}^{4} \sigma^{(q)} \frac{|\widetilde{A}_z^{(q)}(i, j, k)|}{|\widetilde{A}_z(i, j, k)|}$$

averaged according to the conductivities $\sigma^{(q)}$ of the primal grid cells $V^{(q)}$ surrounding $L_z(i, j, k)$ and the surface fractions $\widetilde{A}_z^{(q)}(i, j, k) = V^{(q)} \cap \widetilde{A}_z(i, j, k)$. Analogously, material matrices for ε and ν are obtained, which lead to the discretised material relations

$$\widehat{\widehat{\mathbf{d}}} = \mathbf{M}_\varepsilon \widehat{\mathbf{e}} \, , \qquad \widehat{\widehat{\mathbf{j}}}_\mathrm{c} = \mathbf{M}_\sigma \widehat{\mathbf{e}} \, , \qquad \widehat{\mathbf{h}} = \mathbf{M}_\nu \widehat{\widehat{\mathbf{b}}}$$

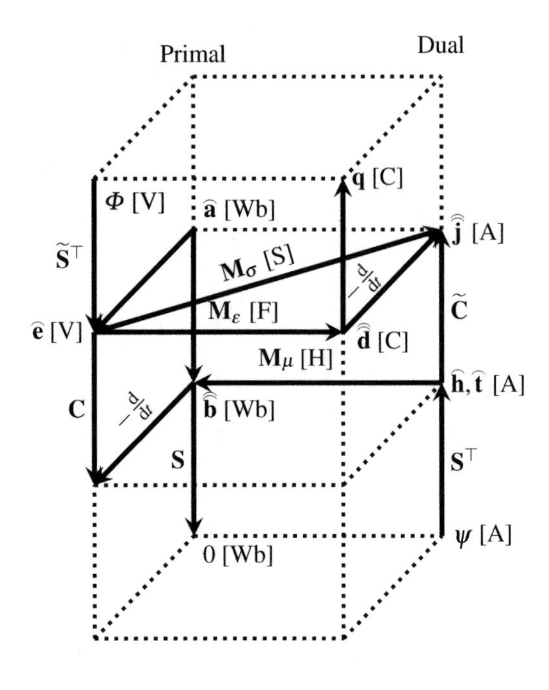

Fig. 5 Maxwell's house after spatial discretisation

and

$$\widehat{\widehat{\mathbf{j}}} = \widehat{\widehat{\mathbf{j}}}_{\mathrm{c}} + \widehat{\widehat{\mathbf{j}}}_{\mathrm{s}}$$

with the source current density $\widehat{\widehat{\mathbf{j}}}_{\mathrm{s}}$, which may be given by the discretisation $\mathbf{X}$ of the winding function (2.7), such that $\widehat{\widehat{\mathbf{j}}}_{\mathrm{s}} = \sum_k \mathbf{X}_k\, i_k$ with currents i_k.

For the material matrices, one can show the following result [84, Appendix A].

Lemma 2 (Material Matrices) *The material matrices $\mathbf{M}_\xi$ are symmetric for all material properties $\xi = \{\sigma, \nu, \varepsilon\}$. If Assumption 2.2 holds, then the matrices $\mathbf{M}_\nu, \mathbf{M}_\varepsilon$ are positive definite whereas $\mathbf{M}_\sigma$ is only positive semidefinite.*

Finally, the discretised version of Maxwell's equations with its corresponding material laws can be visualised by 'Maxwell's house' shown in Fig. 5.

Remark 3 Both Lemmas 1 and 2, as well as Assumption 3.2 hold for Finite Element discretisations with basis functions fulfilling a discrete de Rham sequence.

Remark 4 In many applications the material parameters, e.g. the reluctivity, in (2.5) dependent nonlinearly on the fields. In these cases one may consider the linearised system but since the differential material properties inherit the relevant properties, e.g. [52], the characteristics of the DAE will also remain the same. In particular, there will be no change of nullspaces, see e.g. [7]

4 Differential Algebraic Equations

Starting from Maxwell's grid equations, various discrete time-domain formulations can be obtained. Depending on the choices made according to Definition 3, the resulting system is either static, first or second-order in time. In the dynamic cases, it can be written as a (linear) problem of the form

$$\mathbf{M}\frac{\mathrm{d}}{\mathrm{d}t}\mathbf{x}(t) + \mathbf{K}\mathbf{x}(t) = \mathbf{r}(t), \quad \text{and} \quad \mathbf{x}(t_0) = \mathbf{x}_0 \tag{4.1}$$

where $\mathbf{M}, \mathbf{K} \in \mathbb{R}^{n\times n}$ are matrices, $\mathbf{x} : [t_0, T] \to \mathbb{R}^n$ contains the time-dependent degrees of freedom and $\mathbf{r} : [t_0, T] \to \mathbb{R}^n$ is an input.

Definition 1 (DAE) Equation (4.1) is called a system of differential-algebraic equations (DAE) if $\mathbf{M}$ is singular.

There are many options how to perform time-discretisation ('integration') of a DAE (4.1), see for example [47]. We suggest the simplest approach: implicit Euler's method, i.e.,

$$(\mathbf{M}/\Delta t + \mathbf{K})\,\mathbf{x}_{n+1} = \mathbf{r}(t_{n+1}) + \mathbf{M}/\Delta t\ \mathbf{x}_n \tag{4.2}$$

where $\mathbf{x}_n \doteq \mathbf{x}(t_n)$ and $\Delta t = t_{n+1} - t_n$ is the time step. DAEs are commonly classified according to their *index*. Intuitively, it can be seen as a measure of the equations' sensitivity to perturbations of the input and the numerical difficulties when integrating. There are several competing index concepts. They essentially agree in the case of regular, linear problems, see [65] for detailed discussion. Therefore, we employ the simplest concept

Definition 2 (Differential Index [25]) If solvable and the right-hand-side is smooth enough, then the DAE (4.1) has differential index-ϑ if ϑ is the minimal number of analytical differentiations with respect to the time t that are necessary to obtain an ODE for $d\mathbf{x}/dt$ as a continuous function in $\mathbf{x}$ and t by algebraic manipulations only.

For $\vartheta \geq 2$ the time-integration becomes difficult. Let us consider the classical educational index-2 problem to motivate analytically the sensitivity with respect to perturbations. The problem is described by

$$\frac{\mathrm{d}}{\mathrm{d}t}x_1 = x_2 \quad \text{and} \quad x_1 = \sin(t) + \delta(t) \tag{4.3}$$

where $\delta(t) = 10^{-k}\sin(10^{2k}t)$ is a small perturbation with $k \gg 1$. The solution $x_2 = \cos(t) + 10^k\cos(10^{2k}t)$ is easily obtained by the product and chain rules. It shows that a very small perturbation in an index-2 system (at a high frequency) can have a serious impact (in the order of 10^k) on the solution when compared to the original solution $x_2 = \cos(t)$ of the unperturbed problem where $\delta = 0$.

Remark 3 For the index analysis in the following sections we assume that the right-hand sides are smooth enough.

Furthermore, DAEs are known for the fact that solutions have to fulfil certain constraints. One of the difficult parts in solving DAEs numerically is to determine a consistent set of initial conditions in order to start the integration [9, 43, 63].

Remark 4 ([60]) A vector $\mathbf{x}_0 \in \mathbb{R}^n$ is called a consistent initial value if there is a solution of (4.1) through $\mathbf{x}_0$ at time t_0.

The problems discussed in the following will have at most (linear) index-2 components. For this case it has be shown that if we are not interested in a consistent initialisation at time t_0 but accept a solution satisfying the DAE only after the first step, then one may apply the implicit Euler method starting with an operating point and still obtain the same solution after $t > t_0$ that one would have obtained using a particular consistent value [8, 9].

The aim of this paper is to study the index of the systems obtained with different formulations and approximations according to Definition 3.

5 Full-Wave Formulation

On first sight it seems optimal to analyse high-frequency electromagnetic phenomena, e.g. the radiation of antennas, in frequency domain. The right-hand-sides can often be assumed to vary sinusoidally and for a given frequency, the equations are linear as the materials are rather frequency than field-dependent. However, the solution of problems in frequency domain requires the resolution of very large systems of equations and becomes inconvenient if one is interested in many frequencies (*broadband solution*). Therefore, often time-domain simulations are carried out with right-hand-sides that excite a large frequency spectrum.

5.1 First-Order Formulation Time-Stepped by Leapfrog

When solving Maxwell's grid equations for lossless ($\sigma \equiv 0$) wave propagation problems in time domain, a problem formulation based on the electric and magnetic field is commonly proposed. Assuming that the initial conditions fulfil the divergence relations of System (2.2), one starts with Faraday's and Ampère's laws

$$\frac{\partial \mathbf{B}}{\partial t} + \nabla \times \mathbf{E} = 0 \quad \text{and} \quad \frac{\partial \mathbf{D}}{\partial t} - \nabla \times \mathbf{H} = \mathbf{J}_{\mathrm{s}} .$$

After inserting the material laws, the system becomes

$$\mu \frac{\partial \mathbf{H}}{\partial t} + \nabla \times \mathbf{E} = 0 \quad \text{and} \quad \boldsymbol{\varepsilon} \frac{\partial \mathbf{E}}{\partial t} - \nabla \times \mathbf{H} = \mathbf{J}_{\mathrm{s}} ,$$

with $\boldsymbol{\nu} = \boldsymbol{\mu}^{-1}$. Using Maxwell's grid equations (3.1), the semi-discrete initial value problem (IVP) has the form of Eq. (4.1) with unknown voltages $\mathbf{x}^\top := [\widehat{\mathbf{h}}^\top, \widehat{\mathbf{e}}^\top]$, right-hand-side $\mathbf{r}^\top := [0, \widehat{\widehat{\mathbf{j}}}_s^\top]$ and matrices

$$\mathbf{M} := \begin{bmatrix} \mathbf{M}_\nu^{-1} & 0 \\ 0 & \mathbf{M}_\varepsilon \end{bmatrix} \quad \text{and} \quad \mathbf{K} := \begin{bmatrix} 0 & \mathbf{C} \\ -\widetilde{\mathbf{C}} & 0 \end{bmatrix}. \tag{5.1}$$

If Assumptions 2.2 and 3.2 holds, all superfluous degrees of freedom are removed and the material matrices $\mathbf{M}_\nu$ and $\mathbf{M}_\varepsilon$ have full rank. With FIT the matrices are furthermore diagonal and thus easily inverted. A transformation by the matrices $\mathbf{M}_\nu^{-1/2}$ and $\mathbf{M}_\varepsilon^{1/2}$ allows us to rewrite (4.1) as

$$\frac{\mathrm{d}}{\mathrm{d}t}\bar{\mathbf{x}}(t) = \bar{\mathbf{K}}\bar{\mathbf{x}}(t) + \bar{\mathbf{r}}(t) \qquad \bar{\mathbf{x}}(t_0) = \bar{\mathbf{x}}_0 \tag{5.2}$$

in the new unknowns $\bar{\mathbf{x}}^\top = [(\mathbf{M}_\nu^{-1/2}\widehat{\mathbf{h}})^\top, (\mathbf{M}_\varepsilon^{1/2}\widehat{\mathbf{e}})^\top]$ with the skew-symmetric stiffness matrix

$$\bar{\mathbf{K}} = \begin{bmatrix} 0 & -\mathbf{M}_\nu^{1/2}\mathbf{C}\mathbf{M}_\varepsilon^{-1/2} \\ \mathbf{M}_\varepsilon^{-1/2}\widetilde{\mathbf{C}}\mathbf{M}_\nu^{1/2} & 0 \end{bmatrix}. \tag{5.3}$$

and right-hand-side $\bar{\mathbf{r}}^\top = [0, (\mathbf{M}_\varepsilon^{-1/2}\widehat{\widehat{\mathbf{j}}}_s)^\top]$. Let us conclude this by the following result.

Theorem 5.1 *Let Assumptions 2.1, 2.2 and 3.2 hold. Then, the semidiscrete full-wave Maxwell equations expressed in the field strengths, i.e.,* (5.2) *are an explicit system of ordinary differential equations.*

The resulting IVP could be readily solved by the implicit Euler method (4.2) or any method that is tailored for second order differential equations. However, as explained above FIT allows to efficiently invert the mass matrix $\mathbf{M}$ and thus explicit methods become interesting. Typically the leapfrog scheme (or equivalently Störmer-Verlet) are used [91]. The restriction on the time step size related to the Courant-Friedrichs-Lewy-condition (CFL) is tolerable if the dynamics of the right-hand-side are in a similar order of magnitude. Leapfrog is second-order accurate and symplectic, which is particularly interesting if there is no damping, i.e., no conductors present ($\boldsymbol{\sigma} \equiv 0$). Furthermore it can be shown that space and time errors are well balanced when using the leapfrog scheme with the a time step size close to the CFL limit ("magic time step") [90, Chapters 2.4 and 4].

Let the initial conditions be

$$\widehat{\mathbf{e}}^{(0)} = \widehat{\mathbf{e}}_0 \quad \text{and} \quad \widehat{\mathbf{h}}^{(\frac{1}{2})} = \widehat{\mathbf{h}}_{1/2},$$

then the update equations for the leapfrog scheme read [90, 97, 99]

$$\widehat{\mathbf{e}}^{(m+1)} = \widehat{\mathbf{e}}^{(m)} + \Delta t \mathbf{M}_{\varepsilon}^{-1} \left(\widetilde{\mathbf{C}} \widehat{\mathbf{h}}^{(m+\frac{1}{2})} - \widehat{\widehat{\mathbf{j}}}^{(m+\frac{1}{2})} \right),$$

$$\widehat{\mathbf{h}}^{(m+\frac{3}{2})} = \widehat{\mathbf{h}}^{(m+\frac{1}{2})} - \Delta t \mathbf{M}_{\nu} \widetilde{\mathbf{C}} \widehat{\mathbf{e}}^{(m+1)}$$

for the electric and magnetic voltages $\widehat{\mathbf{e}}^{(m)}$, $\widehat{\mathbf{h}}^{(m+\frac{1}{2})}$ at time instants t_m and $t_{m+\frac{1}{2}}$ with step size Δt. For equidistant grids, the resulting scheme is (up to scaling and interpretation) equivalent to Yee's FDTD scheme [100].

Remark 5.2 In practice, one may choose to violate Assumption 3.2. Instead one imposes the boundary conditions by setting the corresponding entries in the material matrices $\mathbf{M}_{\varepsilon}^{-1}$ and $\mathbf{M}_{\nu}$ to zero. In this case the system (5.1) comes with additional (trivial) equations when compared to a system that is projected to the lower dimensional subspace containing the boundary conditions. However, this preserves a simpler structure of the equation system and the topological grid operators, e.g. the discrete curl matrix $\mathbf{C}$, keep their banded structure.

Benchmark 5.2 *In [11] a spiral inductor model with coplanar lines located on a substrate layer with an air bridge was proposed as a benchmark example for high-frequency problems. The CST Microwave tutorial discusses the same model to advocate the usage of 3D field simulation instead of circuit models [34]. A slightly simplified geometry is illustrated in Fig. 6. The dimensions of the layer are* $7 \cdot 10^{-4}\,\text{m} \times 4.75 \cdot 10^{-4}\,\text{m} \times 2.5 \cdot 10^{-5}\,\text{m}$ *and Fig. 7 illustrates the dimensions of the coil.*

Fig. 6 Spiral inductor model with coplanar lines located on a substrate layer with an air bridge (Benchmark 5.2)

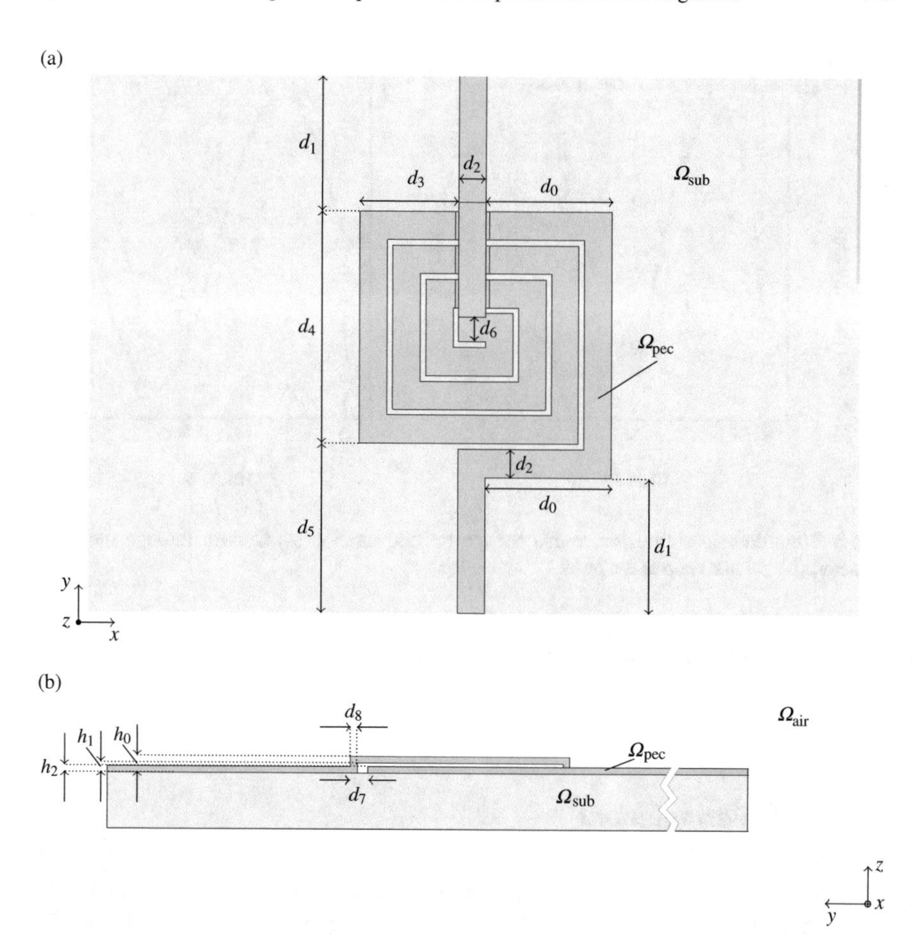

Fig. 7 Model of a spiral inductor of Benchmark 5.2. The distances are $d_0 = 1.15 \cdot 10^{-4}$ m, $d_1 = 1.2 \cdot 10^{-4}$ m, $d_2 = 2.5 \cdot 10^{-5}$ m, $d_3 = 9 \cdot 10^{-5}$ m, $d_4 = 2.05 \cdot 10^{-4}$ m, $d_5 = 1.5 \cdot 10^{-4}$ m, $d_6 = 2.2 \cdot 10^{-5}$ m, $d_7 = 9 \cdot 10^{-6}$ m, $d_8 = 3 \cdot 10^{-6}$ m, $h_0 = 8 \cdot 10^{-6}$ m, $h_1 = 5 \cdot 10^{-6}$ m and $h_2 = 3 \cdot 10^{-6}$ m. (**a**) x–y cross section of spiral inductor, (**b**) y–z cross section of spiral inductor

The bottom of the substrate layer is constrained by ebc and the other five boundaries are by mbc. On each side of the bridge the coil is connected by a straight line of perfect conductor with the ebc bottom plane. One side is excited by a discrete port which is given by a current source $i(t) = \sin(2\pi f t)\mathrm{A}$ *with* $f = 50 \cdot 10^9$ 1/s. *The coil* (Ω_{pec}) *is assumed to be a perfect conductor, i.e., modelled by homogeneous electric boundary conditions, the substrate* (Ω_{sub}) *is given a relative permittivity of* $\varepsilon_r = 12$, *in the air region* Ω_{air} $\varepsilon_r = 1$ *and vacuum permeability* $\mu = 4\pi \cdot 1 \cdot 10^{-7}$ H/m *is assumed everywhere else.*

The structure is discretised using FIT with 4,06,493 *mesh cells and* 1,283,040 *degrees of freedom. Leapfrog is used with a time step of* $\Delta t = 3.4331 \cdot 10^{-15}$ s *based on the CFL condition and zero initial condition, see Fig. 8. The performance*

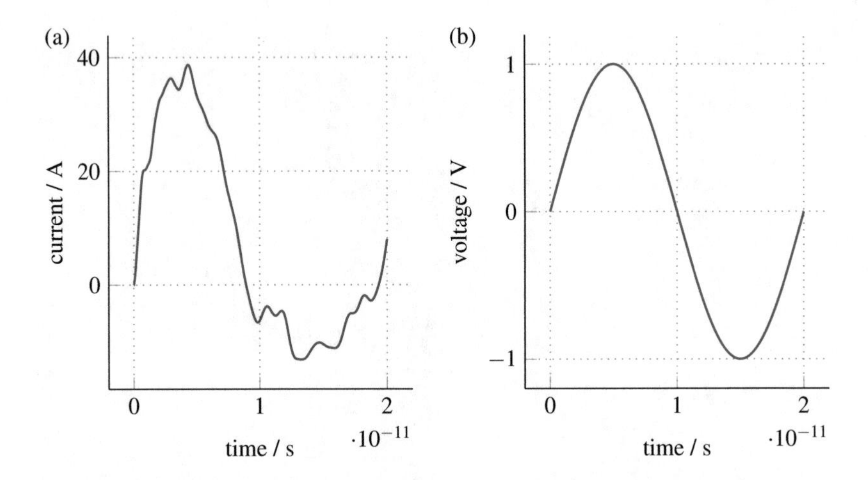

Fig. 8 Time domain simulation results for the Benchmark 5.2. (**a**) Current through the spiral inductor, (**b**) voltage drop at the ports

of leapfrog and exponential integrators for this model was recently discussed in [66].

5.2 A–Φ Formulations

If there are small geometric features, slowly varying excitations, conducting or semiconducting materials [27], the leapfrog scheme becomes inefficient. An alternative formulation is obtained if one rewrites Maxwell's equations as a second-order partial differential equation by combining Faraday's law, Ampère's law and the material equations complemented by Gauss's law, i.e.,

$$\boldsymbol{\varepsilon}\frac{\partial^2}{\partial t^2}\mathbf{E} + \boldsymbol{\sigma}\frac{\partial}{\partial t}\mathbf{E} + \nabla \times \nu \nabla \times \mathbf{E} = \frac{\partial}{\partial t}\mathbf{J}_{\mathrm{s}}. \tag{5.4}$$

$$-\nabla \cdot \boldsymbol{\varepsilon}\mathbf{E} = \rho \tag{5.5}$$

The inconvenience of a time-differentiated source current density can be mitigated by exploiting the potentials as defined in (2.15)

$$\boldsymbol{\varepsilon}\frac{\partial^2}{\partial t^2}\mathbf{A} + \frac{\partial}{\partial t}\boldsymbol{\varepsilon}\nabla\phi + \boldsymbol{\sigma}\frac{\partial}{\partial t}\mathbf{A} + \boldsymbol{\sigma}\nabla\phi + \nabla \times \nu \nabla \times \mathbf{A} = \mathbf{J}_{\mathrm{s}} \tag{5.6}$$

$$-\nabla \cdot \boldsymbol{\varepsilon}\frac{\partial}{\partial t}\mathbf{A} - \nabla \cdot \boldsymbol{\varepsilon}\nabla\phi = \rho. \tag{5.7}$$

There is an ambiguity of the electromagnetic potentials since $\mathbf{A}$ is only fixed up to a gradient field [54]. To this end, several gauging techniques have been introduced. For example grad-div formulations that are based on the Coulomb gauge, have been introduced for low frequencies [22, 30, 32] and high-frequency applications [46].

Let us define a regularisation of the electrodynamic potentials by the following gauge condition

$$\xi_1 \nabla \cdot \mathbf{A} + \xi_2 \phi + \xi_3 \frac{\partial}{\partial t} \phi = 0 \tag{5.8}$$

which yields for $\xi_1 = 1$, $\xi_2 = \xi_3 = 0$ the *Coulomb* gauge and for $\xi_1 = \nu$, $\xi_2 = 0$ and $\xi_3 = \varepsilon$ the *Lorenz* gauge if the considered materials are conducting, uniform, isotropic and linear. In the case $\xi_1 = \nu$, $\xi_2 = \sigma$ and $\xi_3 = \varepsilon$ the curl-curl equation (5.6) can be written as a pair of damped wave equations

$$\left[\Delta - \mu\sigma \frac{\partial}{\partial t} - \mu\varepsilon \frac{\partial^2}{\partial t^2}\right] \mathbf{A} = -\mu \mathbf{J}_\mathrm{s} \tag{5.9}$$

$$\left[\Delta - \mu\sigma \frac{\partial}{\partial t} - \mu\varepsilon \frac{\partial^2}{\partial t^2}\right] \phi = -\frac{\rho}{\varepsilon} \tag{5.10}$$

where Δ denotes the (scalar and vector) Laplace operators. In the undamped case ($\sigma = 0$), this system reduces to the well-known d'Alembert equations [54]. The right-hand-sides are still coupled via the continuity equation (2.3)

$$\nabla \cdot \mathbf{J}_\mathrm{s} + \frac{\sigma}{\varepsilon}\rho + \frac{\partial}{\partial t}\rho = 0 \tag{5.11}$$

where we have again exploited isotropy and homogeneity of σ and ε to obtain $-\nabla \cdot \sigma\mathbf{E} = \frac{\sigma}{\varepsilon}\rho$. When solving the system (5.9)–(5.11) we have to ensure that the (generalised) Lorenz gauge (5.8) is still fulfilled, which requires compatible boundary conditions for $\mathbf{A}$ and ϕ [10].

Now, let us derive a similar semidiscrete formulation based on the spatial discretisation introduced above. We start with the A–ϕ formulation (5.6) using the discretised laws of Ampère (3.1b) and Gauss (3.1d):

$$\tilde{\mathbf{C}}\mathbf{M}_\nu \mathbf{C}\widehat{\mathbf{a}} + \mathbf{M}_\sigma \left[\frac{\mathrm{d}}{\mathrm{d}t}\widehat{\mathbf{a}} + \mathbf{G}\boldsymbol{\Phi}\right] + \mathbf{M}_\varepsilon \left[\frac{\mathrm{d}^2}{\mathrm{d}t^2}\widehat{\mathbf{a}} + \mathbf{G}\frac{\mathrm{d}}{\mathrm{d}t}\boldsymbol{\Phi}\right] = \widehat{\widehat{\mathbf{j}}}_\mathrm{s} \tag{5.12}$$

$$-\tilde{\mathbf{S}}\mathbf{M}_\varepsilon \frac{\mathrm{d}}{\mathrm{d}t}\widehat{\mathbf{a}} + \mathbf{L}_\varepsilon \boldsymbol{\Phi} = \mathbf{q} \tag{5.13}$$

which contains the discrete Laplace operators

$$\mathbf{L}_\varepsilon := -\tilde{\mathbf{S}}\mathbf{M}_\varepsilon \mathbf{G} \quad \text{and} \quad \mathbf{L}_\sigma := -\tilde{\mathbf{S}}\mathbf{M}_\sigma \mathbf{G}\,, \tag{5.14}$$

for permittivity and conductivity, respectively.

Lemma 5.3 (Discrete Laplacians) *Let Assumptions 2.2 and 3.2 hold true, the discrete Laplace operator* $\mathbf{L}_\varepsilon$ *in* (5.14) *is symmetric positive definite and* $\mathbf{L}_\sigma$ *is symmetric positive semidefinite.*

Proof As we assume Dirichlet boundary conditions (ebc) in Assumption 3.2 and $\mathbf{G} = -\tilde{\mathbf{S}}^\top$ due to (3.3), the proof is straight forward. □

Equations (5.12) and (5.13) are coupled by the potentials and right-hand-sides via the continuity equation

$$\tilde{\mathbf{S}}\widehat{\widehat{\mathbf{j}}}_\mathrm{s} + \mathbf{L}_\sigma\mathbf{L}_\varepsilon^{-1}\mathbf{q} + \frac{\mathrm{d}}{\mathrm{d}t}\mathbf{q} = \left[\tilde{\mathbf{S}}\mathbf{M}_\sigma - \mathbf{L}_\sigma\mathbf{L}_\varepsilon^{-1}\tilde{\mathbf{S}}\mathbf{M}_\varepsilon\right]\frac{\mathrm{d}}{\mathrm{d}t}\widehat{\mathbf{a}}\,, \tag{5.15}$$

that is obtained by a left multiplication of Ampère's law by $\tilde{\mathbf{S}}$ and inserting Gauss' law etc. The steps are the same as in the continuous case, e.g., applying the divergence operator. Nonetheless, the discrete continuity equation (5.15) is more general than its continuous counterpart (5.11) as it covers anisotropic and non-homogeneous material distributions.

The ambiguity of the potentials is not yet fixed. The generalised discrete Lorenz gauge (5.8) for a conductive domain in FIT notation is given by

$$\mathbf{M}_\varepsilon\mathbf{G}\mathbf{M}_\mathrm{N}\tilde{\mathbf{S}}\mathbf{M}_\varepsilon\widehat{\mathbf{a}} + \mathbf{M}_\sigma\mathbf{G}\boldsymbol{\Phi} + \mathbf{M}_\varepsilon\mathbf{G}\frac{\mathrm{d}}{\mathrm{d}t}\boldsymbol{\Phi} = 0 \tag{5.16}$$

with a scaling matrix $\mathbf{M}_\mathrm{N}$ which is mainly introduced to guarantee correct units. A consistent but rather inconvenient choice is

$$\mathbf{M}_\mathrm{N} := \mathbf{M}_\varepsilon^{-1/2}\mathbf{M}_\nu^{1/2}\mathbf{L}_\varepsilon^{-1}\mathbf{M}_\nu^{1/2}\mathbf{M}_\varepsilon^{-1/2}.$$

This regularisation is similar to the Lagrange-multiplier formulation for the eddy-current problem [28]. Left-multiplication of (5.16) by $\mathbf{M}_\mathrm{N}^{-1}\mathbf{L}_\varepsilon^{-1}\tilde{\mathbf{S}}$ yields

$$\tilde{\mathbf{S}}\mathbf{M}_\varepsilon\widehat{\mathbf{a}} + \mathbf{M}_\mathrm{N}^{-1}\mathbf{L}_\varepsilon^{-1}\mathbf{L}_\sigma\boldsymbol{\Phi} + \mathbf{M}_\mathrm{N}^{-1}\frac{\mathrm{d}}{\mathrm{d}t}\boldsymbol{\Phi} = 0. \tag{5.17}$$

which simplifies to Coulomb's gauge

$$\tilde{\mathbf{S}}\mathbf{M}_\varepsilon\widehat{\mathbf{a}} = 0 \tag{5.18}$$

with respect to the permittivities if we set $\boldsymbol{\Phi} = 0$.

To obtain a discrete version of the damped wave equation (5.9)–(5.10), we utilise (5.17). Now, using (5.16) and (5.17) the system (5.12)–(5.13) becomes two discrete damped wave equations

$$\mathbf{L}_\nu\widehat{\mathbf{a}} + \mathbf{M}_\sigma\frac{\mathrm{d}}{\mathrm{d}t}\widehat{\mathbf{a}} + \mathbf{M}_\varepsilon\frac{\mathrm{d}^2}{\mathrm{d}t^2}\widehat{\mathbf{a}} = \widehat{\widehat{\mathbf{j}}}_\mathrm{s} \tag{5.19}$$

$$\mathbf{L}_\varepsilon \boldsymbol{\Phi} + \mathbf{M}_\mathrm{N}^{-1}\mathbf{L}_\varepsilon^{-1}\mathbf{L}_\sigma \frac{\mathrm{d}}{\mathrm{d}t}\boldsymbol{\Phi} + \mathbf{M}_\mathrm{N}^{-1}\frac{\mathrm{d}^2}{\mathrm{d}t^2}\boldsymbol{\Phi} = \mathbf{q} \tag{5.20}$$

with $\mathbf{L}_\nu := \widetilde{\mathbf{C}}\mathbf{M}_\nu\mathbf{C} - \mathbf{M}_\varepsilon\mathbf{G}\mathbf{M}_\mathrm{N}\widetilde{\mathbf{S}}\mathbf{M}_\varepsilon$ and given right-hand-sides $\widehat{\widehat{\mathbf{j}}}_\mathrm{s}$ and $\mathbf{q}$ that fulfil the continuity equation (5.15). The resulting problem (5.19)–(5.20) is a system of second-order ordinary differential equations:

Theorem 5.4 *Let Assumptions 2.1, 2.2 and 3.2 hold. Then, the A–Φ-formulation with Lorenz gauge* (5.17) *and known charges* $\mathbf{q}$ *leads to an ordinary differential equation (ODE) system which is given in* (5.19)–(5.20).

5.2.1 Full Maxwell with Lorenz Gauge

Let us now investigate the case where the charges $\mathbf{q}$ are not known. We start from Lorenz' gauge (5.16). Left-multiplication of the equation by $-\tilde{\mathbf{S}}$ yields

$$\mathbf{L}_\varepsilon\mathbf{M}_\mathrm{N}\tilde{\mathbf{S}}\mathbf{M}_\varepsilon\widehat{\mathbf{a}} + \mathbf{L}_\sigma\boldsymbol{\Phi} + \mathbf{L}_\varepsilon\frac{\mathrm{d}}{\mathrm{d}t}\boldsymbol{\Phi} = 0\,.$$

Following the notation of Schoenmaker, e.g. [83], we denote the derivative of the magnetic vector potential by $\widehat{\pi} := \mathrm{d}\widehat{\mathbf{a}}/\mathrm{d}t$. Then, the Eqs. (5.12)–(5.13) can be rearranged as the following system of DAEs

$$\mathbf{L}_\varepsilon\mathbf{M}_\mathrm{N}\tilde{\mathbf{S}}\mathbf{M}_\varepsilon\widehat{\mathbf{a}} + \mathbf{L}_\sigma\boldsymbol{\Phi} + \mathbf{L}_\varepsilon\frac{\mathrm{d}}{\mathrm{d}t}\boldsymbol{\Phi} = 0 \tag{5.21}$$

$$\tilde{\mathbf{C}}\mathbf{M}_\nu\mathbf{C}\widehat{\mathbf{a}} + \mathbf{M}_\sigma[\widehat{\pi} + \mathbf{G}\boldsymbol{\Phi}] + \mathbf{M}_\varepsilon\left[\frac{\mathrm{d}}{\mathrm{d}t}\widehat{\pi} + \mathbf{G}\frac{\mathrm{d}}{\mathrm{d}t}\boldsymbol{\Phi}\right] = \widehat{\widehat{\mathbf{j}}}_\mathrm{s} \tag{5.22}$$

$$\tilde{\mathbf{S}}\mathbf{M}_\varepsilon\widehat{\pi} - \mathbf{L}_\varepsilon\boldsymbol{\Phi} + \mathbf{q} = 0 \tag{5.23}$$

$$\frac{\mathrm{d}}{\mathrm{d}t}\widehat{\mathbf{a}} - \widehat{\pi} = 0 \tag{5.24}$$

with $\mathbf{x}^\top = (\mathbf{q}^\top, \boldsymbol{\Phi}^\top, \widehat{\mathbf{a}}^\top, \widehat{\pi}^\top)$ such that we can write (5.21)–(5.24) in the form of (4.1) with the definitions

$$\mathbf{M} = \begin{bmatrix} 0 & \mathbf{L}_\varepsilon & 0 & 0 \\ 0 & \mathbf{M}_\varepsilon\mathbf{G} & 0 & \mathbf{M}_\varepsilon \\ 0 & 0 & 0 & 0 \\ 0 & 0 & \mathbf{I} & 0 \end{bmatrix}, \quad \mathbf{K} = \begin{bmatrix} 0 & \mathbf{L}_\sigma & \mathbf{L}_\varepsilon\mathbf{M}_\mathrm{N}\tilde{\mathbf{S}}\mathbf{M}_\varepsilon & 0 \\ 0 & \mathbf{M}_\sigma\mathbf{G} & \tilde{\mathbf{C}}\mathbf{M}_\nu\mathbf{C} & \mathbf{M}_\sigma \\ \mathbf{I} & -\mathbf{L}_\varepsilon & 0 & \tilde{\mathbf{S}}\mathbf{M}_\varepsilon \\ 0 & 0 & 0 & -\mathbf{I} \end{bmatrix} \quad \text{and} \quad \mathbf{r} = \begin{bmatrix} 0 \\ \widehat{\widehat{\mathbf{j}}}_\mathrm{s} \\ 0 \\ 0 \end{bmatrix}.$$

Now, any standard time integrator, e.g. the implicit Euler method (4.2), can be applied.

Next we determine the differential index of the system (5.21)–(5.24). Equation (5.21) is an ODE for $\boldsymbol{\Phi}$

$$\frac{\mathrm{d}}{\mathrm{d}t}\boldsymbol{\Phi} = -\mathbf{M}_{\mathrm{N}}\tilde{\mathbf{S}}\mathbf{M}_{\varepsilon}\widehat{\mathbf{a}} - \mathbf{L}_{\varepsilon}^{-1}\mathbf{L}_{\sigma}\boldsymbol{\Phi}. \tag{5.25}$$

Then, we deduce from (5.22) and (5.25) an ODE for $\widehat{\boldsymbol{\pi}}$:

$$\frac{\mathrm{d}}{\mathrm{d}t}\widehat{\boldsymbol{\pi}} = -\mathbf{M}_{\varepsilon}^{-1}\left[\mathbf{L}_{\nu}\widehat{\mathbf{a}}+\mathbf{M}_{\sigma}[\widehat{\boldsymbol{\pi}} + \mathbf{G}\boldsymbol{\Phi}]-\mathbf{M}_{\varepsilon}\mathbf{G}\mathbf{L}_{\varepsilon}^{-1}\mathbf{L}_{\sigma}\boldsymbol{\Phi} - \widehat{\widehat{\mathbf{j}}}_{\mathrm{s}}\right]$$

Finally, only one differentiation with respect to time of (5.23) is needed to obtain an ordinary differential equation for $\mathbf{q}$:

$$\frac{\mathrm{d}}{\mathrm{d}t}\mathbf{q} = \tilde{\mathbf{S}}\mathbf{M}_{\sigma}\widehat{\boldsymbol{\pi}} - \mathbf{L}_{\sigma}\boldsymbol{\Phi} - \tilde{\mathbf{S}}\widehat{\widehat{\mathbf{j}}}_{\mathrm{s}}\,.$$

Hence we conclude the following result [10]

Theorem 5.5 *Let Assumptions 2.1, 2.2 and 3.2 hold. The system* (5.21)–(5.24) *has differential index-*1 *and the initial vector* $\mathbf{x}_0^\top = (\mathbf{q}_0^\top, \boldsymbol{\Phi}_0^\top, \widehat{\mathbf{a}}_0^\top, \widehat{\boldsymbol{\pi}}_0^\top)$ *is a consistent initial value if* $\mathbf{q}_0 = \mathbf{L}_{\varepsilon}\boldsymbol{\Phi}_0 - \tilde{\mathbf{S}}\mathbf{M}_{\varepsilon}\widehat{\boldsymbol{\pi}}_0$ *is fulfilled.*

5.2.2 Full Maxwell with Coulomb Gauge

Instead of augmenting the equations by a Lorenz gauge, one can choose the Coulomb gauge (5.18). Starting by left-multiplying Coulomb's gauge by $\mathbf{M}_{\varepsilon}\mathbf{G}\mathbf{M}_{\mathrm{N}}$, we obtain

$$\mathbf{M}_{\varepsilon}\mathbf{G}\mathbf{M}_{\mathrm{N}}\tilde{\mathbf{S}}\mathbf{M}_{\varepsilon}\widehat{\mathbf{a}} = 0\,. \tag{5.26}$$

Using (5.18) and (5.26), the system (5.12)–(5.13) becomes a semi-discrete damped wave equation accompanied by a Laplace equation, i.e.,

$$\begin{aligned} \mathbf{L}_{\nu}\widehat{\mathbf{a}}+\mathbf{M}_{\sigma}\left[\frac{\mathrm{d}}{\mathrm{d}t}\widehat{\mathbf{a}} + \mathbf{G}\boldsymbol{\Phi}\right]+\mathbf{M}_{\varepsilon}\left[\frac{\mathrm{d}^2}{\mathrm{d}t^2}\widehat{\mathbf{a}} + \mathbf{G}\frac{\mathrm{d}}{\mathrm{d}t}\boldsymbol{\Phi}\right] &= \widehat{\widehat{\mathbf{j}}}_{\mathrm{s}} \\ \mathbf{L}_{\varepsilon}\boldsymbol{\Phi} &= \mathbf{q} \end{aligned}$$

with right-hand-sides that fulfil the continuity equation (5.15) and thus for given $\widehat{\widehat{\mathbf{j}}}_{\mathrm{s}}$ the resulting semi-discrete problem is again a system of DAEs. The Coulomb-gauged system reads

$$\tilde{\mathbf{S}}\mathbf{M}_{\varepsilon}\widehat{\mathbf{a}} = 0 \tag{5.27}$$

$$\tilde{\mathbf{C}}\mathbf{M}_\nu\mathbf{C}\widehat{\mathbf{a}}+\mathbf{M}_\sigma[\widehat{\boldsymbol{\pi}}+\mathbf{G}\boldsymbol{\Phi}]+\mathbf{M}_\varepsilon\left[\frac{\mathrm{d}}{\mathrm{d}t}\widehat{\boldsymbol{\pi}}+\mathbf{G}\frac{\mathrm{d}}{\mathrm{d}t}\boldsymbol{\Phi}\right]=\widehat{\widehat{\mathbf{j}}}_\mathrm{s} \tag{5.28}$$

$$\tilde{\mathbf{S}}\mathbf{M}_\varepsilon\widehat{\boldsymbol{\pi}}-\mathbf{L}_\varepsilon\boldsymbol{\Phi}+\mathbf{q}=0 \tag{5.29}$$

$$\frac{\mathrm{d}}{\mathrm{d}t}\widehat{\mathbf{a}}-\widehat{\boldsymbol{\pi}}=0 \tag{5.30}$$

with $\mathbf{x}^\top=(\mathbf{q}^\top,\boldsymbol{\Phi}^\top,\widehat{\mathbf{a}}^\top,\widehat{\boldsymbol{\pi}}^\top)$. Similarly as before we can identify a first order DAE system of form (4.1) and apply for example the implicit Euler method.

Next, we determine the differential index of the system (5.27)–(5.30). Differentiating (5.27) twice with respect to time and inserting (5.30) leads to

$$\tilde{\mathbf{S}}\mathbf{M}_\varepsilon\frac{\mathrm{d}}{\mathrm{d}t}\widehat{\boldsymbol{\pi}}=0\,. \tag{5.31}$$

This indicates already that the differential index is at least $\vartheta\geq 2$. Left-multiplying (5.28) by $\tilde{\mathbf{S}}$ and applying (5.31) yields:

$$\frac{\mathrm{d}}{\mathrm{d}t}\boldsymbol{\Phi}=-\mathbf{L}_\varepsilon^{-1}\left[\mathbf{L}_\sigma\boldsymbol{\Phi}-\tilde{\mathbf{S}}\mathbf{M}_\sigma\widehat{\boldsymbol{\pi}}+\tilde{\mathbf{S}}\widehat{\widehat{\mathbf{j}}}_\mathrm{s}\right]$$

Furthermore, from (5.28), we obtain:

$$\begin{aligned}\frac{\mathrm{d}}{\mathrm{d}t}\widehat{\boldsymbol{\pi}}=-\mathbf{M}_\varepsilon^{-1}\Big[&(\mathbf{M}_\sigma\mathbf{G}-\mathbf{M}_\varepsilon\mathbf{G}\mathbf{L}_\varepsilon^{-1}\mathbf{L}_\sigma)\boldsymbol{\Phi}+\tilde{\mathbf{C}}\mathbf{M}_\nu\mathbf{C}\widehat{\mathbf{a}}\\&+(\mathbf{M}_\sigma+\mathbf{M}_\varepsilon\mathbf{G}\mathbf{L}_\varepsilon^{-1}\tilde{\mathbf{S}}\mathbf{M}_\sigma)\widehat{\boldsymbol{\pi}}-(\mathbf{I}+\mathbf{M}_\varepsilon\mathbf{G}\mathbf{L}_\varepsilon^{-1}\tilde{\mathbf{S}})\widehat{\widehat{\mathbf{j}}}_\mathrm{s}\Big]\end{aligned}$$

Finally, one differentiation with respect to time of (5.29) results in

$$\frac{\mathrm{d}}{\mathrm{d}t}\mathbf{q}=\tilde{\mathbf{S}}\mathbf{M}_\sigma\widehat{\boldsymbol{\pi}}-\mathbf{L}_\sigma\boldsymbol{\Phi}-\tilde{\mathbf{S}}\widehat{\widehat{\mathbf{j}}}_\mathrm{s}$$

and thus the overall problem has a differential index-2 [10].

Theorem 5.6 *Let Assumptions 2.1, 2.2 and 3.2 hold. The system* (5.27)–(5.30) *has differential index-2 and the initial vector* $\mathbf{x}_0^\top=(\mathbf{q}_0^\top,\boldsymbol{\Phi}_0^\top,\widehat{\mathbf{a}}_0^\top,\widehat{\boldsymbol{\pi}}_0^\top)$ *is a consistent initial value if* $\tilde{\mathbf{S}}\mathbf{M}_\varepsilon\widehat{\mathbf{a}}_0=0$, $\tilde{\mathbf{S}}\mathbf{M}_\varepsilon\widehat{\boldsymbol{\pi}}_0=0$ *and* $\mathbf{q}_0=\mathbf{L}_\varepsilon\boldsymbol{\Phi}_0-\tilde{\mathbf{S}}\mathbf{M}_\varepsilon\widehat{\boldsymbol{\pi}}_0$ *are fulfilled.*

Lorenz' and Coulomb's gauge lead to systems that describe the same phenomena and have eventually, i.e. in the mesh size limit, the same electromagnetic fields (strengths or fluxes) as solutions. On the other hand, the structural properties are different, i.e., the Lorenz gauge yields an index-1 problem whereas the Coulomb gauge gives index-2. Hence, the latter formulation will be much more affected by perturbations and the computation of consistent initial values is more cumbersome. This has been observed in simulations [8, 10].

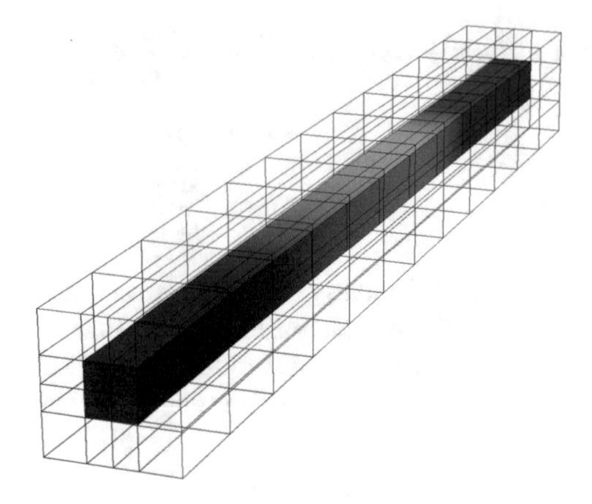

Fig. 9 Copper bar in air, excited by sinusoidal source (Benchmark 5.7)

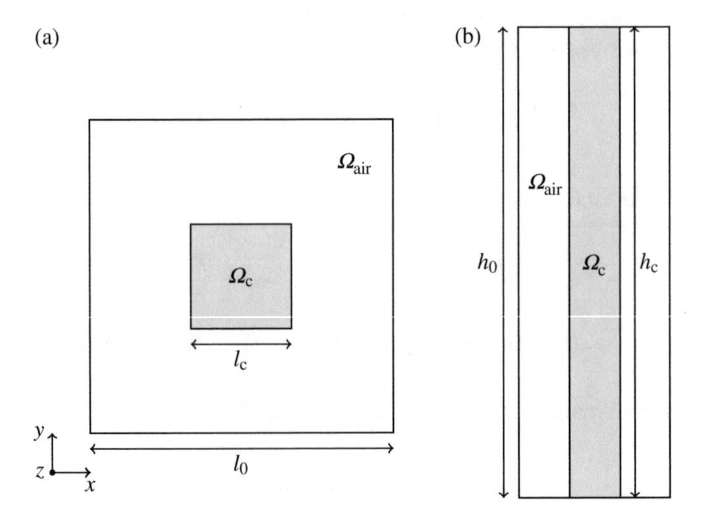

Fig. 10 Copper bar with square cross-section surrounded by air. The distances are $l_{\text{c}} = (1+\sqrt{1.5})\cdot 10^{-3}$ m, $l_0 = (3+\sqrt{1.5})\cdot 10^{-3}$ m, and $h_0 = h_{\text{c}} = 3$ m. (**a**) Squared cross-section of the copper bar at the $z = 0$ plane. (**b**) Cross-section of the copper bar at the $x = 0$ plane

Benchmark 5.7 *The benchmark example Fig. 9 was proposed in [10] to numerically analyse the DAE index of the two gauged A–ϕ formulations. The model is a copper bar with a cross-sectional area of* $0.25\,mm^2$ *surrounded by air and discretised by FIT. A detailed characterisation of the dimensions can be seen in Fig. 10.*

On the copper bar Ω_{c}*, a conductivity of* $\sigma_{\text{c}} = 5.7\cdot 10^7$ S/m *is set and on the air region* $\sigma_{\text{air}} = 0\,\text{S/m}$*. Vacuum permeability* $\mu = 4\pi\cdot 10^{-7}$ H/m *and relative permittivity* $\varepsilon_r = 1$ *is assumed in the entire domain.*

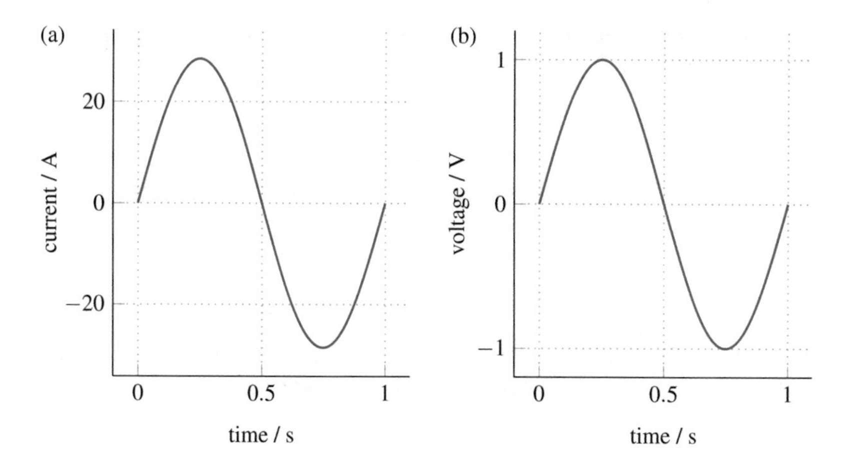

Fig. 11 Time domain simulation results for the Benchmark 5.7 with Lorenz gauge $\Delta t = 1 \cdot 10^{-4}$ s. (**a**) Current through the bar, (**b**) voltage drop at the ports

One contact is excited by a sinusoidal voltage $v = \sin(2\pi t)\mathrm{V}$*, the other contact is grounded (ebc) and the remaining boundary is set to mbc.*

The structure is discretised using FIT with 325 *mesh cells and* 845 *degrees of freedom. The implicit Euler method is applied with a time step of* $\Delta t = 1 \cdot 10^{-4}\,\mathrm{s}$ *and zero initial condition, see Fig. 11.*

6 Quasistatic Maxwell's Equations

In the case of slowly time-varying fields, certain time-derivatives of Maxwell's equations can be disregarded with respect to other phenomena, see Definition 3. This is convenient to simplify the numerical treatment. However, the resulting (quasi-) static approximations have different structural properties and their differential algebraic index is studied next.

6.1 *Electroquasistatic Maxwell's Equations*

We start with the index study of the electroquasistatic approximation, that is given in Definition 3(b). Maxwell's equations can be rewritten as

$$\nabla \times \mathbf{E} = 0\,, \qquad \nabla \times \mathbf{H} = \frac{\partial \mathbf{D}}{\partial t} + \mathbf{J}\,, \qquad \nabla \cdot \mathbf{D} = \rho\,, \qquad \nabla \cdot \mathbf{B} = 0\,.$$

As the curl of electric field $\mathbf{E}$ vanishes, it can be described as the gradient of the electric scalar potential ϕ

$$\mathbf{E} = -\nabla\phi\ , \tag{6.1}$$

i.e., the magnetic vector potential's contribution to $\mathbf{E}$ in (2.15) is negligible.

6.1.1 Electric Scalar Potential ϕ-Formulation

Using Eq. (6.1), Maxwell's equations for electroquasistatic fields and the material laws, the following potential equation can be obtained to compute ϕ

$$\nabla \cdot \boldsymbol{\sigma} \nabla\phi + \frac{\partial}{\partial t} \nabla \cdot \boldsymbol{\varepsilon} \nabla\phi = 0\ .$$

Eventually, spatial discretisation leads to a system of DAEs

$$\widetilde{\mathbf{S}}\mathbf{M}_{\sigma}\widetilde{\mathbf{S}}^{\top}\boldsymbol{\Phi} + \widetilde{\mathbf{S}}\mathbf{M}_{\varepsilon}\widetilde{\mathbf{S}}^{\top}\frac{\mathrm{d}}{\mathrm{d}t}\boldsymbol{\Phi} = 0\ , \tag{6.2}$$

where $\boldsymbol{\Phi}$ contains the degrees of freedom of our problem, i.e. the electric scalar potential on the nodes of the primal grid. However, if boundary conditions are properly set, then one can show

Theorem 6.1 *The system* (6.2) *under Assumptions 2.1, 2.2 and 3.2 is an ODE.*

Theorem 6.1 follows immediately from Lemma 2.

Benchmark 6.2 *In a DC high-voltage cable, the insulation between the inner high-voltage electrode and the outer shielding layer carries a large electric field strength. At the end of the cable (Ω_{cbl}), the voltage has to drop along the surface of the insulation layer (Ω_{ins}) with a substantially smaller electric field strength. This necessitates the design of a so-called cable termination with field-shaping capability. A sketch of the domain and its distances can be seen in Fig. 12.*

The computational domain is $\Omega = \Omega_{\mathrm{air}} \cup \Omega_{\mathrm{ins}}$ and the rest of the domain is modelled via boundary conditions and thus not considered by the discretisation. In the air region Ω_{air} the conductivity σ is set to zero and the permittivity of vacuum ε_0 =8.85 $\cdot\, 10^{-12}$ F/m is assumed. The insulating domain Ω_{ins} has conductivity 1 S/m *and permittivity $6\varepsilon_0$.*

Due to symmetry reasons, only an axisymmetric cross-section (i.e. the right half of the domain sketched in Fig. 12) is simulated. The cable endings Ω_{cbl} are modelled by zero Dirichlet boundary conditions (ebc) on the boundary $\Gamma_{\mathrm{cbl}} = \partial\Omega_{\mathrm{cbl}} \cap \left(\overline{\Omega}_{\mathrm{ins}} \cup \overline{\Omega}_{\mathrm{air}}\right)$. Similarly, the domain Ω_{c} is modelled by non-homogeneous Dirichlet boundary conditions that set the potential ϕ to a time dependent value $f(t)$ on $\Gamma_{\mathrm{c}} = \partial\Omega_{\mathrm{ins}} \cap \overline{\Omega}_{\mathrm{c}}$, see Fig. 13a. At the rest of the boundary zero Neumann boundary conditions (mbc) are set.

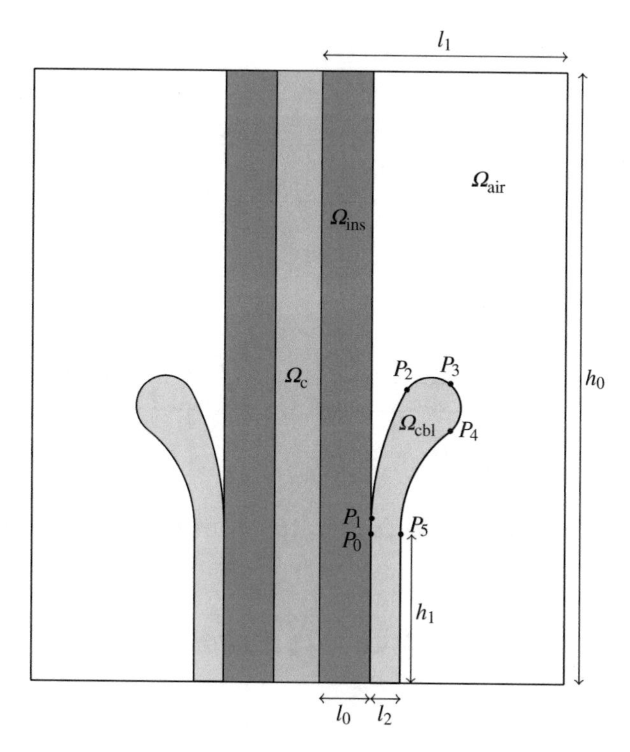

Fig. 12 Sketch of electroquasistatic benchmark domain. The distances are $l_0 = 1.2 \cdot 10^{-2}$ m, $l_1 = 2.5 \cdot 10^{-2}$ m, $l_2 = 2 \cdot 10^{-3}$ m, $h_0 = 4 \cdot 10^{-2}$ m and $h_1 = 1 \cdot 10^{-2}$ m. To describe the arc segments, points $P_0 = (12,\ 10)$, $P_1 = (12.03,\ 11)$, $P_2 = (14.24,\ 18.94)$, $P_3 = (17.3,\ 19.52)$, $P_4 = (17.2,\ 16,4)$ and $P_5 = (14,\ 10)$ are defined. The first arc segment from point P_2 to P_1 has an angle of 25.06° and is described by straight segments with a grid spacing of 3.02°. Both arc segments from P_3 to P_2 as well as from P_4 to P_3 have 102.53° and 28.65° spacing. The last one from P_4 to P_5 has 53.13° and 7.16° spacing

Using the Finite Element Method yields 2,892 number of nodes and 2,078 degrees of freedom. Time integration is carried out with the implicit Euler method from time $t_0 = 0$ s *to* $t_{\mathrm{end}} = 2 \cdot 10^{-3}$ s *with step size* $\Delta t = 1 \cdot 10^{-5}$ s. *The steady state solution is set as initial condition. Figure 13 shows the excitation function* $f(t)$ *and the electric energy* $E_{\mathrm{elec}} = \frac{1}{2}\int_{\Omega} \mathbf{E}\cdot\mathbf{D}\,\mathrm{d}\Omega \approx \frac{1}{2}\widehat{\mathbf{e}}^{\top}\mathbf{M}_{\varepsilon}\widehat{\mathbf{e}}$ *over time.*

For the electroquasistatic problem other formulations are not common as (6.2) has convenient properties, e.g. a low-number of degrees of freedom, since no vectorial fields are needed and ordinary differential character. Rarely, a mixed charge/potential formulation [74]

$$\begin{aligned} \widetilde{\mathbf{S}}\mathbf{M}_{\sigma}\widetilde{\mathbf{S}}^{\top}\boldsymbol{\Phi} + \frac{\mathrm{d}}{\mathrm{d}t}\mathbf{q} &= 0\,, \\ \mathbf{q} - \widetilde{\mathbf{S}}\mathbf{M}_{\varepsilon}\widetilde{\mathbf{S}}^{\top}\boldsymbol{\Phi} &= 0\,. \end{aligned}$$

is employed which is a simple and easy to solve DAE index-1 system.

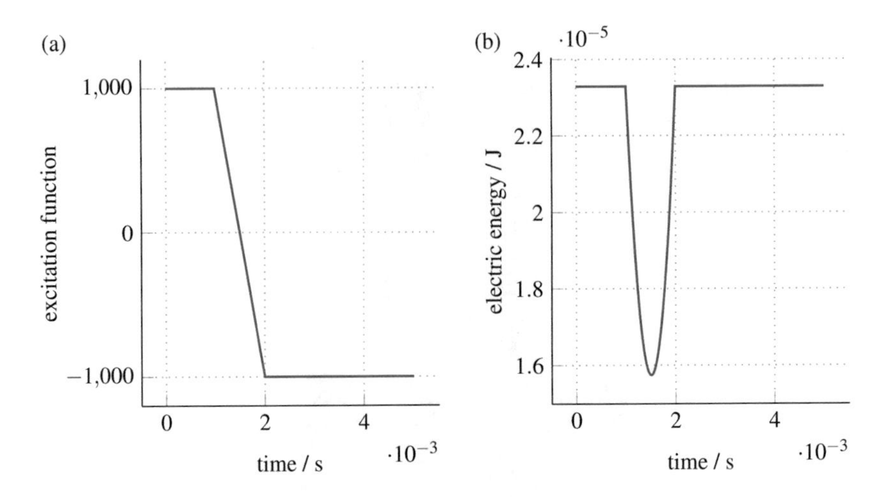

Fig. 13 Electric energy and excitation function of benchmark 6.2. (**a**) Excitation function $f(t)$. (**b**) Electric energy

6.2 *Magnetoquasistatic Maxwell's Equations*

Now the magnetoquasistatic case is studied. Following Definition 3(c) Maxwell's equations take the form

$$\nabla \times \mathbf{E} = -\frac{\partial \mathbf{B}}{\partial t}\,, \qquad \nabla \times \mathbf{H} = \mathbf{J}\,, \qquad \nabla \cdot \mathbf{D} = \rho\,, \qquad \nabla \cdot \mathbf{B} = 0\,.$$

Due to the non-vanishing curl of the electric field strength, a scalar potential formulation is no longer possible. Several competing vector potential formulations are common, see e.g. [15].

6.2.1 Magnetic Vector Potential A-formulations

Using the definition of the magnetic vector and the electric scalar potentials **A** and ϕ (see Eq. 2.15) and inserting the material laws and magnetoquasistatic equations into each other, one finds the curl-curl equation

$$\sigma\left(\frac{\partial \mathbf{A}}{\partial t} + \nabla\phi\right) + \nabla \times (\boldsymbol{\nu}\nabla \times \mathbf{A}) = \mathbf{J}_{\mathrm{s}}\,.$$

In a three dimensional domain, a gauge condition is necessary to ensure uniqueness of solution, due to the kernel of the curl-operator ($\mathbf{C}\widetilde{\mathbf{S}}^{\top} = 0$). The so-called A* formulation exploits this freedom of choice and assumes that the gradient of the electric scalar potential is zero ($\nabla\phi = 0$). Applying it yields the spatially discretised magnetoquasistatic curl-curl equation

$$\mathbf{M}_\sigma \frac{\mathrm{d}}{\mathrm{d}t}\widehat{\mathbf{a}} + \mathbf{C}^\top \mathbf{M}_\nu \mathbf{C}\widehat{\mathbf{a}} = \widehat{\widehat{\mathbf{j}}}_\mathrm{s} \tag{6.3}$$

with $\widehat{\widehat{\mathbf{j}}}_\mathrm{s}$ known and $\widehat{\mathbf{a}}$ containing the degrees of freedom (the magnetic vector potential integrated on the edges of the primal grid). Alternatively, mixed formulations incorporating a gauging conditions have been proposed, e.g. [28]

$$\begin{pmatrix} \mathbf{M}_\sigma & 0 \\ 0 & 0 \end{pmatrix} \frac{\mathrm{d}}{\mathrm{d}t} \begin{pmatrix} \widehat{\mathbf{a}} \\ \boldsymbol{\Phi} \end{pmatrix} + \begin{pmatrix} \mathbf{C}^\top \mathbf{M}_\nu \mathbf{C} & \mathbf{M}_1 \widetilde{\mathbf{S}}^\top \\ -\widetilde{\mathbf{S}}\mathbf{M}_1 & -\mathbf{M}_\mathrm{N}^{-1} \end{pmatrix} \begin{pmatrix} \widehat{\mathbf{a}} \\ \boldsymbol{\Phi} \end{pmatrix} = \begin{pmatrix} \widehat{\widehat{\mathbf{j}}}_\mathrm{s} \\ 0 \end{pmatrix}$$

where $\mathbf{M}_1$ is a regularised version of $\mathbf{M}_\sigma$ and $\mathbf{M}_\mathrm{N}$ is a regular matrix to ensure the correct physical units as in (5.16). Using the Schur complement, one derives a grad-div regularisation, e.g. [30], where

$$\mathbf{Z}_\sigma = \mathbf{M}_1 \widetilde{\mathbf{S}}^\top \mathbf{M}_\mathrm{N} \widetilde{\mathbf{S}} \mathbf{M}_1 .$$

can be used to finally arrive at

Assumption 6.3 *Let us assume that system* (6.3) *is rewritten as*

$$\mathbf{M}_\sigma \frac{\mathrm{d}}{\mathrm{d}t}\widehat{\mathbf{a}} + \mathbf{K}_\nu \widehat{\mathbf{a}} = \widehat{\widehat{\mathbf{j}}}_\mathrm{s}, \tag{6.4}$$

where $\mathbf{K}_\nu = \mathbf{C}^\top \mathbf{M}_\nu \mathbf{C} + \mathbf{Z}_\sigma$, *provided* $\mathbf{Z}_\sigma$ *is a positive semidefinite matrix that enforces the matrix pencil* $\lambda \mathbf{M}_\sigma + \mathbf{K}_\nu$ *to be positive definite for* $\lambda > 0$.

Theorem 6.4 *Under Assumptions 2.1, 2.2, 3.2 and 6.3, system* (6.4) *has Kronecker and tractability index 1.*

The proof of the *Kronecker index* of system (6.4) has been originally given in [72]. More recently, [57] obtained the same result for the tractability index and [7] used the tractability index concept to analyse the DAE index of this formulation with an attached network description.

Instead of using $\mathbf{Z}_\sigma$, various gauging techniques have been proposed for this formulation, such as the tree-cotree gauge [69] or the weak gauging property of iterative linear solvers [29]. Due to the simple structure of $\mathbf{M}_\sigma$ and as long as the gauge leads to a positive definite matrix pencil, the index can be derived analogously as before.

Benchmark 6.5 *A common benchmark for magnetoquasistatic models are inductors with a metal core. The example in Fig. 14 from [33] features an aluminium core, i.e.* $\sigma = 35 \cdot 10^6$ S/m *in* Ω_c *with a copper coil* Ω_str *surrounded by air* Ω_air. *The coil is given by the stranded conductor model consisting of 120 turns with conductivity* $\sigma = 1 \cdot 10^6$ S/m. *The conductivity in* Ω_air *is zero and disregarded in* Ω_str

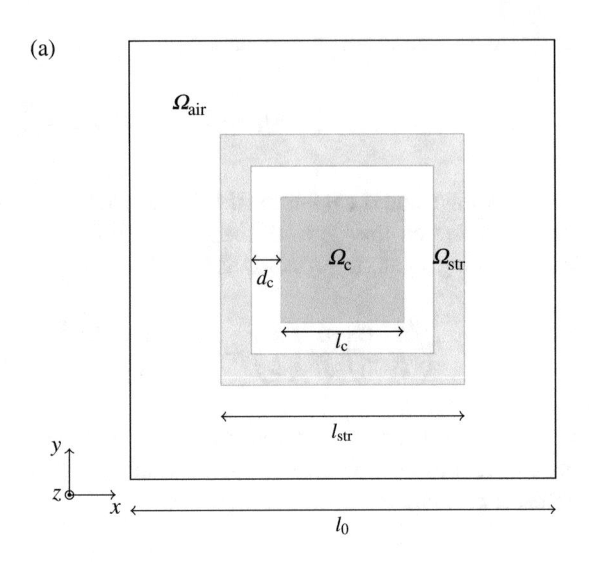

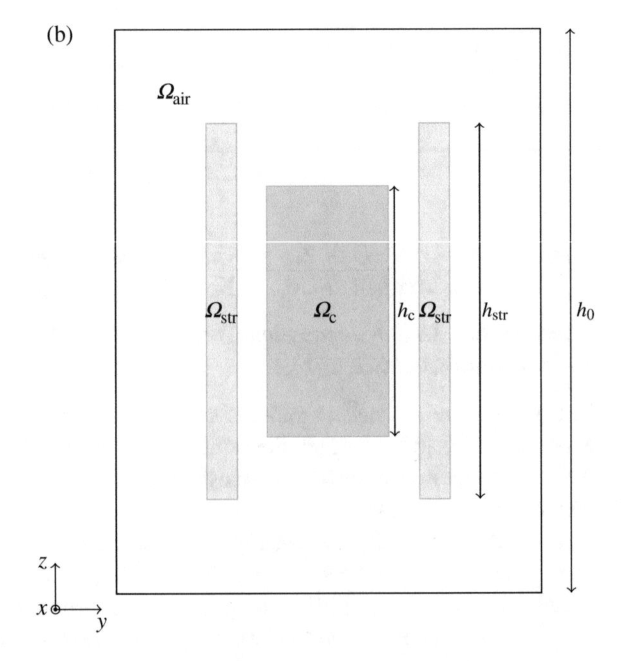

Fig. 14 Magnetoquasistatic model of an inductor with metal core. The distances are $l_0 = 1.4 \cdot 10^{-2}$ m, $l_{\text{str}} = 8 \cdot 10^{-3}$ m $l_{\text{c}} = 4 \cdot 10^{-3}$ m, $d_{\text{c}} = 1 \cdot 10^{-3}$ m, $h_0 = 1.8 \cdot 10^{-2}$ m, $l_{\text{str}} = 1.2 \cdot 10^{-2}$ m and $l_{\text{c}} = 8 \cdot 10^{-3}$ m. (**a**) Cross-section of inductor at the $z = 0$ plane. (**b**) Cross-section of inductor at the $x = 0$ plane

in the construction of $\mathbf{M}_\sigma$ *as eddy currents are assumed negligible in the windings. Vacuum permeability* $\mu = 4\pi \cdot 1 \cdot 10^{-7}$ H/m *is assumed everywhere* (Ω) *and electric boundary conditions are enforced* $\Gamma = \Gamma_{ebc}$, *cf.* (2.4). *The FIT discretisation uses an equidistant hexahedral grid with step size* 10^{-3}, *which leads to* 3,528 *elements.*

The discretisation of the winding function for the A* *formulation can be visualised in Fig. 15. As no gauging is performed,* 9,958 *degrees of freedom arise.*

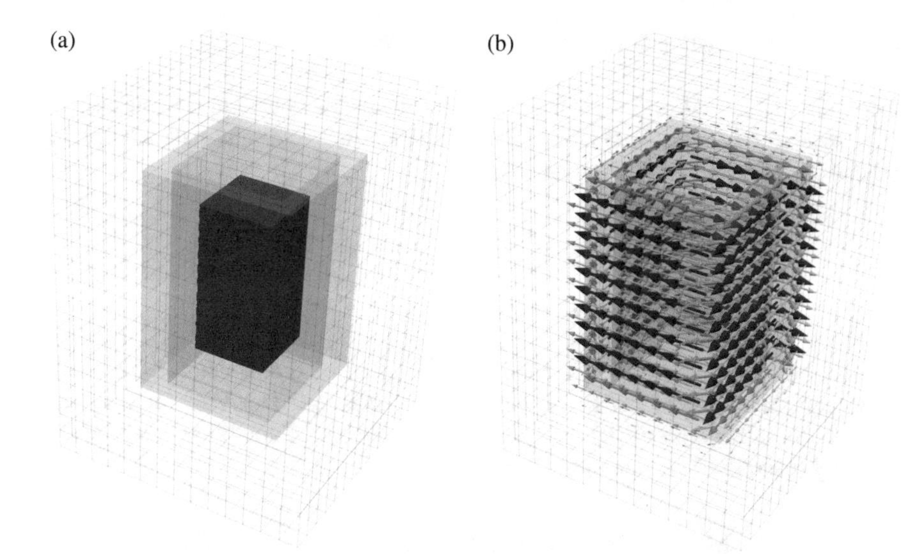

Fig. 15 Simple magnetoquasistatic model of an inductor with metal core. Coil is given by the stranded conductor model (Benchmark 6.5). (**a**) Iron core (red) with surrounding coil (transparent grey). (**b**) Source current density given by winding function $\mathbf{J}_\mathrm{s} = \boldsymbol{\chi}_\mathrm{s} i_\mathrm{s}$

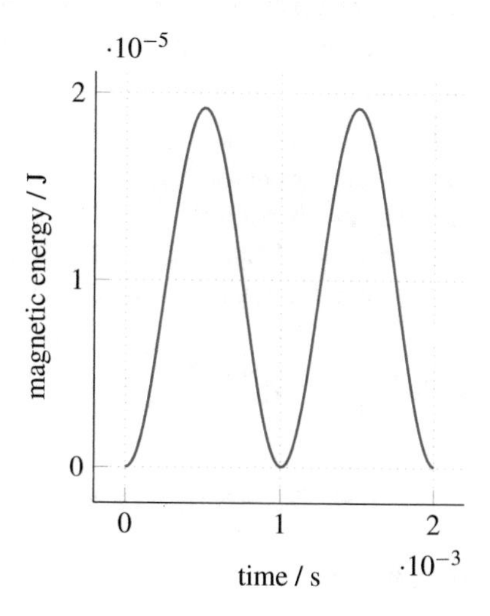

Fig. 16 Magnetic energy of the inductor in Fig. 14 simulated with the A* formulation

For the simulation, the source current density is $\widehat{\widehat{\mathbf{j}}}_\mathrm{s} = \mathbf{X}_\mathrm{s}\mathbf{i}_\mathrm{s}$*, where* $\mathbf{X}_\mathrm{s}$ *is the discretisation of the winding function in Fig. 15,* $\mathbf{i}_\mathrm{s} = \sin(2\pi f_\mathrm{s} t)$ *is the source current and the frequency* f_s *is* 500 Hz*. Time integration is performed with the implicit Euler method with step size* $\Delta t = 2 \cdot 10^{-5}\,\mathrm{s}$ *from time* $t_0 = 0\,\mathrm{s}$ *to* $t_\mathrm{end} = 2 \cdot 10^{-3}\,\mathrm{s}$ *and zero initial condition. Figure 16 shows the magnetic energy* $E_\mathrm{mag} = \frac{1}{2}\int_\Omega \mathbf{H} \cdot \mathbf{B}\,\mathrm{d}\Omega \approx \frac{1}{2}\widehat{\widehat{\mathbf{b}}}^\top \mathbf{M}_\nu \widehat{\widehat{\mathbf{b}}}$ *over time.*

6.2.2 Electric Vector Potential T–Ω-formulation

Maxwell's equations and their material laws in magnetoquasistatics can also be expressed in the T–Ω formulation. Using the electric vector and magnetic scalar potentials from Eq. (2.16), the system reads

$$\nabla \times (\boldsymbol{\rho} \nabla \times \mathbf{T}) + \boldsymbol{\mu} \frac{\partial \mathbf{T}}{\partial t} - \boldsymbol{\mu} \nabla \frac{\partial \psi}{\partial t} = -\boldsymbol{\mu} \frac{\partial \mathbf{H}_\mathrm{s}}{\partial t} - \nabla \times \mathbf{E}_\mathrm{s}$$
$$\nabla \cdot (\boldsymbol{\mu} \mathbf{T}) - \nabla \cdot (\boldsymbol{\mu} \nabla \psi) = -\nabla \cdot (\boldsymbol{\mu} \mathbf{H}_\mathrm{s}),$$

with $\mathbf{E}_\mathrm{s}$ being the source electric field strength when solid conductors are present. Here, $\boldsymbol{\rho}$ is the electrical resistivity, which corresponds to the inverse of $\boldsymbol{\sigma}$ wherever $\boldsymbol{\sigma} \neq 0$. As in the case of the A–ϕ formulation, the gauge condition is set for the spatially discretised version of our system

$$\mathbf{C}^\top \mathbf{M}_\rho \mathbf{C} \widehat{\mathbf{t}} + \mathbf{M}_\mu \frac{\mathrm{d}}{\mathrm{d}t} \widehat{\mathbf{t}} + \mathbf{M}_\mu \widetilde{\mathbf{S}}^\top \frac{\mathrm{d}}{\mathrm{d}t} \Psi = -\mathbf{M}_\mu \frac{\mathrm{d}}{\mathrm{d}t} \widehat{\mathbf{h}}_\mathrm{s} - \mathbf{C} \widehat{\mathbf{e}}_\mathrm{s}$$
$$\widetilde{\mathbf{S}} \mathbf{M}_\mu \widehat{\mathbf{t}} + \widetilde{\mathbf{S}} \mathbf{M}_\mu \widetilde{\mathbf{S}}^\top \Psi = -\widetilde{\mathbf{S}} \mathbf{M}_\mu \widehat{\mathbf{h}}_\mathrm{s},$$

with degrees of freedom $\widehat{\mathbf{t}}$, which contains the on the dual grid's edges integrated electric vector potential and Ψ, which consists of the magnetic scalar potential on the dual grid's nodes.

Definition 6.5 For gauging, a tree T is generated on the dual grid's edges inside the conducting region Ω_c. We define a projector $\tilde{\mathbf{P}}_\mathrm{t}$ onto the cotree of T and truncate it by deleting all the linearly dependent columns to obtain $\mathbf{P}_\mathrm{t}$. As gauging condition we set $\mathbf{Q}_\mathrm{t} \widehat{\mathbf{t}} = 0$, where $\mathbf{Q}_\mathrm{t}$ spans the kernel of $\mathbf{P}_\mathrm{t}^\top$. This corresponds to setting to zero the values of the electric vector potential on the edges T.

$$\mathbf{P}_\mathrm{t}^\top \mathbf{C}^\top \mathbf{M}_\rho \mathbf{C} \mathbf{P}_\mathrm{t} \widehat{\mathbf{t}} + \mathbf{P}_\mathrm{t}^\top \mathbf{M}_\mu \mathbf{P}_\mathrm{t} \frac{\mathrm{d}}{\mathrm{d}t} \widehat{\mathbf{t}} + \mathbf{P}_\mathrm{t}^\top \mathbf{M}_\mu \widetilde{\mathbf{S}}^\top \frac{\mathrm{d}}{\mathrm{d}t} \psi = -\mathbf{P}_\mathrm{t}^\top \mathbf{M}_\mu \frac{\mathrm{d}}{\mathrm{d}t} \widehat{\mathbf{h}}_\mathrm{s} - \mathbf{P}_\mathrm{t}^\top \mathbf{C} \widehat{\mathbf{e}}_\mathrm{s} \tag{6.5}$$

$$\widetilde{\mathbf{S}} \mathbf{M}_\mu \mathbf{P}_\mathrm{t} \widehat{\mathbf{t}} + \widetilde{\mathbf{S}} \mathbf{M}_\mu \widetilde{\mathbf{S}}^\top \psi = -\widetilde{\mathbf{S}} \mathbf{M}_\mu \widehat{\mathbf{h}}_\mathrm{s}. \tag{6.6}$$

Property 6.6 The matrix $\mathbf{P}_\mathrm{t}$ fulfils

1. $\det \left(\mathbf{P}_\mathrm{t}^\top \mathbf{C}^\top \mathbf{M}_\rho \mathbf{C} \mathbf{P}_\mathrm{t} \right) \neq 0$,
2. $\operatorname{im} \mathbf{P}_\mathrm{t} \cap \operatorname{im} \widetilde{\mathbf{S}}^\top = \emptyset$.

Property 6.6(1) is a consequence of the tree-cotree gauge (see [69]) and Property 6.6(2) follows from Property 6.6(1) and the fact that $\operatorname{im} \widetilde{\mathbf{S}}^\top \subseteq \ker \mathbf{C}$ (Lemma 1).

Property 6.7 ([33]) Every $\mathbf{x} \in \mathbb{R}^n$ can be written as $\mathbf{x} = \mathbf{M}_\mu^{1/2} \widetilde{\mathbf{S}}^\top \mathbf{x}_1 + \mathbf{M}_\mu^{-1/2} \mathbf{W}^\top \mathbf{x}_2$, where $n := \operatorname{rank} \mathbf{M}_\mu$ and $\mathbf{W}$ is the matrix whose columns span $\ker \widetilde{\mathbf{S}}$.

Proof As $\mathbf{M}_\mu$ has full rank and is symmetric, $\text{rank}(\mathbf{M}_\mu^{1/2}\widetilde{\mathbf{S}}^\top) = \text{rank}\,\widetilde{\mathbf{S}}^\top$ and $\text{rank}(\mathbf{M}_\mu^{-1/2}\mathbf{W}^\top) = \text{rank}\,\mathbf{W}^\top$. Using the rank-nullity theorem together with the fact that both subspaces are orthogonal and thus linear independent, we obtain that their direct sum spans $\mathbb{R}^n$. □

Theorem 6.8 ([33]) *Under Assumptions 2.1, 2.2 and 3.2, the system of DAEs* (6.5)–(6.6) *has differential index-1.*

Proof As the system has an algebraic constraint, it has at least index-1. Equation (6.6) is differentiated and $\frac{\mathrm{d}}{\mathrm{d}t}\Psi$ is extracted as

$$\frac{\mathrm{d}}{\mathrm{d}t}\Psi = -(\widetilde{\mathbf{S}}\mathbf{M}_\mu\widetilde{\mathbf{S}}^\top)^{-1}\widetilde{\mathbf{S}}\mathbf{M}_\mu\mathbf{P}_\mathrm{t}\frac{\mathrm{d}}{\mathrm{d}t}\mathbf{t} - (\widetilde{\mathbf{S}}\mathbf{M}_\mu\widetilde{\mathbf{S}}^\top)^{-1}\widetilde{\mathbf{S}}\mathbf{M}_\mu\widehat{\mathbf{h}}_\mathrm{s}.$$

This can be inserted into Eq. (6.5) and now it is sufficient to see that $\det(\mathbf{P}_\mathrm{t}\mathbf{Z}\mathbf{P}_\mathrm{t}) \neq 0$ for

$$\mathbf{Z} = (\mathbf{M}_\mu - \mathbf{M}_\mu\widetilde{\mathbf{S}}^\top(\widetilde{\mathbf{S}}\mathbf{M}_\mu\widetilde{\mathbf{S}}^\top)^{-1}\widetilde{\mathbf{S}}\mathbf{M}_\mu).$$

We can write $\mathbf{M}_\mu^{1/2}\mathbf{P}_\mathrm{t}\mathbf{x} = \mathbf{M}_\mu^{1/2}\widetilde{\mathbf{S}}^\top\mathbf{x}_1 + \mathbf{M}_\mu^{-1/2}\mathbf{W}^\top\mathbf{x}_2$ (Property 6.7). As $\mathbf{M}_\mu^{1/2}$ is invertible, $\mathbf{P}_\mathrm{t}\mathbf{x} = \widetilde{\mathbf{S}}^\top\mathbf{x}_1 + \mathbf{M}_\mu^{-1}\mathbf{W}^\top\mathbf{x}_2$ and, as $\mathbf{P}_\mathrm{t}\mathbf{x} \neq \widetilde{\mathbf{S}}^\top\mathbf{x}_1$ (Property 6.6), $\mathbf{M}_\mu^{-1}\mathbf{W}^\top\mathbf{x}_2 \neq 0$. Thus

$$\mathbf{x}^\top\mathbf{P}_\mathrm{t}^\top\mathbf{Z}\mathbf{P}_\mathrm{t}\mathbf{x} = \mathbf{x}_2^\top\mathbf{W}\mathbf{M}_\mu^{-1}\mathbf{W}^\top\mathbf{x}_2 > 0,$$

as long as $\mathbf{x} \neq 0$. We conclude that $\mathbf{P}_\mathrm{t}^\top\mathbf{Z}\mathbf{P}_\mathrm{t}$ is positive definite. □

Benchmark 6.9 *The physical specifications and discretisation of the benchmark example of the T–Ω formulation is equivalent to Benchmark 6.5.*

The construction of the winding function is different and it is depicted in Fig. 17. This time a tree-cotree gauge is performed, which yield only 4,610 *degrees of freedom.*

Again for the simulation the source current is set to $\mathbf{i}_\mathrm{s} = \sin(2\pi f_\mathrm{s} t)$ *with frequency* $f_\mathrm{s} = 500\,\mathrm{Hz}$. *The source magnetic field is* $\widehat{\mathbf{h}}_\mathrm{s} = \mathbf{Y}_\mathrm{s}\mathbf{i}_\mathrm{s}$, *with* $\mathbf{Y}_\mathrm{s}$ *being the discretisation of the winding function in Fig. 17. Like in Benchmark 6.5, time integration is performed with implicit Euler with step size* $\Delta t = 2\cdot 10^{-5}\,\mathrm{s}$ *from time* $t_0 = 0\,\mathrm{s}$ *to* $t_\mathrm{end} = 2\cdot 10^{-3}\,\mathrm{s}$ *and with zero initial condition. The resulting magnetic energy is depicted in Fig. 18.*

Remark 6.9 Note that the magnetic energy obtained with the A* formulation in Fig. 16 and the one obtained with the T–Ω one in Fig. 18 differ. Both formulations are dual to each other, i.e., their degrees of freedom are on dual sides of Maxwell's House in Fig. 5. They converge to the unique physical solution from below and above. This property can be used in order to study the error of the spatial discretisation (see [1]).

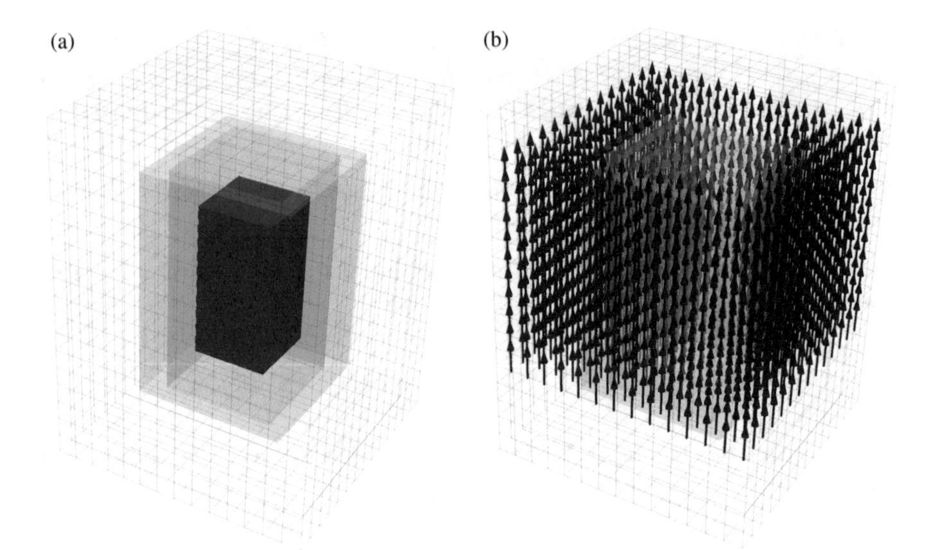

Fig. 17 Simple magnetoquasistatic model of an inductor with metal core. Coil is given by the stranded conductor model (Benchmark 6.9). (**a**) Iron core (red) with surrounding coil (transparent grey). (**b**) Source magnetic field strength $\mathbf{H}_s$ such that $\nabla \times \mathbf{H}_s = \chi_s i_s$

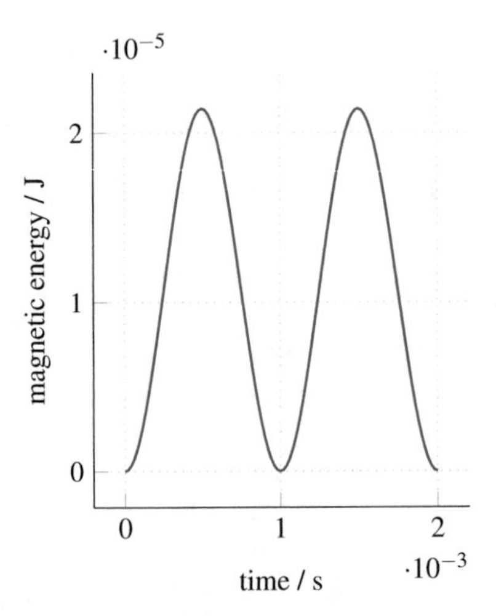

Fig. 18 Magnetic energy of the inductor in Fig. 14 simulated with the T–Ω formulation

6.3 Darwin Model

In many situations, phenomena related to electric energy, magnetic energy and Joule losses coincide, while at the same time, for the considered operating frequencies, the wave lengths are much larger than the model size, which indicates that wave propagation effects can be neglected. An example thereof is a filter design with concentrated elements, i.e., coils and capacitors connected by strip lines, arranged on a printed circuit board [88]. In such models, resonances between the discrete elements and Joule losses in all conducting parts need to be considered, whereas for the considered frequency range, cross-talk due to electromagnetic radiation is irrelevant. For such configurations, a combination of electroquasistatics and magnetoquasistatics in the form of the *Darwin* approximation of the Maxwell equations is the appropriate formulation.

The Darwin approximation starts from the decomposition of the electric field strength $\mathbf{E} = \mathbf{E}_{\mathrm{irr}} + \mathbf{E}_{\mathrm{rem}}$ into an *irrotational part* $\mathbf{E}_{\mathrm{irr}}$ and a *remainder part* $\mathbf{E}_{\mathrm{rem}}$. This decomposition is not unique, in contrast to the Helmholtz decomposition, which can be considered as a special case enforcing $\mathbf{E}_{\mathrm{rem}}$ to be solenoidal. In this paper, the non-uniqueness of the decomposition will be resolved when choosing a gauge condition below. The Darwin approximation consists of removing the displacement currents related to $\mathbf{E}_{\mathrm{rem}}$ from the law of Ampère-Maxwell. This affects the set of Maxwell equations in the sense that, in the formulations derived below, second derivatives with respect to time vanish and the overall PDE looses its hyperbolic character, which is equivalent to neglecting wave propagation effects. The set of relevant equations is

$$\nabla \times \mathbf{E}_{\mathrm{rem}} = -\frac{\partial \mathbf{B}}{\partial t} \quad ; \tag{6.7}$$

$$\nabla \times (\boldsymbol{\nu}\mathbf{B}) = \mathbf{J}_{\mathrm{s}} + \boldsymbol{\sigma}\mathbf{E}_{\mathrm{irr}} + \boldsymbol{\sigma}\mathbf{E}_{\mathrm{rem}} + \frac{\partial}{\partial t}(\varepsilon\mathbf{E}_{\mathrm{irr}}) \quad ; \tag{6.8}$$

$$\nabla \cdot \mathbf{B} = 0 \quad ; \tag{6.9}$$

$$\nabla \cdot (\varepsilon\mathbf{E}_{\mathrm{irr}}) + \nabla \cdot (\varepsilon\mathbf{E}_{\mathrm{rem}}) = \rho \quad . \tag{6.10}$$

Almost all publications on the Darwin approximation are limited to the case with homogeneous materials. They choose the Helmholtz decomposition, i.e., $\nabla \cdot \mathbf{E}_{\mathrm{rem}} = 0$, which causes the term $\nabla \cdot (\varepsilon\mathbf{E}_{\mathrm{rem}})$ in (6.10) to vanish [39, 61]. In the more general case considered here, however, this term is important as it models the charges accumulating at material interfaces due to the induced electric field strength [59].

An ambiguity arises when expressing the continuity equation $\nabla \cdot \mathbf{J} + \frac{\partial \rho}{\partial t} = 0$. When inserting Gauss law, the result reads

$$\nabla \cdot \mathbf{J}_{\mathrm{s}} + \nabla \cdot (\boldsymbol{\sigma}\mathbf{E}_{\mathrm{irr}}) + \nabla \cdot (\boldsymbol{\sigma}\mathbf{E}_{\mathrm{rem}}) + \frac{\partial}{\partial t}(\nabla \cdot \varepsilon\mathbf{E}_{\mathrm{irr}}) + \frac{\partial}{\partial t}(\nabla \cdot \varepsilon\mathbf{E}_{\mathrm{rem}}) = 0 \quad , \tag{6.11}$$

whereas when applying the divergence operator to (6.8), the last term in (6.11) is missing. Hence, the Darwin model introduces an anomaly in the continuity equation, which can only be alleviated by also neglecting the displacement currents due to $\mathbf{E}_{\mathrm{rem}}$ in the continuity equation.

6.3.1 ϕ–A–ψ Formulation of the Darwin Model

The irrotational part in of the electric field strength is represented by an electric scalar potential ϕ, i.e., $\mathbf{E}_{\text{irr}} = -\nabla\phi$. The magnetic Gauss law (6.9) is resolved by the definition of the magnetic vector potential $\mathbf{A}$, i.e., $\mathbf{B} = \nabla\times\mathbf{A}$, whereas Faraday's law (6.7) is fulfilled by the definition of an additional scalar potential ψ such that $\mathbf{E}_{\text{rem}} = -\frac{\partial\mathbf{A}}{\partial t} - \nabla\psi$. The law of Ampère-Darwin, Gauss' law and the continuity equation become

$$\nabla\times(\nu\nabla\times\mathbf{A}) + \boldsymbol{\sigma}\frac{\partial\mathbf{A}}{\partial t} + \boldsymbol{\sigma}\nabla\psi + \boldsymbol{\sigma}\nabla\phi + \varepsilon\nabla\frac{\partial\phi}{\partial t} = \mathbf{J}_{\text{s}} \quad ; \tag{6.12}$$

$$-\nabla\cdot\left(\varepsilon\frac{\partial\mathbf{A}}{\partial t}\right) - \nabla\cdot(\varepsilon\nabla\psi) - \nabla\cdot(\varepsilon\nabla\phi) = \rho \quad ; \tag{6.13}$$

$$-\nabla\cdot\left(\boldsymbol{\sigma}\frac{\partial\mathbf{A}}{\partial t}\right) - \nabla\cdot(\boldsymbol{\sigma}\nabla\psi) - \nabla\cdot(\boldsymbol{\sigma}\nabla\phi) - \nabla\cdot\left(\varepsilon\nabla\frac{\partial\phi}{\partial t}\right) = 0 \quad . \tag{6.14}$$

The introduced potentials ϕ, $\mathbf{A}$ and ψ lead to too many degrees of freedom. The electric field strength is decomposed as

$$\mathbf{E} = \underbrace{\overbrace{-\nabla\phi}^{\mathbf{E}_\phi}}_{\mathbf{E}_{\text{irr}}} \underbrace{\overbrace{-\frac{\partial\mathbf{A}}{\partial t}}^{\mathbf{E}_\mathbf{A}} \overbrace{-\nabla\psi}^{\mathbf{E}_\psi}}_{\mathbf{E}_{\text{rem}}} \tag{6.15}$$

Irrotational parts of the electric field strength can be attributed to $\mathbf{E}_\phi$, $\mathbf{E}_\mathbf{A}$ or $\mathbf{E}_\psi$. However, the splitting determines which displacement currents are considered in the law of Ampère-Maxwell and which are only considered in the electric law of Gauss. All components of $\mathbf{E}$ contribute to the Joule losses and contribute in the electric Gauss law, whereas only $\mathbf{E}_\phi$ contributes to the displacement currents. The component $\mathbf{E}_\psi$ can be fully integrated into $\mathbf{E}_\mathbf{A}$.

The choice of ψ determines the fraction of the displacement currents which is neglected in the Darwin model. In the case of a homogeneous material distribution, the choice $\psi = 0$ implies a Helmholtz splitting of $\mathbf{E}$ into the irrotational part $\mathbf{E}_{\text{irr}}$ and a solenoidal part $\mathbf{E}_{\text{rem}}$. Then, the Darwin approximation amounts to neglecting the displacement currents related to the solenoidal part of the electric field strength. Other choices for ψ lead to other Darwin models yielding different results. Hence, Darwin's approximation to the Maxwell equations is not gauge-invariant [61].

6.3.2 ψ–A*-formulation of the Darwin Model

A straightforward choice for ψ is $\psi = 0$ in which case, it makes sense to combine (6.12) and (6.13), i.e.,

$$\nabla \times (\nu \nabla \times \mathbf{A}) + \sigma \frac{\partial \mathbf{A}}{\partial t} + \sigma \nabla \phi + \varepsilon \nabla \frac{\partial \phi}{\partial t} = \mathbf{J}_\mathrm{s} \quad ; \tag{6.16}$$

$$-\nabla \cdot \left(\varepsilon \frac{\partial \mathbf{A}}{\partial t} \right) - \nabla \cdot (\varepsilon \nabla \phi) = \rho \quad . \tag{6.17}$$

The drawback of this formulation is that the charge density must be prescribed, which may be cumbersome in the presence of metallic parts at floating potentials.

6.3.3 Alternative ψ–A*-formulation of the Darwin Model

Another possible formulation [58] arises with $\psi = 0$, but by combining (6.12) and (6.14), i.e.,

$$\nabla \times (\nu \nabla \times \mathbf{A}) + \sigma \frac{\partial \mathbf{A}}{\partial t} + \sigma \nabla \phi + \varepsilon \nabla \frac{\partial \phi}{\partial t} = \mathbf{J}_\mathrm{s} \quad ; \tag{6.18}$$

$$-\nabla \cdot \left(\sigma \frac{\partial \mathbf{A}}{\partial t} \right) - \nabla \cdot (\sigma \nabla \phi) - \nabla \cdot \left(\varepsilon \nabla \frac{\partial \phi}{\partial t} \right) = 0 \quad . \tag{6.19}$$

Here, however, the magnetic vector potential **A** is not uniquely defined in the non-conductive model parts. Then, a gauge is necessary is fix the irrotational part of **A** in the non-conductive model parts. During post-processing, one should discard the irrotational part of **A** when calculating the electric field strength or displacement current density in the non-conducting parts.

6.3.4 Discretisation of the Darwin Model

The discrete counterpart of (6.16) and (6.17) reads

$$\mathbf{C}^\top \mathbf{M}_\nu \mathbf{C} \widehat{\mathbf{a}} + \mathbf{M}_\sigma \frac{\mathrm{d}}{\mathrm{d}t} \widehat{\mathbf{a}} - \mathbf{M}_\sigma \widetilde{\mathbf{S}}^\top \Phi - \mathbf{M}_\varepsilon \widetilde{\mathbf{S}}^\top \frac{\mathrm{d}}{\mathrm{d}t} \Phi = \widehat{\widehat{\mathbf{j}}}_\mathrm{s} \quad ; \tag{6.20}$$

$$\widetilde{\mathbf{S}} \mathbf{M}_\varepsilon \frac{\mathrm{d}}{\mathrm{d}t} \widehat{\mathbf{a}} + \widetilde{\mathbf{S}} \mathbf{M}_\varepsilon \widetilde{\mathbf{S}}^\top \Phi = \mathbf{q} \quad , \tag{6.21}$$

whereas the discrete counterpart of (6.18) and (6.19) reads

$$\mathbf{C}^\top \mathbf{M}_\nu \mathbf{C} \widehat{\mathbf{a}} + \mathbf{M}_\sigma \frac{\mathrm{d}}{\mathrm{d}t} \widehat{\mathbf{a}} - \mathbf{M}_\sigma \widetilde{\mathbf{S}}^\top \Phi - \mathbf{M}_\varepsilon \widetilde{\mathbf{S}}^\top \frac{\mathrm{d}}{\mathrm{d}t} \Phi = \widehat{\widehat{\mathbf{j}}}_\mathrm{s} \quad ; \tag{6.22}$$

$$\widetilde{\mathbf{S}} \mathbf{M}_\sigma \frac{\mathrm{d}}{\mathrm{d}t} \widehat{\mathbf{a}} + \widetilde{\mathbf{S}} \mathbf{M}_\sigma \widetilde{\mathbf{S}}^\top \Phi + \widetilde{\mathbf{S}} \mathbf{M}_\varepsilon \widetilde{\mathbf{S}}^\top \frac{\mathrm{d}}{\mathrm{d}t} \Phi = 0 \quad , \tag{6.23}$$

with all matrices defined as above. The second formulation needs a gauge for the non-conducting model parts as previously discussed in Assumption 6.3.

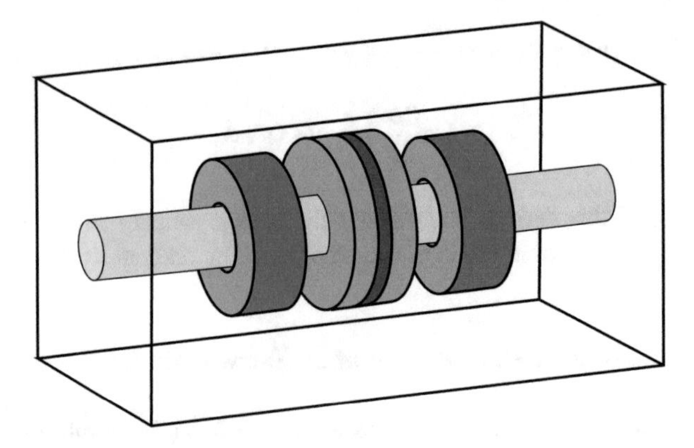

Fig. 19 Benchmark example for Darwin model, originally proposed in [59], see Benchmark 6.10

In [59], simulation results for the Darwin approximation, the electroquasistatic approximation and the magnetoquasistatic approximation are compared to results obtained by a full-wave solver, serving as a reference. This comparison clearly shows the limitations (the neglected wave propagation effects) and the advantages (more accurate than electroquasistatics and magnetoquasistatics and faster than a full-wave solver) of the Darwin model.

A DAE index analysis of the systems (6.20)–(6.21) and (6.22)–(6.23) is an open research question. Also, the following benchmark is merely a literature reference.

Benchmark 6.10 *The Darwin benchmark example shown in Fig. 19 is taken from [59]. It is axisymmetric and consists of conductive, dielectric and magnetic pieces. A solid conductor is connected to two plates which form a capacitor in combination with the dielectric material layer in between. Two solid ferrite rings surround the conductor. The problem is excited by potentials* ϕ_1 *and* ϕ_0 *at the ends of the conductor, the remaining boundaries are mbc. Geometry specifications and material data can be found in [59]. Koch et al. propose to regularise the model by an artificial conductivity.*

7 Conclusions

This paper has discussed various formulation for low and high-frequency problems in computational electromagnetics. In contrast to electric circuit simulation, e.g. [92], most formulations are rather harmless, i.e., they lead to a systems of low DAE index (≤ 2). More precisely, the only index-2 problem arises in the case of considering Maxwell's equations in A–ϕ-potential formulation with a Coulomb gauge. It has been shown that this can be mitigated by a reformulation based on Lorenz' gauge. Obviously, when coupling various formulations with each other or with electric circuits, the situation becomes more complex.

References

1. Albanese, R., Coccorese, E., Martone, R., Miano, G., Rubinacci, G.: On the numerical solution of the nonlinear three-dimensional eddy current problem. IEEE Trans. Magn. **27**(5), 3990–3995 (1991). https://doi.org/10.1109/20.104976
2. Alonso Rodríguez, A., Raffetto, M.: Unique solvability for electromagnetic boundary value problems in the presence of partly lossy inhomogeneous anisotropic media and mixed boundary conditions. Math. Models Methods Appl. Sci. **13**(04), 597–611 (2003). https://doi.org/10.1142/S0218202503002672
3. Alonso Rodríguez, A., Valli, A.: Eddy Current Approximation of Maxwell Equations. Modeling, Simulation and Applications, vol. 4. Springer, Heidelberg (2010). https://doi.org/10.1007/978-88-470-1506-7
4. Alotto, P., De Cian, A., Molinari, G.: A time-domain 3-D full-Maxwell solver based on the cell method. IEEE Trans. Magn. **42**(4), 799–802 (2006). https://doi.org/10.1109/tmag.2006.871381
5. Assous, F., Ciarlet, P., Labrunie, S.: Mathematical Foundations of Computational Electromagnetism. Springer, Cham (2018)
6. Außerhofer, S., Bíró, O., Preis, K.: Discontinuous Galerkin finite elements in time domain eddy-current problems. IEEE Trans. Magn. **45**(3), 1300–1303 (2009)
7. Bartel, A., Baumanns, S., Schöps, S.: Structural analysis of electrical circuits including magnetoquasistatic devices. Appl. Numer. Math. **61**, 1257–1270 (2011). https://doi.org/10.1016/j.apnum.2011.08.004
8. Baumanns, S.: Coupled electromagnetic field/circuit simulation: modeling and numerical analysis. Ph.D. thesis, Universität zu Köln, Köln (2012)
9. Baumanns, S., Selva Soto, M., Tischendorf, C.: Consistent initialization for coupled circuit-device simulation. In: Roos, J., Costa, L.R.J. (eds.) Scientific Computing in Electrical Engineering SCEE 2008. Mathematics in Industry, vol. 14, pp. 297–304. Springer, Berlin (2010). https://doi.org/10.1007/978-3-642-12294-1_38
10. Baumanns, S., Clemens, M., Schöps, S.: Structural aspects of regularized full Maxwell electrodynamic potential formulations using FIT. In: Manara, G. (ed.) Proceedings of 2013 URSI International Symposium on Electromagnetic Theory (EMTS), pp. 1007–1010. IEEE, New York (2013)
11. Becks, T., Wolff, I.: Analysis of 3-d metallization structures by a full-wave spectral-domain technique. IEEE Trans. Microwave Theory Tech. **40**(12), 2219–2227 (1992). https://doi.org/10.1109/22.179883
12. Bedrosian, G.: A new method for coupling finite element field solutions with external circuits and kinematics. IEEE Trans. Magn. **29**(2), 1664–1668 (1993). https://doi.org/10.1109/20.250726
13. Bíró, O., Preis, K.: On the use of the magnetic vector potential in the finite-element analysis of three-dimensional eddy currents. IEEE Trans. Magn. **25**(4), 3145–3159 (1989). https://doi.org/10.1109/20.34388
14. Bíró, O., Preis, K.: Finite element analysis of 3-d eddy currents. IEEE Trans. Magn. **26**(2), 418–423 (1990). https://doi.org/10.1109/20.106343
15. Bíró, O., Preis, K., Richter, K.R.: Various FEM formulations for the calculation of transient 3d eddy currents in nonlinear media. IEEE Trans. Magn. **31**(3), 1307–1312 (1995). https://doi.org/10.1109/20.376269
16. Boffi, D.: Finite element approximation of eigenvalue problems. Acta. Numer. **19**, 1–120 (2010). https://doi.org/10.1017/S0962492910000012
17. Bondeson, A., Rylander, T., Ingelström, P.: Computational Electromagnetics. Texts in Applied Mathematics. Springer, Berlin (2005). https://doi.org/10.1007/b136922
18. Bossavit, A.: Whitney forms: a class of finite elements for three-dimensional computations in electromagnetism. IEE Proc. **135**(8), 493–500 (1988). https://doi.org/10.1049/ip-a-1:19880077

19. Bossavit, A.: Differential geometry for the student of numerical methods in electromagnetism. Technical Report, Électricité de France (1991)
20. Bossavit, A.: Computational Electromagnetism: Variational Formulations, Complementarity, Edge Elements. Academic Press, San Diego (1998)
21. Bossavit, A.: On the geometry of electromagnetism. (4): 'Maxwell's house'. J. Jpn. Soc. Appl. Electromagn. Mech. **6**(4), 318–326 (1999)
22. Bossavit, A.: Stiff problems in eddy-current theory and the regularization of Maxwell's equations. IEEE Trans. Magn. **37**(5), 3542–3545 (2001). https://doi.org/0018-9464/01\protect\T1\textdollar10.00
23. Bossavit, A., Kettunen, L.: Yee-like schemes on a tetrahedral mesh, with diagonal lumping. Int. J. Numer. Modell. Electron. Networks Devices Fields **12**(1-2), 129–142 (1999). https://doi.org/10.1002/(SICI)1099-1204(199901/04)12:1/2<129::AID-JNM327>3.0.CO;2-G
24. Bossavit, A., Kettunen, L.: Yee-like schemes on staggered cellular grids: a synthesis between FIT and FEM approaches. IEEE Trans. Magn. **36**(4), 861–867 (2000). https://doi.org/10.1109/20.877580
25. Brenan, K.E., Campbell, S.L., Petzold, L.R.: Numerical Solution of Initial-Value Problems in Differential-Algebraic Equations. SIAM, Philadelphia (1995)
26. Carpenter, C.J.: Comparison of alternative formulations of 3-dimensional magnetic-field and eddy-current problems at power frequencies. IEE Proc. B Electr. Power Appl. **127**(5), 332 (1980). https://doi.org/10.1049/ip-b:19800045
27. Chen, Q., Schoenmaker, W., Chen, G., Jiang, L., Wong, N.: A numerically efficient formulation for time-domain electromagnetic-semiconductor cosimulation for fast-transient systems. IEEE Trans. Comput. Aided. Des. Integrated Circ. Syst. **32**(5), 802–806 (2013). https://doi.org/10.1109/TCAD.2012.2232709
28. Clemens, M.: Large systems of equations in a discrete electromagnetism: formulations and numerical algorithms. IEE. Proc. Sci. Meas. Tech. **152**(2), 50–72 (2005). https://doi.org/10.1049/ip-smt:20050849
29. Clemens, M., Weiland, T.: Transient eddy-current calculation with the FI-method. IEEE Trans. Magn. **35**(3), 1163–1166 (1999). https://doi.org/10.1109/20.767155
30. Clemens, M., Weiland, T.: Regularization of eddy-current formulations using discrete grad-div operators. IEEE Trans. Magn. **38**(2), 569–572 (2002). https://doi.org/10.1109/20.996149
31. Clemens, M., Wilke, M., Weiland, T.: Linear-implicit time-integration schemes for error-controlled transient nonlinear magnetic field simulations. IEEE Trans. Magn. **39**(3), 1175–1178 (2003). https://doi.org/10.1109/TMAG.2003.810221
32. Clemens, M., Schöps, S., De Gersem, H., Bartel, A.: Decomposition and regularization of nonlinear anisotropic curl-curl DAEs. Int. J. Comput. Math. Electr. Electron. Eng. **30**(6), 1701–1714 (2011). https://doi.org/10.1108/03321641111168039
33. Cortes Garcia, I., De Gersem, H., Schöps, S.: A structural analysis of field/circuit coupled problems based on a generalised circuit element (2018, submitted). arXiv:1801.07081
34. CST AG: CST STUDIO SUITE 2016 (2016). https://www.cst.com
35. De Gersem, H., Hameyer, K.: A finite element model for foil winding simulation. IEEE Trans. Magn. **37**(5), 3472–3432 (2001). https://doi.org/10.1109/20.952629
36. De Gersem, H., Weiland, T.: Field-circuit coupling for time-harmonic models discretized by the finite integration technique. IEEE Trans. Magn. **40**(2), 1334–1337 (2004). https://doi.org/10.1109/TMAG.2004.824536
37. De Gersem, H., Hameyer, K., Weiland, T.: Field-circuit coupled models in electromagnetic simulation. J. Comput. Appl. Math. **168**(1-2), 125–133 (2004). https://doi.org/10.1016/j.cam.2003.05.008
38. Deschamps, G.A.: Electromagnetics and differential forms. Proc. IEEE **69**(6), 676–696 (1981). https://doi.org/dx.doi.org/10.1109/PROC.1981.12048
39. Dirks, H.K.: Quasi-stationary fields for microelectronic applications. Electr. Eng. **79**(2), 145–155 (1996). https://doi.org/10.1007/BF01232924
40. Dutiné, J.S., Richter, C., Jörgens, C., Schöps, S., Clemens, M.: Explicit time integration techniques for electro- and magneto-quasistatic field simulations. In: Graglia, R.D. (ed.)

Proceedings of the International Conference on Electromagnetics in Advanced Applications (ICEAA) 2017. IEEE, New York (2017). https://doi.org/10.1109/ICEAA.2017.8065562
41. Dyck, D.N., Webb, J.P.: Solenoidal current flows for filamentary conductors. IEEE Trans. Magn. **40**(2), 810–813 (2004). https://doi.org/10.1109/TMAG.2004.824594
42. Eller, M., Reitzinger, S., Schöps, S., Zaglmayr, S.: A symmetric low-frequency stable broadband Maxwell formulation for industrial applications. SIAM J. Sci. Comput. **39**(4), B703–B731 (2017). https://doi.org/10.1137/16M1077817
43. Estévez Schwarz, D.: Consistent initialization of differential-algebraic equations in circuit simulation. Technical Report 99-5, Humboldt Universität Berlin, Berlin (1999)
44. Gödel, N., Schomann, S., Warburton, T., Clemens, M.: GPU accelerated Adams-Bashforth multirate discontinuous Galerkin FEM simulation of high-frequency electromagnetic fields. IEEE Trans. Magn. **46**(8), 2735–2738 (2010)
45. Griffiths, D.F.: Introduction to Electrodynamics. Prentice-Hall, Upper Saddle River (1999)
46. Hahne, P., Weiland, T.: 3d eddy current computation in the frequency domain regarding the displacement current. IEEE Trans. Magn. **28**(2), 1801–1804 (1992). https://doi.org/10.1109/20.124056
47. Hairer, E., Nørsett, S.P., Wanner, G.: Solving Ordinary Differential Equations II: Stiff and Differential-Algebraic Problems. Springer Series in Computational Mathematics, 2 edn. Springer, Berlin (2002)
48. Harrington, R.F.: Field Computation by Moment Methods. Wiley-IEEE, New York (1993)
49. Haus, H.A., Melcher, J.R.: Electromagnetic Fields and Energy. Englewood Cliffs, Prentice-Hall (1989)
50. Heaviside, O.: On the forces, stresses, and fluxes of energy in the electromagnetic field. Proc. R. Soc. Lond. Ser. I **50**, 126–129 (1891)
51. Hehl, F.W., Obukhov, Y.N.: Foundations of Classical Electrodynamics – Charge, Flux, and Metric. Progress in Mathematical Physics. Birkhäuser, Basel (2003)
52. Heise, B.: Analysis of a fully discrete finite element method for a nonlinear magnetic field problem. SIAM J. Numer. Anal. **31**(3), 745–759 (1994)
53. Hesthaven, J.S., Warburton, T.: Nodal Discontinuous Galerkin Methods: Algorithms, Analysis, and Applications. Texts in Applied Mathematics. Springer, New York (2008)
54. Jackson, J.D.: Classical Electrodynamics, 3rd edn. Wiley, New York (1998)
55. Jochum, M.T., Farle, O., Dyczij-Edlinger, R.: A new low-frequency stable potential formulation for the finite-element simulation of electromagnetic fields. IEEE Trans. Magn. **51**(3), 7402,304 (2015). https://doi.org/10.1109/TMAG.2014.2360080
56. Kameari, A.: Calculation of transient 3D eddy-current using edge elements. IEEE Trans. Magn. **26**(5), 466–469 (1990). https://doi.org/10.1109/20.106354
57. Kerler-Back, J., Stykel, T.: Model reduction for linear and nonlinear magneto-quasistatic equations. Int. J. Numer. Methods Eng. **111**(13), 1274–1299 (2017). https://doi.org/10.1002/nme.5507
58. Koch, S., Weiland, T.: Time domain methods for slowly varying fields. In: URSI International Symposium on Electromagnetic Theory (EMTS 2010), pp. 291–294 (2010). https://doi.org/10.1109/URSI-EMTS.2010.5636991
59. Koch, S., Weiland, T.: Different types of quasistationary formulations for time domain simulations. Radio Sci. **46**(5) (2011). https://doi.org/10.1029/2010RS004637
60. Lamour, R., März, R., Tischendorf, C.: Differential-Algebraic Equations: A Projector Based Analysis. Differential-Algebraic Equations Forum. Springer, Heidelberg (2013). https://doi.org/10.1007/978-3-642-27555-5
61. Larsson, J.: Electromagnetics from a quasistatic perspective. Am. J. Phys. **75**(3), 230–239 (2007). https://doi.org/10.1119/1.2397095
62. Manges, J.B., Cendes, Z.J.: Tree-cotree decompositions for first-order complete tangential vector finite elements. Int. J. Numer. Methods Eng. **40**(9), 1667–1685 (1997). https://doi.org/10.1002/(SICI)1097-0207(19970515)40:9<1667::AID-NME133>3.0.CO;2-9
63. März, R.: Differential algebraic systems with properly stated leading term and MNA equations. In: Anstreich, K., Bulirsch, R., Gilg, A., Rentrop, P. (eds.) Modelling, Simulation and Optimization of Integrated Circuits, pp. 135–151. Birkhäuser, Berlin (2003)

64. Maxwell, J.C.: A dynamical theory of the electromagnetic field. Phil. Trans. R. Soc. London **CLV**, 459–512 (1864)
65. Mehrmann, V.: Index Concepts for Differential-Algebraic Equations, pp. 676–681. Springer, Berlin (2015). https://doi.org/10.1007/978-3-540-70529-1_120
66. Merkel, M., Niyonzima, I., Schöps, S.: Paraexp using leapfrog as integrator for high-frequency electromagnetic simulations. Radio Sci. **52**(12), 1558–1569 (2017). https://doi.org/10.1002/2017RS006357
67. Monk, P.: Finite Element Methods for Maxwell's Equations. Oxford University Press, Oxford (2003)
68. Monk, P., Süli, E.: A convergence analysis of Yee's scheme on nonuniform grids. SIAM J. Numer. Anal. **31**(2), 393–412 (1994). https://doi.org/10.1137/0731021
69. Munteanu, I.: Tree-cotree condensation properties. ICS Newsl. (International Compumag Society) **9**, 10–14 (2002). http://www.compumag.org/jsite/images/stories/newsletter/ICS-02-09-1-Munteanu.pdf
70. Nagel, L.W., Pederson, D.O.: Simulation program with integrated circuit emphasis. Technical Report, University of California, Berkeley, Electronics Research Laboratory, ERL-M382 (1973)
71. Nédélec, J.C.: Mixed finite elements in r^3. Numer. Math. **35**(3), 315–341 (1980). https://doi.org/10.1007/BF01396415
72. Nicolet, A., Delincé, F.: Implicit Runge-Kutta methods for transient magnetic field computation. IEEE Trans. Magn. **32**(3), 1405–1408 (1996). https://doi.org/0.1109/20.497510
73. Ostrowski, J., Hiptmair, R., Krämer, F., Smajic, J., Steinmetz, T.: Transient full Maxwell computation of slow processes. In: Michielsen, B., Poirier, J.R. (eds.) Scientific Computing in Electrical Engineering SCEE 2010. Mathematics in Industry, vol. 16, pp. 87–95. Springer, Berlin (2012). https://doi.org/10.1007/978-3-642-22453-9_10
74. Ouédraogo, Y., Gjonaj, E., Weiland, T., De Gersem, H., Steinhausen, C., Lamanna, G., Weigand, B., Preusche, A., Dreizler, A., Schremb, M.: Electrohydrodynamic simulation of electrically controlled droplet generation. Int. J. Heat Fluid Flow **64**, 120–128 (2017)
75. Petzold, L.R.: Differential/algebraic equations are not ODE's. SIAM J. Sci. Stat. Comput. **3**(3), 367–384 (1982). https://doi.org/10.1137/0903023
76. Rapetti, F., Rousseaux, G.: On quasi-static models hidden in Maxwell's equations. Appl. Numer. Math. **79**, 92–106 (2014). https://doi.org/10.1016/j.apnum.2012.11.007
77. Rautio, J.C.: The long road to Maxwell's equations. IEEE Spectr. **51**(12), 36–56 (2014). https://doi.org/10.1109/MSPEC.2014.6964925
78. Raviart, P.A., Thomas, J.M.: Primal hybrid finite element methods for 2nd order elliptic equations. Math. Comput. **31**(138), 391–413 (1977)
79. Ruehli, A.E.: Equivalent circuit models for three-dimensional multiconductor systems. IEEE Trans. Microwave Theory Tech. **22**(3), 216–221 (1974)
80. Ruehli, A.E., Antonini, G., Jiang, L.: The Partial Element Equivalent Circuit Method for Electro-Magnetic and Circuit Problems. Wiley, Hoboken (2015)
81. Schilders, W.H.A., Ciarlet, P., ter Maten, E.J.W. (eds.): Handbook of Numerical Analysis. Numerical Methods in Electromagnetics. Handbook of Numerical Analysis, vol. 13. North-Holland, Amsterdam (2005)
82. Schmidt, K., Sterz, O., Hiptmair, R.: Estimating the eddy-current modeling error. IEEE Trans. Magn. **44**(6), 686–689 (2008). https://doi.org/10.1109/TMAG.2008.915834
83. Schoenmaker, W.: Computational Electrodynamics. River Publishers Series in Electronic Materials and Devices. River Publishers, Delft (2017)
84. Schöps, S.: Multiscale modeling and multirate time-integration of field/circuit coupled problems. Dissertation, Bergische Universität Wuppertal & Katholieke Universiteit Leuven, Düsseldorf (2011). VDI Verlag. Fortschritt-Berichte VDI, Reihe 21
85. Schöps, S., Bartel, A., Clemens, M.: Higher order half-explicit time integration of eddy current problems using domain substructuring. IEEE Trans. Magn. **48**(2), 623–626 (2012). https://doi.org/10.1109/TMAG.2011.2172780

86. Schöps, S., De Gersem, H., Weiland, T.: Winding functions in transient magnetoquasistatic field-circuit coupled simulations. Int. J. Comput. Math. Electr. Electron. Eng. **32**(6), 2063–2083 (2013). https://doi.org/10.1108/COMPEL-01-2013-0004
87. Schuhmann, R., Weiland, T.: Conservation of discrete energy and related laws in the finite integration technique. Prog. Electromagn. Res. **32**, 301–316 (2001). https://doi.org/10.2528/PIER00080112
88. Schuhmacher, S., Klaedtke, A., Keller, C., Ackermann, W., De Gersem, H.: Optimizing the inductance cancellation behavior in an EMI filter design with the help of a sensitivity analysis. In: EMC Europe. Angers, France (2017)
89. Steinmetz, T., Kurz, S., Clemens, M.: Domains of validity of quasistatic and quasistationary field approximations. Int. J. Comput. Math. Electr. Electron. Eng. **30**(4), 1237–1247 (2011). https://doi.org/10.1108/03321641111133154
90. Taflove, A.: Computational Electrodynamics: The Finite-Difference Time-Domain-Method. Artech House, Dedham (1995)
91. Taflove, A.: A perspective on the 40-year history of FDTD computational electrodynamics. Appl. Comput. Electromagn. Soc. J. **22**(1), 1–21 (2007)
92. Tischendorf, C.: Topological index calculation of DAEs in circuit simulation. Technical Report 3-4, Humboldt Universität Berlin, Berlin (1999)
93. Tonti, E.: On the formal structure of physical theories. Technical Report, Politecnico di Milano, Milano, Italy (1975)
94. Tsukerman, I.A.: Finite element differential-algebraic systems for eddy current problems. Numer. Algorithms **31**(1), 319–335 (2002). https://doi.org/10.1023/A:1021112107163
95. Webb, J.P., Forghani, B.: The low-frequency performance of $h - \phi$ and $t - \omega$ methods using edge elements for 3d eddy current problems. IEEE Trans. Magn. **29**(6), 2461–2463 (1993). https://doi.org/10.1109/20.280983
96. Weeks, W., Jimenez, A., Mahoney, G., Mehta, D., Qassemzadeh, H., Scott, T.: Algorithms for ASTAP – a network-analysis program. IEEE Trans. Circuit Theory **20**(6), 628–634 (1973). https://doi.org/10.1109/TCT.1973.1083755
97. Weiland, T.: A discretization method for the solution of Maxwell's equations for six-component fields. Int. J. Electron. Commun. (AEU) **31**, 116–120 (1977)
98. Weiland, T.: On the unique numerical solution of Maxwellian eigenvalue problems in three dimensions. Part. Accel. **17**(227–242) (1985)
99. Weiland, T.: Time domain electromagnetic field computation with finite difference methods. Int. J. Numer. Modell. Electron. Networks Devices Fields **9**(4), 295–319 (1996). https://doi.org/10.1002/(SICI)1099-1204(199607)9:4<295::AID-JNM240>3.0.CO;2-8
100. Yee, K.S.: Numerical solution of initial boundary value problems involving Maxwell's equations in isotropic media. IEEE Trans. Antennas Propag. **14**(3), 302–307 (1966). https://doi.org/10.1109/TAP.1966.1138693

Gas Network Benchmark Models

Peter Benner, Sara Grundel, Christian Himpe, Christoph Huck, Tom Streubel, and Caren Tischendorf

Abstract The simulation of gas transportation networks becomes increasingly more important as its use-cases broaden to more complex applications. Classically, the purpose of the gas network was the transportation of predominantly natural gas from a supplier to the consumer for long-term scheduled volumes. With the rise of renewable energy sources, gas-fired power plants are often chosen to compensate for the fluctuating nature of the renewables, due to their on-demand power generation capability. Such an only short-term plannable supply and demand setting requires sophisticated simulations of the gas network prior to the dispatch to ensure the supply of all customers for a range of possible scenarios and to prevent damages to the gas network. In this work we describe the modeling of gas networks and present benchmark systems to test implementations and compare new or extended models.

Keywords Flow network · Gas network · Gas transport · Isothermal Euler equation · Natural gas · Pipeline

Mathematics Subject Classification (2010) 76N15, 68U20, 35L60

Electronic supplementary material The online version of this article (https://doi.org/10.1007/11221_2018_5) contains supplementary material, which is available to authorized users.

P. Benner · S. Grundel (✉) · C. Himpe
Max Planck Institute for Dynamics of Complex Technical Systems, Magdeburg, Germany
e-mail: benner@mpi-magdeburg.mpg.de; grundel@mpi-magdeburg.mpg.de; himpe@mpi-magdeburg.mpg.de

C. Huck · C. Tischendorf
Department of Mathematics, Humboldt-Universität zu Berlin, Berlin, Germany
e-mail: christoph.huck@math.hu-berlin.de; tischendorf@math.hu-berlin.de

T. Streubel
Department of Optimization at Zuse Institute Berlin, Berlin, Germany

Department of Mathematics, Humboldt-Universität zu Berlin, Berlin, Germany
e-mail: streubel@zib.de

S. Campbell et al. (eds.), *Applications of Differential-Algebraic Equations: Examples and Benchmarks*, Differential-Algebraic Equations Forum,
https://doi.org/10.1007/11221_2018_5

1 Introduction

The simulation of gas transport over large pipeline networks is essential to a safe and timely dispatch and delivery of contracted denominations. The modeling and verified implementation of gas transport is a prerequisite for the reliable simulation of gas transportation scenarios. Beyond the basic network of pipelines, further (active) components such as compressors have to be included into realistic models. To this end, a modeling approach for gas networks including these components is presented in this work together with four benchmark examples and associated reference solutions allowing the test of gas network simulation implementations.

The basis for gas network models are the Euler equations as introduced in [25], which describe the transient behavior in terms of conservation of momentum, conservation of mass and the gas state. Discretizations of this gas network model given by partial differential algebraic equations have been investigated in [1, 6]. An index reduction of this differential-algebraic to a purely differential model in a model order reduction setting has been investigated in [10–12]. The modeling of complex network elements such as compressors in the context of gas networks is described in [7, 13], while verification of this model has been conducted for example in [2, 26]. Modeling based on practical engineering considerations can be found in [8, 21].

In this work, we will present a modular gas network model based on the isothermal Euler equations. The modularity rests upon factor approximations which in different regimes are chosen accordingly. The focus of the modeling effort is hereby directed towards transient simulations of the gas transport. Beyond basic pipeline networks, the following modeling approach includes gas network elements like valves, resistors or compressors, and allows the extension with new elements. Additionally, certain benchmark networks are outlined together with respective scenarios, describing the transient boundary value behavior in order to provide testable discretized model instances.

In Sect. 2 we describe the model for a single pipe, which is extended to a network of pipes and additional components in Sect. 3. Section 4 details the partial discretization of the network model and finally, Sect. 5 describes four benchmark networks with increasing degree of complexity.

2 Pipe Physics

2.1 The Isothermal Euler Equations

The flow of (a real) gas is modeled by the Euler equations, which describe the conservation of mass (2.1a), conservation of momentum (2.1b), and inherent state of the gas (2.1c). In the following, we discuss the analytic modeling, assumptions and simplifications of these partial differential equations (PDEs).

First of all, we assume that the temperature variations throughout the network have negligible effects on the dynamic behavior of the gas transport. This may seem unrealistic, but for strictly on-shore gas networks this is actually a sensible assumption [21, Ch. 45] and greatly reduces the complexity of the model. Hence, we fix the temperature to a constant value T_0.

For the transport of gas in a network of pipes, we first model a single pipe of length L by the one-dimensional isothermal Euler equations over the spatial domain $x \in [0, L]$ and time $t \in \mathbb{R}^+$:

$$\frac{\partial}{\partial t}\rho = -\frac{\partial}{\partial x}\varphi, \tag{2.1a}$$

$$\frac{\partial}{\partial t}\varphi = -\frac{\partial}{\partial x}p - \frac{\partial}{\partial x}(\rho v^2) - g\rho\frac{\partial}{\partial x}h - \frac{\hat{\lambda}(\varphi)}{2D}\rho v|v|, \tag{2.1b}$$

$$p = \gamma(T)z(p, T)\rho. \tag{2.1c}$$

This system of coupled (PDEs) in space and time consists of the variables: density $\rho = \rho(x, t)$, flow rate $\varphi = \varphi(x, t)$, pressure $p = p(x, t)$, velocity $v = v(x, t)$ and pipe elevation $h = h(x)$. Note, that the Euler equations are nonlinear (2.1b) and of hyperbolic nature [19].

The remaining components are: the gravity constant g, pipe diameter D, gas state $\gamma(T)$, friction factor $\hat{\lambda}(\varphi)$ and the compressibility factor $z(p, T)$. The latter two functions will be discussed in Sects. 2.2 and 2.3. Also, as the temperature T is assumed constant, the temperature dependency of the gas state and compressibility factor is fixed to $T \equiv T_0$. See Table 1 for a list of all symbols and their associated units.

Subsequently we will transform this model, based on physical laws, to a representation which contains the measurable quantities as solution variables, to a more convenient form with respect to the numerical simulation. To this end we introduce the mass flow $q = q(x, t) := m\varphi(x, t)$ and the pipe's cross-sectional area

Table 1 List of symbols

Symbol	Meaning	SI-unit	Symbol	Meaning	SI-unit
ρ	Density	$[\frac{\text{kg}}{\text{m}^3}]$	R_s	Specific gas constant	$[\frac{\text{m}^2}{\text{s}^2\,\text{K}}]$
p	Pressure	$[\frac{\text{kg}}{\text{s}^2\,\text{m}}]$	γ	Gas state	$[\frac{\text{m}^2}{\text{s}^2}]$
φ	Flow-rate	$[\frac{\text{m}^3}{\text{s}}]$	z	Compressibility factor	[1]
q	Mass-flow	$[\frac{\text{kg}}{\text{s}}]$	S	Cross-sectional area	$[\text{m}^2]$
v	Velocity	$[\frac{\text{m}}{\text{s}}]$	D	Pipe diameter	[m]
g	Gravity constant	$[\frac{\text{m}}{\text{s}^2}]$	L	Pipe length	[m]
h	Pipe elevation	[m]	T	Temperature	[K]
λ	Friction factor	[1]	c	Speed of sound	$[\frac{\text{m}}{\text{s}}]$
k	Roughness of pipe wall	[m]	P	Power	[W]

$$\text{Pressure (Continuity)} \left\{ \frac{1}{\gamma_0} \frac{\partial}{\partial t} \frac{p}{z_0(p)} = -\frac{1}{S} \frac{\partial}{\partial x} q, \right.$$

$$\text{Mass Flux (Momentum)} \left\{ \frac{1}{S} \frac{\partial}{\partial t} q = -\frac{\partial}{\partial x} p \underbrace{- \frac{\gamma_0}{S^2} \frac{\partial}{\partial x} q^2 \frac{z_0(p)}{p}}_{\text{Inertia Term}} \underbrace{- \frac{g}{\gamma_0} \frac{p}{z_0(p)} \frac{\partial}{\partial x} h}_{\text{Gravity Term}} \underbrace{- \frac{\lambda(q)\gamma_0}{2DS^2} \frac{q|q|}{\left(\frac{p}{z_0(p)}\right)}}_{\text{Friction Term}} \right.$$

Fig. 1 Term-wise highlighted PDE model with respect to physical meaning

$S := \frac{1}{4}\pi D^2$, over which the gas flow in the pipe is averaged, to replace the velocity by mass flux (mass flow per area) over density $v = \frac{1}{S}\frac{q}{\rho}$ and obtain:

$$\frac{\partial}{\partial t}\rho = -\frac{1}{S}\frac{\partial}{\partial x}q, \tag{2.2a}$$

$$\frac{1}{S}\frac{\partial}{\partial t}q = -\frac{\partial}{\partial x}p - \frac{1}{S^2}\frac{\partial}{\partial x}\frac{q^2}{\rho} - g\rho\frac{\partial}{\partial x}h - \frac{\lambda(q)}{2DS^2}\frac{q|q|}{\rho}, \tag{2.2b}$$

$$p = \gamma(T)z(p,T)\rho. \tag{2.2c}$$

To match the change in variables, the friction factor is also adapted to the representation $\lambda(q) := \hat{\lambda}(S\varphi)$.

Using Boyle's Law and given the specific gas constant R_s, the gas state is constant $\gamma_0 := \gamma(T_0) = R_s T_0$ and we define $z_0(p) := z(p, T_0)$ due to the isothermality assumption. Finally, we substitute the pressure relation (2.2c) into (2.2a) and (2.2b) to obtain the following formulation of the isothermal Euler equations:

$$\frac{1}{\gamma_0}\frac{\partial}{\partial t}\frac{p}{z_0(p)} = -\frac{1}{S}\frac{\partial}{\partial x}q, \tag{2.3a}$$

$$\frac{1}{S}\frac{\partial}{\partial t}q = -\frac{\partial}{\partial x}p - \frac{\gamma_0}{S^2}\frac{\partial}{\partial x}q^2\frac{z_0(p)}{p} - \frac{g}{\gamma_0}\frac{p}{z_0(p)}\frac{\partial}{\partial x}h - \frac{\lambda(q)\gamma_0}{2DS^2}\frac{q|q|}{\left(\frac{p}{z_0(p)}\right)}; \tag{2.3b}$$

see also Fig. 1.

The inertia (or kinematic) term of the mass-flux equation in the pipe gas flow model evolves on a much smaller scale compared to the other components [25]. This is justified by comparing the coupling term and the inertia term (first two right-hand side components in (2.3b)) after factoring the spatial derivative operator $\frac{\partial}{\partial x}$:

$$\left|\frac{\gamma_0}{S^2}\frac{q^2 z_0(p)}{p}\right| = \left|p\frac{v^2}{z_0(p)\gamma_0}\right| \ll |p|, \quad \text{for} \quad z_0(p)\gamma_0 \gg v^2,$$

Table 2 Boundary values and quantities of interest

	Boundary	QoI
Pressure	$p(0,t)$	$p(L,t)$
Mass-flux	$q(L,t)$	$q(0,t)$

based on the velocity-mass-flux relation used in (2.2b). Since the speed of sound (in the medium) $c \approx \sqrt{z_0(p)\gamma_0}$ typically exceeds the transport velocity v, the inertia term is discarded in several works, as i.e. [10–12, 14]. We will follow this approach and similarly exclude the inertia term from the model, which then leads to:

$$\begin{aligned} \frac{\partial}{\partial t}\frac{p}{\gamma_0 z_0(p)} &= -\frac{1}{S}\frac{\partial}{\partial x}q, \\ \frac{\partial}{\partial t}q &= -S\frac{\partial}{\partial x}p - Sg\frac{p}{\gamma_0 z_0(p)}\frac{\partial}{\partial x}h - \frac{\lambda(q)}{2DS}\frac{q|q|}{\left(\frac{p}{\gamma_0 z_0(p)}\right)}. \end{aligned} \tag{2.4}$$

The nonlinearity in the friction term $\frac{q|q|}{p}$ may be treated as quadratic, i.e. $\frac{q^2}{p}$, only if the flow does not change direction. Since not only pipelines, but cyclic networks of pipes are considered, a flow direction may change throughout the course of a simulation. Some works [1, 29] linearize the friction term around the steady state of a given scenario, which is not considered in this work to preserve accuracy. Yet, the linearized equations may be used to obtain an approximate steady state given some boundary condition.

In terms of boundary conditions, the pressure and mass flow in the pipe at time $t = 0$ as well as the pressure at the inflow boundary $p_l(t) := p(0,t)$ and mass flow at the outflow $q_r(t) := q(L,t)$ are given. With this set up, the aim is the computation of the pressure at the outflow boundary $p_r(t) := p(L,t)$ and the mass flow at the inflow boundary $q_l(t) := q(0,t)$. Table 2 summarizes this relation of given boundary quantities and sought quantities of interest (QoI).

It remains to be specified how the friction and compressibility factor are included into the model. As these factors are typically derived from formulas determined by experimental measurements, we will not specify which formula to use, but instead keep the model modular in this regard and present different popular choices for the aforementioned factors in the following.

2.2 *Friction Factor*

The friction factor $\lambda(q)$ scales the (nonlinear) friction term and depends on the Reynolds number $\text{Re}(q)$, which in turn depends on the mass flow variable q for a flow in a pipe, and, depending on the approximation method, on the pipe roughness k and pipe diameter D. We will present two sets of approximation formulas for the friction factor (for turbulent flows): The first is predominately used in European countries, while the second is preferably used in the Commonwealth of Independent States (CIS) [21, Ch. 28]. In both regions, for a laminar flow (Reynolds numbers $\mathrm{Re} < 2100$), the well-known Hagen-Poisseuille formula is used to approximate the friction factor:

$$\lambda_{\mathrm{HP}}(q) := \frac{64}{\mathrm{Re}(q)}.$$

For Reynolds numbers $\mathrm{Re} > 4000$ a flow is considered turbulent. In the European region, the Colebrook-White formula [4], also known as Prandtl-Colebrook formula, is the most accurate approximation of the friction factor [30]:

$$\frac{1}{\sqrt{\lambda_{\mathrm{CW}}(q)}} = -2\log_{10}\Big(\frac{2.51}{\mathrm{Re}(q)\sqrt{\lambda_{CW}(q)}} + \frac{k}{3.71D}\Big),$$

yet of implicit nature. An explicit variant of the Colebrook-White formula is given by the Hofer approximation [15]:

$$\lambda_{\mathrm{H}}(q) := \Big(-2\log_{10}\Big(\frac{4.518}{\mathrm{Re}(q)}\log_{10}\Big(\frac{\mathrm{Re}(q)}{7}\Big) + \frac{k}{3.71D}\Big)\Big)^{-2},$$

which is of sufficient accuracy for transient gas network simulations. The Nikuradse formula [24] results from the Hofer formula for $\mathrm{Re} \to \infty$:

$$\lambda_{\mathrm{N}}(q) := \Big(-2\log_{10}\Big(\frac{k}{3.71D}\Big)\Big)^{-2}.$$

In the CIS region, approximations based on the Altschul formula [23, Ch. 7.26] are favored:

$$\lambda_{\mathrm{A}}(q) := 0.11\Big(\frac{68}{\mathrm{Re}(q)} + \frac{k}{D}\Big)^{\frac{1}{4}}.$$

Similarly, for $\mathrm{Re} \to \infty$, a simplified formula by Schifrinson [21] exists:

$$\lambda_{\mathrm{S}}(q) := 0.11\Big(\frac{k}{D}\Big)^{\frac{1}{4}}.$$

Lastly, a simple yet commonly used approximation [20] of the friction factor for turbulent flows is given by the Chodanovich-Odischarija formula [3]:

$$\lambda_{\mathrm{CO}}(q) := 0.067\Big(\frac{158}{\mathrm{Re}(q)} + \frac{2k}{D}\Big)^{2}.$$

2.3 Compressibility Factor

The inner state of the gas is described by (2.2c) and relates pressure, volume and temperature. To account for medium specific behavior deviating from an ideal gas, the compressibility factor $z(p, T)$ is utilized. For an ideal gas, the compressibility factor is given independent from pressure and temperature by the unit constant:

$$z_1(p, T) := 1.$$

Typically, the compressibility factor is approximated using the Virial expansion:

$$z(p, T) = 1 + \sum_{k=1}^{\infty} B_k p^k,$$

for real gases. Usually, this expansion is truncated after the first terms, and the associated coefficients B_k are estimated heuristically. The AGA8-DC92 and SGERG [26] approximations are assembled in this fashion; see also [8]. Yet, the partial derivatives of the compressibility factor in (2.4) induce a root-finding problem due to the higher-order terms in the truncated series for these formulas. To avoid this additional complexity, we allow coarser but explicit approximations to the compressibility factor. Such explicit formulas for the compressibility factor are given first, by the AGA88 formula [18]:

$$z_2(p, T) := 1 + 0.257\frac{p}{p_c} - 0.533\frac{pT_c}{p_c T},$$

which is valid for pressures $p < 70$ bar, and second, by the Papay formula [27]:

$$z_3(p, T) := 1 - 3.52\frac{p}{p_c}\,\mathrm{e}^{-2.26\frac{T}{T_c}} + 0.274\left(\frac{p}{p_c}\right)^2 \mathrm{e}^{-1.878\frac{T}{T_c}}.$$

The latter is valid up to $p < 150$ bar and hence should be preferred due to the higher accuracy. The symbols p_c and T_c refer to the critical pressure and critical temperature, respectively. Since the temperature is assumed constant in this work, the compressibility factor formula z_2 is a linear and z_3 is a quadratic polynomial.

3 Gas Network

The abstract gas transportation network is described by a directed graph:

$$\mathscr{G} = (\mathscr{N}, \mathscr{E}),$$

consisting of a tuple: A set of nodes $\mathcal{N}$ and a set of oriented edges $\mathcal{E}$. The edges embody a (possibly large) number of pipes (P), as well as short pipes (S), valves (V), compressors (C), resistors (R), and controlled valves (CV). We introduce an index set $\mathcal{I} = \{P, S, V, C, R, CV\}$ to represent these components. Similarly, the nodes are divided into pressure nodes (p) and flux nodes (q), and we create an index set for those as well $\mathcal{J} = \{p, q\}$. In total this means we can write the set of edges as a union over the different components by using the corresponding index set. Likewise, this can be done for the set of nodes:

$$\mathcal{E} = \mathcal{E}_P \cup \mathcal{E}_S \cup \mathcal{E}_V \cup \mathcal{E}_C \cup \mathcal{E}_R \cup \mathcal{E}_{CV} = \bigcup_{i \in \mathcal{I}} \mathcal{E}_i,$$

$$\mathcal{N} = \mathcal{N}_p \cup \mathcal{N}_q = \bigcup_{j \in \mathcal{J}} \mathcal{N}_j.$$

In the following we will repeat the set of equations used on each pipe and introduce the set of equations used for all other components.

3.1 Pipes

The Euler equation (2.4) for the pressure p and the mass flux q, depend on a set of function and parameters that are listed again in Eq. (3.1) for an individual pipe:

$$\{\gamma_0, z_0, h, \lambda, D, S\}. \tag{3.1}$$

As specified before, γ_0 is the constant gas state, the function z_0 is the compressibility of the gas dependent on the pressure p, while the friction factor of the pipe λ depends on the mass flux q. The diameter of the pipe D, is utilized for the cross-sectional area $S = \frac{1}{2} D\pi^2$ of the pipe. Each pipe has furthermore a pipe elevation function h, which depends on the local elevation, so $\partial h/\partial x = 0$ for level pipes.

In the modeling of gas dynamics over an entire network we have a pressure function p and a flux function q in each pipe, which we label by its pipe edge indices $e \in \mathcal{E}_P$: $p_e(x, t)$ and $q_e(x, t)$. Similarly, the above parameters and functions from (3.1) are indexed this way, as they vary for each pipe within a network. The equations are then given by:

$$\begin{aligned} \frac{\partial}{\partial t} d_{\text{pipe},e}(p_e) &= -\frac{1}{S_e} \frac{\partial}{\partial x} q_e, \quad \forall e \in \mathcal{E}_P \\ \frac{\partial}{\partial t} q_e &= -S_e \frac{\partial}{\partial x} p_e + f_{\text{pipe},e}(p_e, q_e), \quad \forall e \in \mathcal{E}_P \end{aligned} \tag{3.2}$$

where we also introduce the two nonlinear functions $d_{\text{pipe},e}$ and $f_{\text{pipe},e}$ for a simplified notation:

$$\begin{aligned} d_{\text{pipe},e}(p_e) &:= \frac{p_e}{\gamma_0^e z_0^e(p_e)}, \\ f_{\text{pipe},e}(p_e, q_e) &:= -S_e g_e d_{\text{pipe},e}(p_e)\frac{\partial}{\partial x}h_e - \frac{\lambda_e(q_e)}{2D_e S_e}\frac{q_e|q_e|}{d_{\text{pipe},e}(p_e)}. \end{aligned} \tag{3.3}$$

3.2 Non-pipe Edge Components

We will give a short description of all the components used in the given benchmark models, namely the ones defined above: short pipe, valves, compressors, resistors and controlled valves. For a comprehensive description of these and further gas network components, such as reservoirs or heaters, see also [8] and [21]. In order to describe these components we need four variables for each component, namely: $p_{e,r}(t), p_{e,l}(t), q_{e,r}(t), q_{e,l}(t)$, referring to the left and right pressure and the left and right flux for the edge $e \in \mathscr{E}$, meaning we work on a model that is discrete in space. For the sake of readability we will drop the time dependency in our notation (e.g. $p_{e,l}$ instead of $p_{e,l}(t)$) for the remainder of this section.

3.2.1 Short Pipe

First, we introduce a short pipe element, which is an idealized network element with neither friction nor pressure loss due to height differences. The model is simply given by:

$$\begin{aligned} q_{e,r} - q_{e,l} &= 0, \\ p_{e,l} - p_{e,r} &= 0, \quad \forall e \in \mathscr{E}_S. \end{aligned} \tag{3.4}$$

3.2.2 Valves

Valves are gas network components, which connect two junctions and have two modes of operation: open and close. The open state means that the valve component acts as a short pipe, while the closed state of the valve causes a disconnection of the junctions. Hence, the topology of the network can be changed if the valve is toggled between its two states, and thus significantly alter the behavior of the network for example by disconnecting a part of the graph or introducing cycles. A model for valves is given as follows:

$$\begin{cases} q_{e,r} = q_{e,l}, \quad p_{e,r} = p_{e,l} & \text{open valve} \\ q_{e,r} = q_{e,l} = 0 & \text{closed valve} \quad \forall e \in \mathscr{E}_V. \end{cases} \tag{3.5}$$

According to this model, the adjacent nodes are topologically disconnected when the valve is closed, this is formally written by having the flow zero at the left and right end of the valve. The pressure at the left and right end of the valve is not influenced at all by the closed valve. In case of the open valve we have exactly the same two equations as for the short pipes.

3.2.3 Compressor/Ideal Compressor Unit

Compressors are complex gas network components which connect two junctions, and increase the energy (pressure) along the selected path. A basic model is given by the ideal compressor [7]:

$$q_{e,r} = q_{e,l}, \tag{3.6a}$$

$$\frac{p_{e,r}}{p_{e,l}} = \alpha_{C,e}(p_{e,l}, p_{e,r}, q_{e,l}, q_{e,r}, t), \tag{3.6b}$$

where $\alpha : \mathbb{R}^5 \to [1, \infty[$. The model of the idealized compressor unit coincides with that of the short pipe whenever $\alpha \equiv 1$. So this may also serve as the minimum compression ratio provided by the idealized unit. We might also introduce more technical limitations to the capabilities of the ideal unit, e.g. by choosing a maximum compression ratio $\alpha_{\text{cmp,max}}$ (e.g. $\alpha_{\text{cmp,max}} = 80\,\text{bar}/60\,\text{bar}$). A further possibility might be a bound for power consumption. To this end, we solve the power equation from [8, eq. (2.43)] for the compression ratio $p_{e,r}/p_{e,l} \equiv \alpha$ and substitute the power P by the maximum consumption allowed $P_{\max}$:

$$\alpha_{P,\max} \equiv \left[\frac{\eta \cdot P_{\max}}{q_{e,r} \cdot R_s T_0 \cdot z_0(p_{e,l})} \cdot \frac{\gamma - 1}{\gamma} + 1\right]^{\frac{\gamma}{\gamma-1}}.$$

Here $\eta \in\,]0, 1[$ is a unit specific efficiency factor and γ is the isentropic expansion factor or isentropic exponent. The isentropic exponent corresponds to the ratio of specific heats for constant pressure and volume, and is the basis for isentropic processes such as idealized compression of (ideal) gas, which is based on the relation of pressure and volume before and after the compression $p_{e,l} V_{e,l}^{\gamma} = p_{e,r} V_{e,r}^{\gamma}$ [8][1]. By introducing target values $p_{r,\text{set}}$ and $p_{l,\text{set}}$ for the pressures, we can model two modes for α:

$$\alpha_{C,e}(p_{e,l}, p_{e,r}, q_{e,l}, q_{e,r}, t)$$
$$\equiv \begin{cases} \max\left(1, \min\left(\alpha_{\text{cmp,max}}, \alpha_{P,\max}, \frac{p_{r,\text{set}}}{p_{e,l}}\right)\right) & p_{r,\text{set}} \text{ mode}, \\ \max\left(1, \min\left(\alpha_{\text{cmp,max}}, \alpha_{P,\max}, \frac{p_{e,r}}{p_{l,\text{set}}}\right)\right) & p_{l,\text{set}} \text{ mode}. \end{cases}$$

[1] In [8] an approximation of $\gamma = 1.296$ is used.

In either mode the idealized unit will try to keep the corresponding pressure value close to the *target value* ($p_{r,\text{set}}$ or $p_{l,\text{set}}$) w.r.t. to the modeled technical limitations.

Sometimes compressors consume some of the gas from the network to power themselves. Accordingly to [22, Chapter 4.2] and [5], the fuel consumption can be modeled by replacing formula (3.6a) with:

$$q_{e,r} = q_{e,l} - d_c z(p_{e,l}) \cdot q_{e,l} \left(\alpha_{C,e}(p_{e,l}, p_{e,r}, q_{e,l}, q_{e,r}, t)^{\frac{\gamma-1}{\gamma}} - 1 \right), \tag{3.7}$$

where d_c is a compressor specific constant.

3.2.4 Resistor

There is no existing infrastructure with the intended purpose of generating resistance. So *resistors* are virtual elements which resemble and substitute very local microscopic structures in our macroscopic view on a gas network. The following model of resistors is a simplified pipe model. This means we use some of the parameters used for the pipe as well. Here the friction and length parameter is replaced by a so called drag factor ξ. Height differences are neglected and time derivatives are set to zero:

$$q_{e,r} - q_{e,l} = 0, \qquad p_{e,r} - p_{e,l} = -\xi \frac{R_s T_0 \cdot z_0(p_{e,r})}{2S_e^2} \frac{q_{e,l}|q_{e,l}|}{p_{e,r}}. \tag{3.8}$$

A very similar model is proposed in [8]. However the pressure on the right hand side of (3.8) is evaluated at the right boundary instead of the left. The reason for this subtle difference is that the resistor model (3.8) is derived in a way such that it corresponds to the spatial discretization introduced later in Sect. 4.

3.2.5 Control Valve

Control valves can be derived from the model (3.8) of resistors, but with a variable diameter. To that end we introduce a factor $\alpha : \mathbb{R}^5 \to [0, 1]$:

$$q_{e,r}(t) - q_{e,l}(t) = 0, \tag{3.9a}$$

$$\alpha_{CV,e}(p_{e,l}, p_{e,r}, q_{e,l}, q_{e,r}, t) \cdot (p_{e,r} - p_{e,l}) = -\xi \frac{R_s T_0 \cdot z_0(p_{e,r})}{2S_e^2} \frac{q_{e,l}|q_{e,l}|}{p_{e,r}}. \tag{3.9b}$$

Once again we may introduce target values for the in-going and out-going pressure and so we can model two modes via the degree of openness α:

$$\alpha_{CV,e}(p_{e,l}, p_{e,r}, q_{e,l}, q_{e,r}, t)$$
$$\equiv \begin{cases} \max\left(0, \min\left(1, -\xi \frac{R_s T_0 \cdot z_0(p_{r,\text{set}})}{2S_e^2 \cdot (p_{r,\text{set}} - p_{e,l})} \frac{q_{e,l}|q_{e,l}|}{p_{r,\text{set}}}\right)\right) & p_{r,\text{set}} \text{ mode}, \\ \max\left(0, \min\left(1, -\xi \frac{R_s T_0 \cdot z_0(p_{e,r})}{2S_e^2 \cdot (p_{e,r} - p_{l,\text{set}})} \frac{q_{e,l}|q_{e,l}|}{p_{e,r}}\right)\right) & p_{l,\text{set}} \text{ mode}. \end{cases}$$

3.2.6 Summary of Non-pipe Components

For each $i \in \mathscr{I} \backslash p$ we introduce a function $f_{i,e}$, where $e \in \mathscr{E}_i$:

$$f_{S,e}(p_{e,l}, p_{e,r}, q_{e,l}, t) = p_{e,r} - p_{e,l},$$
$$f_{V,e}(p_{e,l}, p_{e,r}, q_{e,l}, t) = \chi_e(t)(p_{e,r} - p_{e,l}) + (1 - \chi_e(t))q_{e,l},$$
$$f_{C,e}(p_{e,l}, p_{e,r}, q_{e,l}, t) = p_{e,r} - \alpha_{C,e}(p_{e,l}, p_{e,r}, q_e, t) \cdot p_{e,l},$$
$$f_{R,e}(p_{e,l}, p_{e,r}, q_{e,l}, t) = S_e(p_{e,r} - p_{e,l}) + \xi \frac{R_s T_0 z_0(p_{e,r})}{2S_e} \frac{q_{e,l}|q_{e,l}|}{p_{e,r}},$$
$$f_{CV,e}(p_{e,l}, p_{e,r}, q_{e,l}, t) = \alpha_{CV,e}(p_{e,l}, p_{e,r}, q_{e,l}, q_{e,r}, t) \cdot (p_{e,r} - p_{e,l}) + \xi \frac{R_s T_0 \cdot z_0(p_{e,r})}{2S_e^2} \frac{q_{e,l}|q_{e,l}|}{p_{e,r}}.$$

where $\chi_e(t) = 1$ if the valve is open and $\chi_e(t) = 0$ if the valve is closed. This means that we can write the equations for the components as:

$$\begin{aligned} q_{e,r} &= q_{e,l}, \\ 0 &= f_{i,e}(p_{e,l}, p_{e,r}, q_{e,l}, t) \quad e \in \mathscr{E}_i, i \in \mathscr{I} \backslash P. \end{aligned} \tag{3.10}$$

Thus we have a simple way to describe the equation on all edges.

3.3 *Node Conditions*

We have a description of the pipe physics as a partial differential equation for the functions $q_e(x,t)$ and $p_e(x,t)$ for all $e \in \mathscr{E}_P$. Furthermore, we have two algebraic equations for the other components for the four variables $q_{e,l}, q_{e,r}, p_{e,l}, p_{e,r}$. Those four variables exist also for each pipe, namely:

$$q_{e,l}(t) = q_e(0,t),\ q_{e,r}(t) = q_e(L_e,t),\ p_{e,l}(t) = p_e(0,t),\ p_{e,r}(t) = p_e(L_e,t),$$

where L_e is the length of the pipe e. We furthermore introduce the set of pressures p_u belonging to each node $u \in \mathscr{N}$. Describing the graph we distinguish between pressure nodes and flux nodes. At pressure nodes – as the name suggests – a pressure function is given:

$$p_u(t) = p_{\text{set}_u}(t), \qquad u \in \mathscr{N}_p.$$

At the other nodes $u \in \mathscr{N}_q$ we have algebraic constraints given by a set of Kirchhoff-type balance equations:

$$\sum_{e \in \delta^-(u)} q_{e,r} - \sum_{e \in \delta^+(u)} q_{e,l}(t) = q_{\text{set}_u}(t), \qquad u \in \mathscr{N}_q,$$

where $\delta^+(u)$ and $\delta^-(u)$ are sets of edges in which u is a right or left node, respectively. The functions $p_{\text{set}_u}(t)$ and $q_{\text{set}_u}(t)$ are given as time-dependent input functions to the system and are typically encoded in a given scenario. At nodes with neither in- nor outflow the function $q_{\text{set}_u}(t)$ is set to zero. Sometimes the mass flow nodes are separated into ones with identically zero-set flow and those where that is not the case. Besides having pressures at each end of each pipe, we also have variables describing the pressure at each node. However, each end of each pipe is a certain node. We therefore have to make sure that these pressures are the same:

$$p_u = p_{e,l} \text{ and } p_v = p_{e,r} \quad \forall e = (u, v) \in \mathscr{E}.$$

In the following we will only work with the pressure variables on the nodes and replace the others by the corresponding pressure node variable, such that we eliminate this constraint.

3.4 *Partial Differential Algebraic Equation*

The overall so-called Partial Differential Algebraic Equation (PDAE) is given by:

$$\frac{\partial}{\partial t} d_{\text{pipe},e}(p_e) = -\frac{1}{S_e}\frac{\partial}{\partial x} q_e \qquad \forall e \in \mathscr{E}_P, \tag{3.11}$$

$$\frac{\partial}{\partial t} q_e = -S_e \frac{\partial}{\partial x} p_e + f_{\text{pipe},e}(p_e, q_e) \qquad \forall e \in \mathscr{E}_P, \tag{3.12}$$

$$q_{e,l}(t) = q_e(0, t) \quad q_{e,r} = q_e(L_e, t) \tag{3.13}$$

$$p_{e,l}(t) = p_e(0, t) \quad p_{e,r} = p_e(L_e, t) \qquad \forall e \in \mathscr{E}_P, \tag{3.14}$$

$$q_{e,r}(t) = q_{e,l}(t), \qquad \forall e \in \mathscr{E}_i, i \in \mathscr{I} \backslash P, \tag{3.15}$$

$$0 = f_{i,e}(p_{e,l}, p_{e,r}, q_{e,l}, t) \qquad \forall e \in \mathscr{E}_i, i \in \mathscr{I} \backslash P, \tag{3.16}$$

$$p_u(t) = p_{\mathrm{set}_u}(t), \qquad \forall u \in \mathcal{N}_p, \tag{3.17}$$

$$\sum_{e \in \delta^-(u)} q_{e,r}(t) - \sum_{e \in \delta^+(u)} q_{e,l}(t) = q_{\mathrm{set}_u}(t), \qquad \forall u \in \mathcal{N}_q, \tag{3.18}$$

$$p_u(t) = p_{e,l}(t) \text{ and } p_v(t) = p_{e,r}(t) \qquad \forall e \in \mathcal{E},\ e = (u, v),\ u, v \in \mathcal{N} \tag{3.19}$$

In the system above we have the two PDEs for each pipe, describing the flow through a pipe ((3.11), (3.12)), the definition of the boundary values for each pipe ((3.13), (3.14)), and two equations for the nonpipe components. The first equation (3.15) is the same for each component and the second equation (3.16) depends on the component specific function f_i for $i \in \mathcal{I} \backslash P$. Equation (3.17) defines the pressure at the pressure nodes and (3.18) the flow condition at the nodes where no pressure is given. The last equation (3.19) ensures that the different pressure variables at a given node do have the same value.

In order to be able to write this complex system in a concise form, we will first need to introduce the **incidence matrix** of a graph. Given a directed graph with N nodes and M edges, the associated incidence matrix $A \in \mathbb{R}^{N \times M}$ is defined as:

$$A_{ij} := \begin{cases} 1 & \text{edge } j \text{ connects } \textbf{to} \text{ node } i, \\ 0 & \text{edge } j \text{ does } \textbf{not connect} \text{ to node } i, \\ -1 & \text{edge } j \text{ connects } \textbf{from} \text{ node } i. \end{cases}$$

If this graph has a tree structure, meaning it is connected and acyclic, then the associated incidence matrix is of rank $(N - 1)$ [11]. As we distinguish between different type of nodes and different types of edge we can always take only certain edges or nodes of the network and the incidence matrix that corresponds to the subgraph spanned by just those. For example, if we are interested in the incidence matrix for just pipes as edges and just the mass flow condition nodes we denote that matrix by $A_{P,q}$. Furthermore, we may only be interested in the negative part of the matrix or just in the positive part of that matrix. We call the negative matrix A^L, and the positive matrix A^R, since the negative part represents the node-to-edge relationship for the left-hand-side connections, while the positive part corresponds to the right-hand-side connections. We can create submatrices from A^L and A^R as well. In particular we can use only the rows corresponding to nodes that have a mass flow condition, which will then be denoted by A^L_q and A^R_q.

The mass balance Kirchhoff type equation (3.18) can now simply be written as:

$$A^R_q q_r + A^L_q q_l = q_{\mathrm{set}}(t), \tag{3.20}$$

where q_r and q_l is a vector of all left and right fluxes for each edge. Equation (3.19) for the pressure reads:

$$p_l = (A^L)^\intercal p, \quad p_r = (A^R)^\intercal p, \tag{3.21}$$

where the vectors p_r and p_l are as above and p is the vector of all pressures at the individual nodes.

Next, the spatial discretization for this partial differential algebraic equations is described, which then leads to a Differential Algebraic Equation (DAE).

4 Discretization

To perform simulations of the partial differential algebraic equation modeling the gas flow in a network of pipes, the Euler equations in (2.4) need to be discretized. The considered model contains spatial $\frac{\partial}{\partial x}$ and temporal $\frac{\partial}{\partial t}$ derivative operators. We follow the established approach of discretizing first in space to obtain a differential-algebraic equation system, consisting of an ordinary differential equation system (in time) and a set of algebraic constraints. This will lead to an overall representation as an input-output system with the input-output quantities given in Table 2 for each boundary (supply and demand) node.

4.1 Spatial Discretization

4.1.1 Spatial discretization of Pipes

We present a spatial discretization of the pipe model (3.2) yielding index-1 DAEs if the pipes in the network are directed properly. Let $e \in \mathscr{E}_P$ be an arbitrary edge modeling a pipe. As before, we introduce the discrete variables $q_{e,l}(t) = q_e(0,t)$, $q_{e,r}(t) = q_e(L_e,t)$, $p_{e,l}(t) = p_e(0,t)$, $p_{e,r}(t) = p_e(L_e,t)$ and discretize (3.2) spatially as follows:

$$\frac{\mathrm{d}}{\mathrm{d}t} d_{\text{pipe},e}(p_{e,r}(t)) + \frac{1}{S_e L_e}(q_{e,r}(t) - q_{e,l}(t)) = 0, \tag{4.1a}$$

$$\frac{\mathrm{d}}{\mathrm{d}t} q_{e,l}(t) + \frac{S_e}{L_e}(p_{e,r}(t) - p_{e,l}(t)) = f_{\text{pipe},e}(p_{e,r}(t), q_{e,l}(t)), \tag{4.1b}$$

with d_{pipe} and f_{pipe} defined by (3.3). The parameter L_e is the length of the pipe e and S_e is its cross-sectional area. In practice it is often useful to apply a finer spatial discretization by introducing artificial nodes and perform the spatial discretization on the subpipes of shorter length. This will be demonstrated in Sect. 5.

4.1.2 Network DAE

We consider gas networks with network elements described in Sect. 3.2. Using the pipe discretization, we obtain a differential algebraic equation (DAE) of the form:

$$E \frac{\mathrm{d}}{\mathrm{d}t} d(x(t)) + b(x(t), t) = 0.$$

Let $p = (p_P, p_q)$ with p_P the pressure vector of the nodes with pressure conditions and p_q the ones without. Furthermore, $q_{P,l}$ and $q_{P,r}$ are the vectors of all pipe flows at left and right nodes of the pipes. Since the left and right flows of all non-pipe element models are equal, it is sufficient to consider only one flow per non-pipe arc, which are collected in a vector $q_{\mathscr{A}}$. Then the DAE is derived from the PDAE by replacing the first four Eqs. (3.11)–(3.14) with (4.1), removing (3.15) by introducing a single variable for it and removing (3.19) by replacing p_l and p_r everywhere by Eq. (3.21). Equations (3.16)–(3.18) are written more concisely to get the following DAE:

$$\frac{\mathrm{d}}{\mathrm{d}t} d_{\mathrm{pipe}}(A_{P,r}^{\top} p(t)) = D_S^{-1} D_L^{-1}(q_{P,l}(t) - q_{P,r}(t)), \tag{4.2a}$$

$$\frac{\mathrm{d}}{\mathrm{d}t} q_{P,l}(t) = -D_S D_L^{-1}(A_{P,r}^{\top} + A_{P,l}^{\top})p(t) - f_{\mathrm{pipe}}(A_{P,r}^{\top} p(t), q_{P,l}(t)), \tag{4.2b}$$

$$0 = f_{\mathscr{A}}(p(t), q_{\mathscr{A}}(t), t), \tag{4.2c}$$

$$0 = A_{P,r} q_{P,r}(t) + A_{P,l} q_{P,l}(t) + A_{\mathscr{A}} q_{\mathscr{A}}(t) - q_{\mathrm{set}}(t), \tag{4.2d}$$

$$0 = p_P(t) - p_{\mathrm{set}}(t). \tag{4.2e}$$

where d_{pipe} and f_{pipe} are vector-valued functions defined component-wise: i.e. $(d_{\mathrm{pipe}}(x))_e = (d_{\mathrm{pipe},e})(x_e)$. For a concise notation we also introduce the constant diagonal matrices D_S and D_L:

$$D_S = \mathrm{diag}\{S_e,\ e \in \mathscr{E}_P\} \quad D_L = \mathrm{diag}\{L_e,\ e \in \mathscr{E}_P\}.$$

The algebraic element descriptions are given by:

$$f_{\mathscr{A}}(p, q_{\mathscr{A}}, t) = [f_S, f_V, f_C, f_R, f_{CV}]^{\top} (A^L p, A^R p, q_{\mathscr{A}}, t),$$

with edge-wise defined functions $f_i = (f_{i,e})_{e \in \mathscr{E}_i},\ i \in \mathscr{I} \backslash P$.

In order to obtain a DAE system of index 1 for networks with a spatial pipe discretization of the form (4.1), one has to adapt the direction of pipes in the network to their topological location with respect to nodes with pressure and flow conditions.

Assumption 4.1 *Let a gas network with pipes, resistors and compressors be given and described by a graph $\mathscr{G} = (\mathscr{N}, \mathscr{E})$ with the node set $\mathscr{N}$ and the arc set $\mathscr{E}$. Denote the set of nodes with pressure conditions by $\mathscr{N}_p$ and the set of pipe arcs by $\mathscr{E}_P$. Let $\mathscr{N}_{\mathscr{A}}$ be the set of nodes $u \in \mathscr{N} \backslash \mathscr{N}_p$ that have an arc $e \in \mathscr{E} \backslash \mathscr{E}_P$ directing to u. The graph $\mathscr{G}$ shall fulfill the following conditions:*

1. *Each pipe e^P is connected to a node of $\mathscr{N} \backslash (\mathscr{N}_p \cup \mathscr{N}_{\mathscr{A}})$.*
2. *Each connected component of $\mathscr{G}_{\mathscr{P}} := (\mathscr{N}, \mathscr{E}_P)$ has at least one node in $\mathscr{N}_p \cup \mathscr{N}_{\mathscr{A}}$.*
3. *For each node $u \in \mathscr{N}$, there exists at most one arc in $\mathscr{E} \backslash \mathscr{E}_P$ directing to u.*
4. *No arc of $\mathscr{E} \backslash \mathscr{E}_P$ directs to a node of $\mathscr{N}_p$.*

In this assumption we consider nodes that are connected to an arc that is not a pipe as special nodes and call the set of all these nodes by $\mathscr{N}_{\mathscr{A}}$. We assume that a pipe arc is always connected to at least one node that is not such a special node and also not a supply node. The second point in the assumption is that in each connected component of the network, we have a least one node that is either a supply node or connected to a non-pipe arc. The third point is that every node is the end node for at most one non-pipe arc and a non-pipe arc can never end in a supply node.

In [16] it has been shown that under these assumptions, the pipes of such gas networks can be directed in such a way that the resulting DAE formed by (4.2) has index 1 as the next theorem explains.

Theorem 4.1 *Let $\mathscr{G} = (\mathscr{N}, \mathscr{E})$ be a connected, directed graph describing a gas network that fulfills Assumption 4.1. Then, the pipes in $\mathscr{G}$ can be directed in such a way that:*

1. *No arc directs to a node of $\mathscr{N}_p$,*
2. *for each node $u \in \mathscr{N} \backslash \mathscr{N}_p$, there exists an arc directed to u,*
3. *if an arc $e \in \mathscr{E} \backslash \mathscr{E}_P$ directs to $u \in \mathscr{N} \backslash \mathscr{N}_p$ then none of the arcs of $\mathscr{E}_P$ is directed to u*

and the DAE formed by (4.2) *has index 1.*

Given a network, we set up the directions within the network in such a way that, if a certain node is a supply node (it lies in $\mathscr{N}_p$) all edges connected to it are leaving the node and no edge is entering that node. It is always placed as a left end, meaning that the direction of the orientation of the oriented graph points away from the supply node. This makes sense as normally we assume that a supply node is an inlet into the network. All other nodes have at least one arc that ends in them, meaning they are the right node of at least one edge. Furthermore, if an arc that is not a pipe ends in a node than no other arc ends in that node. Under Assumption 4.1 this is always possible (Theorem 4.1) and creates a DAE of index 1.

5 Benchmark Networks

In this section we present four benchmark networks of different complexities. The first benchmark describes a long pipeline, the second benchmark features a small pipe network including a cycle, the third benchmark has compressor and resistor elements, and the fourth benchmark models a real gas transport network.

5.1 Pipeline Benchmark Model

The first benchmark model is taken from [2] and represents a real pipeline. For this pipeline model, with physical specifications given in Table 3, the single pipe

Table 3 Pipeline benchmark model attributes

Pipeline length	36,300 m
Pipeline diameter	1.422 m
Pipeline roughness	0.000015 m
Reynolds number	5,000.0
Isothermal speed of sound	300.0 $\frac{\text{m}}{\text{s}}$
Steady supply	84.0 bar
Steady demand	463.33 $\frac{\text{kg}}{\text{s}}$
Time horizon	200 h

model can be utilized together with the scenario, given in Fig. 2, to simulate outputs from inputs. The inputs of the model are the pressure at the inlet of the pipe (the supply node) and the mass-flux at the outlet (the demand node). The outputs are then the mass-flux at the inlet and pressure at the outlet. Starting from a steady-state, a scenario is given by the input time series at the inlet and outlet (boundary). In the provided scenario the inlet-pressure is kept constant over time, and the outlet-mass-flux varies over time; see Fig. 2.

We reduced the length of the pipeline to 36.3 km compared to [2] to allow an easier discretization. Practically, the pipeline simulation is realized using 1,000 virtual

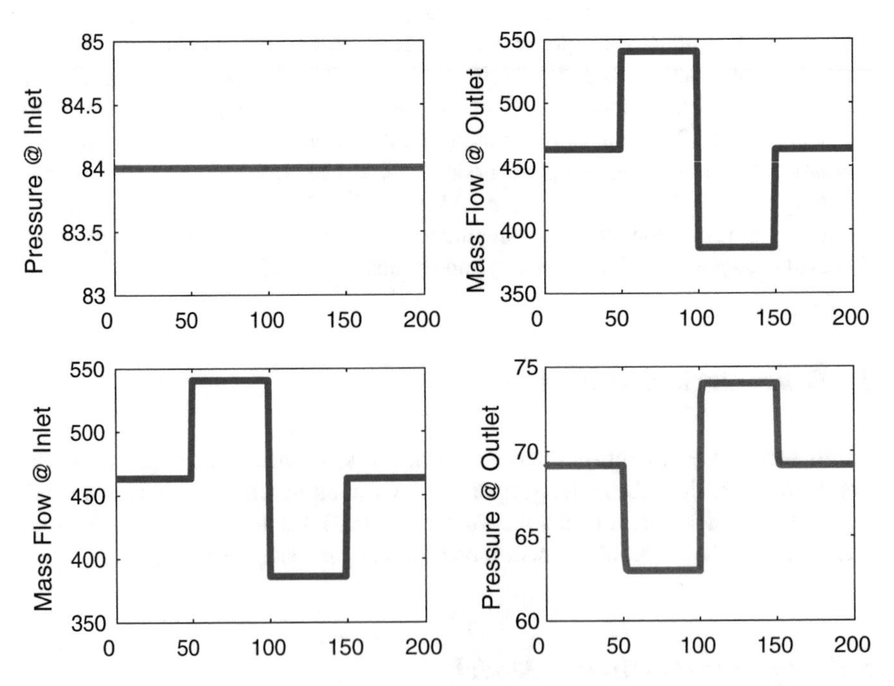

Fig. 2 Input-Output behavior for the pipeline benchmark from Sect. 5.1. Upper left: pressure at inlet [bar], upper right: mass flow at outlet [kg/s], lower left: mass flow at inlet [kg/s], lower right: pressure at outlet [bar]

nodes subdividing the long pipeline into a cascade of shorter sequentially connected pipes. The order of the differential equation is then 2,000. This refinement strategy, also used in [14], relaxes the Courant-Friedrichs-Levy (CFL) number allowing a stable time-stepping. Practically, the first order implicit-explicit method from [9] is utilized to compute the solution. Using the parameters from Table 3 and the aforementioned input scenario, the resulting output quantities over time are depicted in Fig. 2. These results agree with the behavior described in [2].

5.2 *Diamond Network*

This small-scale network is made up of seven pipes, one entry node and five exit-nodes. The topology is given by Fig. 3. Note that:

$$\mathscr{N}_p = \{u_0\}, \qquad \mathscr{N}_q = \{u_1, u_2, u_3, u_4, u_5\}.$$

The gas network *gas_diamond* (see Fig. 3) contains six nodes and seven pipes. The node u_0 is considered to be a source and is modeled by a (constant) pressure condition of 80 bar. The remaining nodes are modeled via flow balance equations, but in our scenario, only at node u_5, gas will exit the network (see Fig. 4). The demand function is given by a piecewise linear function, with a demand between

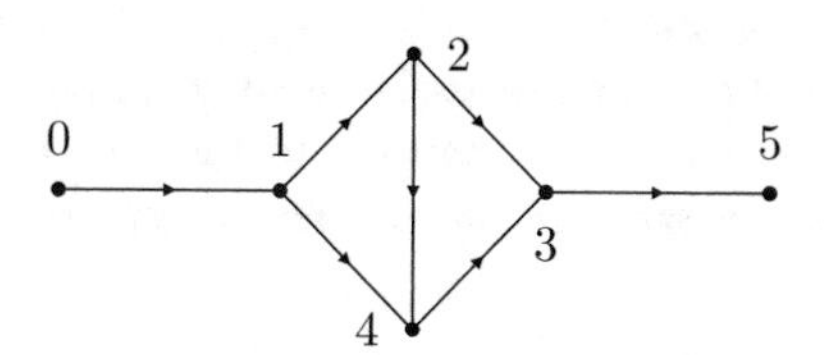

Fig. 3 *gas_diamond* – gas transportation network with *six* nodes and *seven* arcs

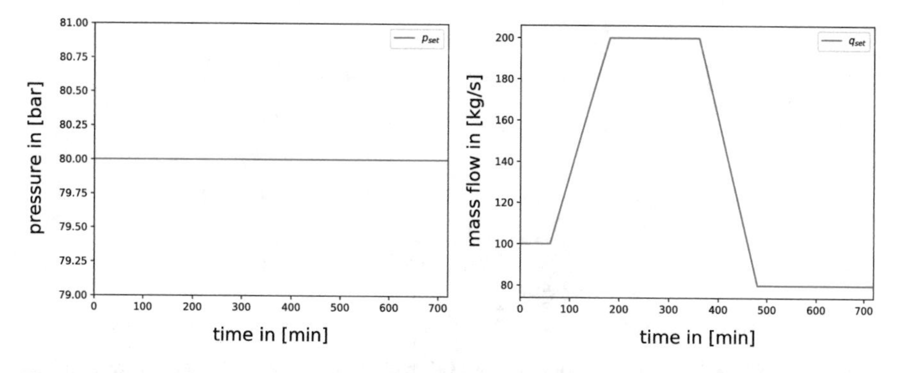

Fig. 4 (Diamond) pressure boundary at the source node u_0 and demand at sink node u_5

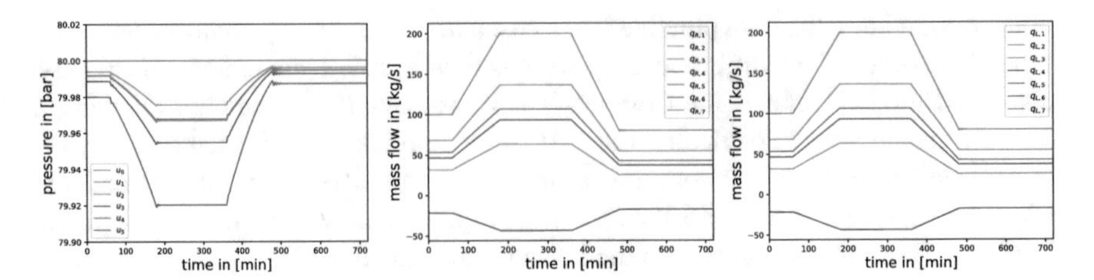

Fig. 5 (Diamond) simulation results for scenario in Fig. 4. Pressures at the nodes (left), mass flows at positions $x = \ell$ (middle) and $x = 0$ (right) for the pipes

80 and 200 $\frac{\text{kg}}{\text{s}}$. A graphical representation of a solution to the scenario described in Fig. 4 can be seen in Fig. 5.

5.3 *Gas Transportation Network* – gas_N23_A24

The gas network *gas_N23_A24* DAE (see Fig. 6) contains 23 nodes, which can be defined as pressure conditions (mostly at sources) or flow conditions (representing other sources, sinks and innodes) depending on a user definable *behavior* property. Per default the nodes N and W are considered to be sources and initialized with pressure conditions. Any other node is considered a flow node, where S, E_1 and E_2 are considered to be the only sinks. Details are listed in Table 4 and Fig. 6 shows the topology. Furthermore, a compressor station belongs to the network in the middle of the 100 km pipeline. The station consists of two resistors, a by-pass valve, a short pipe and a single idealized compressor unit. The scenario including boundary

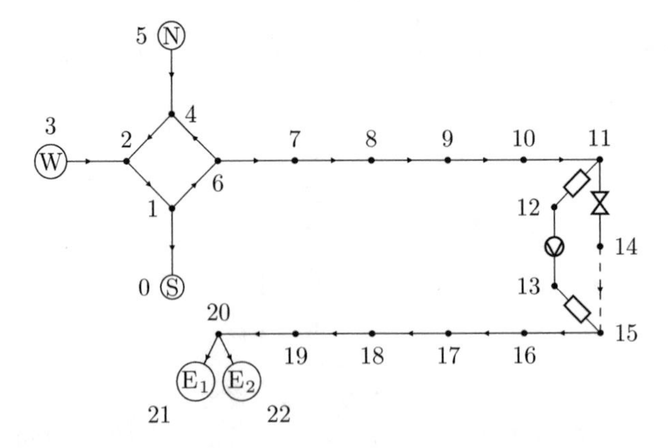

Fig. 6 *gas_N23_A24* – gas transportation network with 23 nodes and 24 arcs. The dashed line represents a short pipe. The node enumeration corresponds to that defined in the benchmark function (for viewing results)

Table 4 *ids* of adjacent nodes and flows of edges from *gas_N23_A24* network, see Fig. 6

Arc	Type	u	v	(Left) Flow	Right flow
e_0	Pipe	1	0	23	24
e_1	Pipe	3	2	25	26
e_2	Pipe	5	4	27	28
e_3	Pipe	2	1	29	30
e_4	Pipe	1	6	31	32
e_5	Pipe	6	4	33	34
e_6	Pipe	4	2	35	36
e_7	Pipe	6	7	37	38
e_8	Pipe	7	8	39	40
e_9	Pipe	8	9	41	42
e_{10}	Pipe	9	10	43	44
e_{11}	Pipe	10	11	45	46
e_{12}	Valve	11	14	47	
e_{13}	Short pipe	14	15	48	
e_{14}	Resistor	11	12	49	
e_{15}	Compressor	12	13	50	
e_{16}	Resistor	13	15	51	
e_{17}	Pipe	15	16	52	53
e_{18}	Pipe	16	17	54	55
e_{19}	Pipe	17	18	56	57
e_{20}	Pipe	18	19	58	59
e_{21}	Pipe	19	20	60	61
e_{22}	Pipe	20	21	62	63
e_{23}	Pipe	20	22	64	65

conditions and target values for the compressor unit is described in an extra file `N23_A24_bconditions.xml` as well as implemented or contained within the python script of this benchmark instance `N23_A24.py`.

Between hour 5 and 7, the compressor unit works at the maximum admissible power and cannot sustain the desired compression ratio until the in-going pressure increases again (Figs. 7, 8, and 9).

5.4 *Gas Transportation Network –* gas_N138_A139 *(Derived from GasLib-134)*

The gas network *gas_N138_A139* DAE (see Fig. 10) contains 138 nodes and is derived from the stationary gas network instance *GasLib-134* (see [17] and [28]). Instead of a macro model for the compressor station and the control valve given in the original network, here, a more detailed model including an in-going and out-going resistor as well as a bypass valve is used. To that end four additional nodes were introduced increasing the number from 134 up to 138. The rest of the network remains unchanged. All nodes and arcs in the Python implementation of

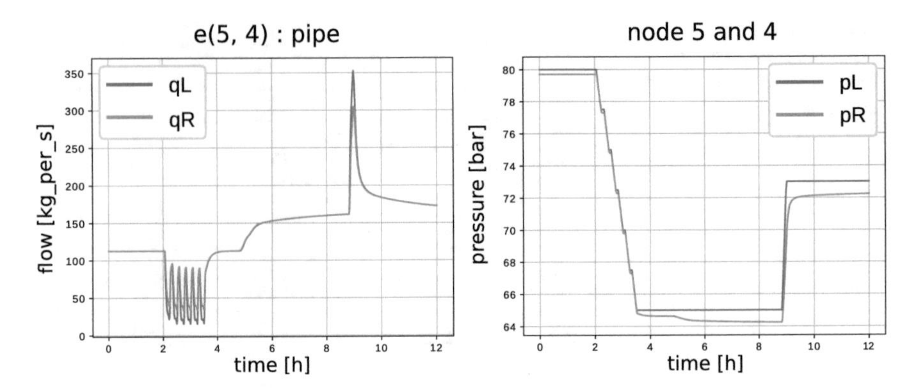

Fig. 7 *gas_N23_A24* – pipe e_2 (node 4 & source node N): flow and pressure curves

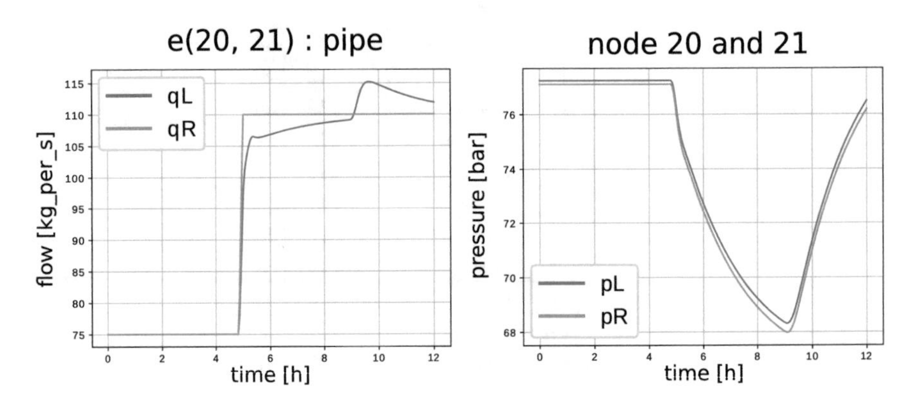

Fig. 8 *gas_N23_A24* – pipe e_{22} (node 20 & exit node E_1): flow and pressure curves

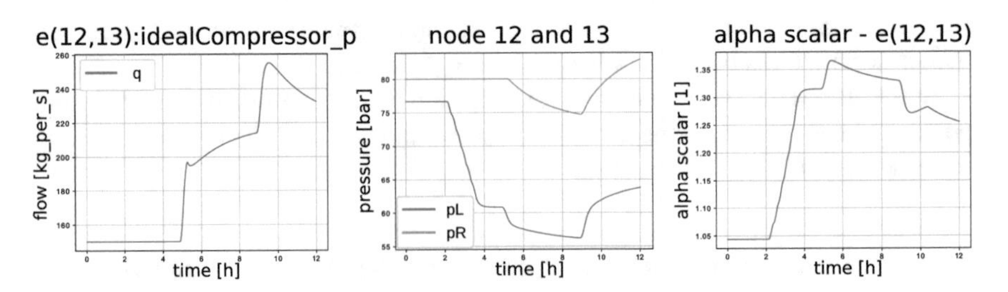

Fig. 9 *gas_N23_A24* – idealized compressor unit (nodes 12 & 13): flow, pressures and compression-factor (α)

this benchmark example are enumerated and do have a name property, too. This example contains an idealized compressor unit and a control valve. The 6 h transient scenario was created on the basis of the daily nominations from the *GasLib-134* instance in that the nominated flows are interpolated piece-wise linear and switched through every 2 h.

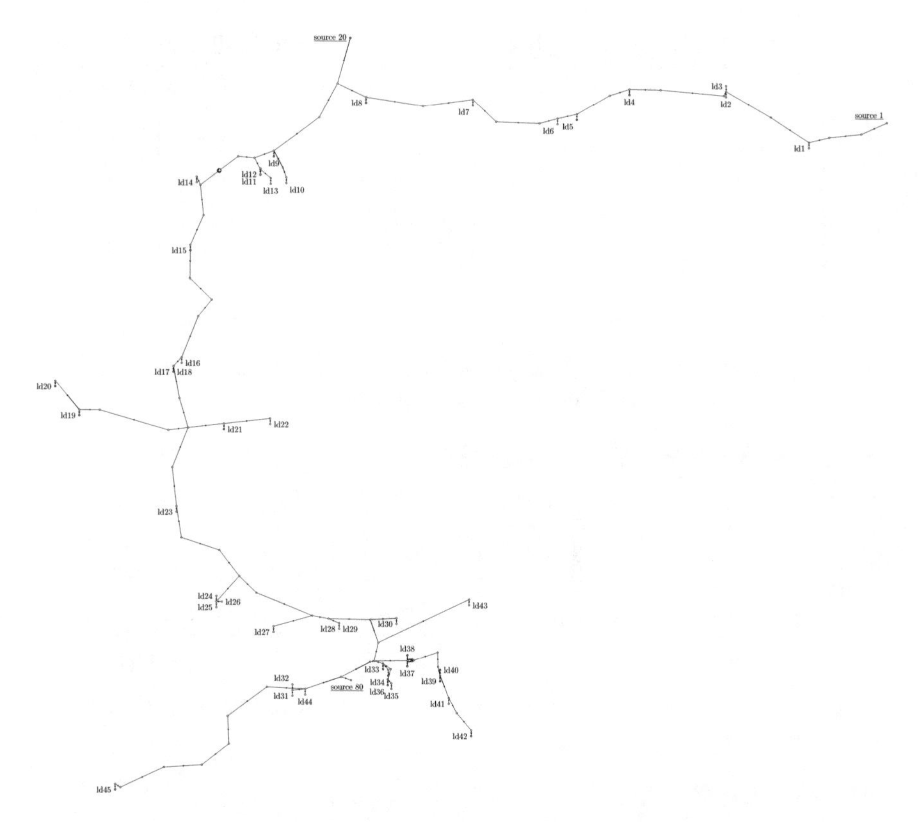

Fig. 10 *gaslib_134* – gas network from http://gaslib.zib.de [17, 28], with 134 nodes and 133 arcs. The node Ids of all 45 sinks and the 3 sources are displayed

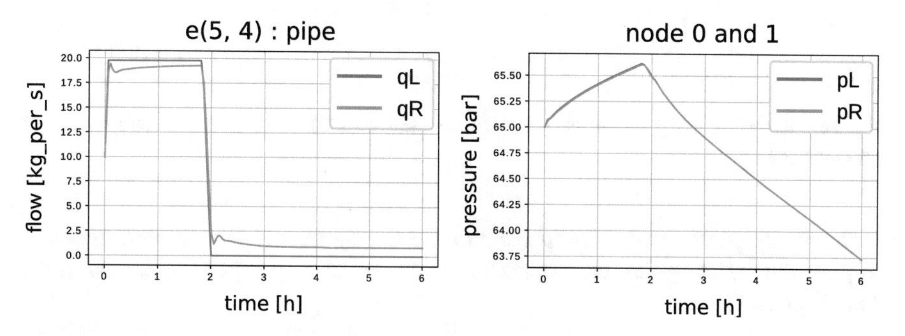

Fig. 11 *gas_N138_A139* – pipe *p_br1* (connecting the source node (name: *node_1*, idx: 0) with innode (name: *node_2*, idx: 1): flow and pressure curves

Between hour 2 and 6 the source node with name node_1 stops providing gas to the network (see Fig. 11). The adjacent pipe with name p_br1 is 14.56 km long such that gas can still be drawn from the other side while the gas pressure and so the gas density decreases It can be observed that the compressor station (see Fig. 13)

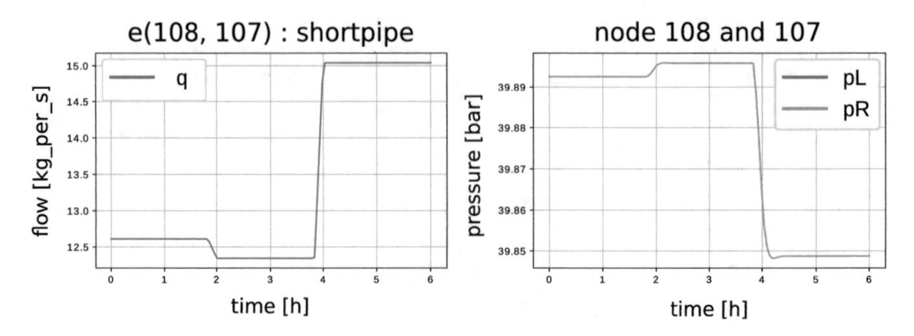

Fig. 12 *gas_N138_A139* – pipe *node_72_ld42* (connecting the innode (name: *node_72*, idx: 108) with sink node (name: *node_ld42*, idx: 107): flow and pressure curves

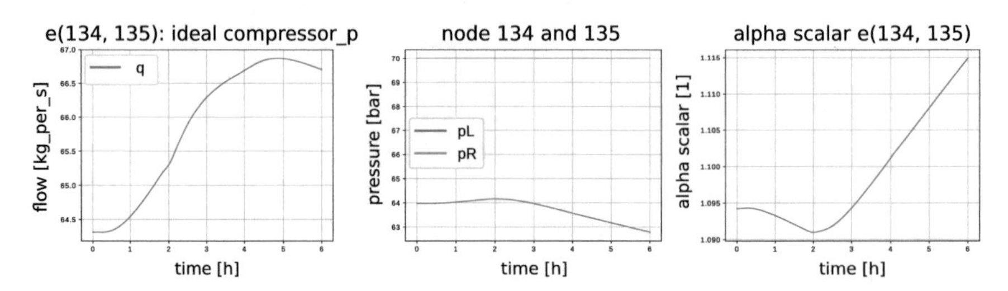

Fig. 13 *gas_N138_A139* – idealised compressor unit (nodes 134 & 135): flow, pressures and compression-factor (α)

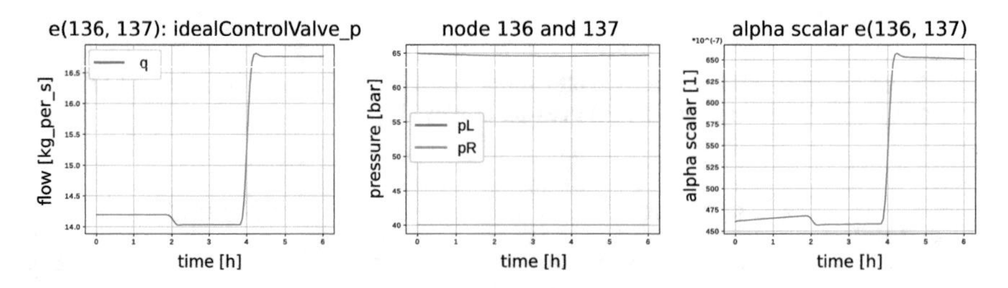

Fig. 14 *gas_N138_A139* – control valves (nodes 136 & 137): flow, pressures and degree of openness (α)

increases its power or the compression ratio α respectively as a counter reaction to preserve the out-going target pressure which is set to $p_{r,\text{set}} \equiv 70$ bar (Figs. 11, 12, 13, and 14).

6 Concluding Remarks

Gas network modeling and numerical solution thereof is at the interface between the real life application and differential algebraic equation research. In the scope of the **MathEnergy** project, the presented models are used for research in large-scale simulations of gas network operation under the influence of volatile renewable energy sources. Within the **Modal GasLab** configurations and recommendations for network operations are solved. They have to satisfy nominations of gas supplies and consumptions at entries and exits of gas networks which have to be announced at least one day in advance. A suitable approach is a back and forth iteration between optimization processes and simulations ranging from single elements and subnetworks up to whole networks. From a numerical point of view, this is also an interesting class of applications due to their manifold challenges, such as nonlinearity or hyperbolicity. In this work we presented a modular gas network model as well as four benchmarks, which enable testing of extensions of this basic model as well as implementations of associated solvers.

7 Code and Data Availability

The code and data used in this work can be obtained as supplementary material. The data is prepared in form of three XML-files for each benchmark example. The `net.xml`-file containing the network topology, the `bconditions.xml`-file providing the boundary conditions (i.e. the in- and out-going flows at sources and sinks as well as fixed pressures and target values for compressors and control valves) and the `result.xml`-file containing our reference solution. XML Schema or *XSD* files for the validation and documentation will be provided alongside. These schema files were created to store transient gas network scenarios and were kindly supplied to us by the SFB Transregio 154 (http://trr154.fau.de). It should be noted, however that the `net.xml` and the `bconditions.xml` files of both the *gas_N23_A24* and the *gas_N138_A139* will not completely validate against the schemes. The formats were intended for a more detailed description of compressor units as so called turbo compressors. In this paper we introduced an idealized description of compressors which are not covered by the schema files. Besides this described data the supplementary material also includes code that creates the differential algebraic equation for each of this networks. For the pipeline this is implemented in MATLAB, and for the others in Python.

Acknowledgements Supported by the German Federal Ministry for Economic Affairs and Energy, in the joint project: "**MathEnergy** – Mathematical Key Technologies for Evolving Energy Grids", sub-project: Model Order Reduction (Grant number: 0324019**B**).

The work for the article has been conducted within the Research Campus MODAL funded by the German Federal Ministry of Education and Research (BMBF) (fund number 05M14ZAM).

We also acknowledge funding through the DFG CRC/Transregio 154 "Mathematical Modelling, Simulation and Optimization using the Example of Gas Networks", Subproject C02.

References

1. Azevedo-Perdicoúlis, T.-P., Jank, G.: Modelling aspects of describing a gas network through a DAE system. In: Proceedings of the 3rd IFAC Symposium on System Structure and Control, vol. 40(20), pp. 40–45 (2007). https://doi.org/10.3182/20071017-3-BR-2923.00007
2. Chaczykowski, M.: Sensitivity of pipeline gas flow model to the selection of the equation of state. Chem. Eng. Res. Des. **87**, 1596–1603 (2009). https://doi.org/10.1016/j.cherd.2009.06.008
3. Chodanovich, I.J., Odischarija, G.E.: Analiž žavisimosti dlja koeffizienta gidravličeskogo soprotivlenija (analysis of the dependency of the pipe friction factor). Gažovaja Promyshlennost **9**(11), 38–42 (1964)
4. Colebrook, C.F.: Turbulent flows in pipes, with particular reference to the transition region between smooth and rough pipe laws. J. Inst. Civ. Eng. **11**, 133–156 (1939)
5. Domschke, P., Geißler, B., Kolb, O., Lang, J., Martin, A., Morsi, A.: Combination of nonlinear and linear optimization of transient gas networks. INFORMS J. Comput. **23**(4), 605–617 (2011). https://doi.org/10.1287/ijoc.1100.0429
6. Dymkou, S., Leugering, G., Jank, G.: Repetitive processes modelling of gas transport networks. In: 2007 International Workshop on Multidimensional (nD) Systems, pp. 101–108 (2007). https://doi.org/10.1109/NDS.2007.4509556
7. Ehrhardt, K., Steinbach, M.C.: Nonlinear optimization in gas networks. In: Bock, H.G., Phu, H.X., Kostina, E., Rannacher, R. (eds.) Modeling, Simulation and Optimization of Complex Processes. Proceedings of the International Conference on High Performance Scientific Computing, pp. 139–148. Springer, Berlin (2005). https://doi.org/10.1007/3-540-27170-8_11
8. Fügenschuh, A., Geißler, B., Gollmer, R., Morsi, A., Pfetsch, M.E., Rövekamp, J., Schmidt, M., Spreckelsen, K., Steinbach, M.C.: Chapter 2: physical and technical fundamentals of gas networks. In: Koch, T., Hiller, B., Pfetsch, M.E., Schewe, L. (eds.) Evaluating Gas Network Capacities. MOS-SIAM Series on Optimization, pp. 17–43. SIAM, Philadelphia (2015). https://doi.org/10.1137/1.9781611973693.ch2
9. Grundel, S., Jansen, L.: Efficient simulation of transient gas networks using IMEX integration schemes and MOR methods. In: 54th IEEE Conference on Decision and Control (CDC), pp. 4579–4584 (2015). https://doi.org/10.1109/CDC.2015.7402934
10. Grundel, S., Hornung, N., Klaassen, B., Benner, P., Clees, T.: Computing surrogates for gas network simulation using model order reduction. In: Surrogate-Based Modeling and Optimization. Applications in Engineering, pp. 189–212. Springer, New York (2013). https://doi.org/10.1007/978-1-4614-7551-4_9
11. Grundel, S., Jansen, L., Hornung, N., Clees, T., Tischendorf, C., Benner, P.: Model order reduction of differential algebraic equations arising from the simulation of gas transport networks. In: Progress in Differential-Algebraic Equations. Differential-Algebraic Equations Forum, pp. 183–205. Springer, Berlin (2014). https://doi.org/10.1007/978-3-662-44926-4_9
12. Grundel, S., Hornung, N., Roggendorf, S.: Numerical aspects of model order reduction for gas transportation networks. In: Simulation-Driven Modeling and Optimization. Proceedings in Mathematics & Statistics, vol. 153, pp. 1–28. Springer, Basel (2016). https://doi.org/10.1007/978-3-319-27517-8_1
13. Herty, M.: Modeling, simulation and optimization of gas networks with compressors. Netw. Heterog. Media **2**(1), 81–97 (2007). https://doi.org/10.3934/nhm.2007.2.81
14. Herty, M., Mohring, J., Sachers, V.: A new model for gas flow in pipe networks. Math. Methods Appl. Sci. **33**, 845–855 (2010). https://doi.org/10.1002/mma.1197
15. Hofer, P.: Beurteilung von Fehlern in Rohrnetzberechnungen (error evaluation in calculation of pipelines). GWF–Gas/Erdgas **114**(3), 113–119 (1973)
16. Huck, C., Tischendorf, C.: Topology motivated discretization of hyperbolic PDAEs describing flow networks. Technical report, Humboldt-Universität zu Berlin (2017). Available from: https://opus4.kobv.de/opus4-trr154/frontdoor/index/index/docId/137

17. Humpola, J., Joormann, I., Kanelakis, N., Oucherif, D., Pfetsch, M.E., Schewe, L., Schmidt, M., Schwarz, R., Sirvent, M. GasLib – a library of gas network instances. Technical report, Mathematical Optimization Society (2017). Available from: http://www.optimization-online.org/DB_HTML/2015/11/5216.html
18. Králik, J., Stiegler, P., Vostrý, Z., Záworka, J.: Dynamic Modeling of Large-Scale Networks with Application to Gas Distribution. Automation and Control, vol. 6. Elsevier, New York (1988)
19. LeVeque, R.J.: Nonlinear conservation laws and finite volume methods. In: Steiner, O., Gautschy, A. (eds.) Computational Methods for Astrophysical Fluid Flow. Saas-Fee Advanced Courses, vol. 27, pp. 1–159. Springer, Berlin (1997). https://doi.org/10.1007/3-540-31632-9_1
20. LIWACOM Informationstechnik GmbH and SIMONE research group s. r. o., Essen. SIMONE Software. Gleichungen und Methoden (2004). Available from: https://www.liwacom.de
21. Mischner, J., Fasold, H.G., Heymer, J. (eds.): gas2energy.net. Edition gas for energy. DIV (2016). Available from: https://www.di-verlag.de/de/gas2energy.net2
22. Moritz, S.: A mixed integer approach for the transient case of gas network optimization. Ph.D. thesis, Technische Universität, Darmstadt (2007). Available from: http://tuprints.ulb.tu-darmstadt.de/785/
23. Nekrasov, B.: Hydraulics for Aeronautical Engineers. Peace Publishers, Moscow (1969). Available from: https://archive.org/details/in.ernet.dli.2015.85993
24. Nikuradse, J.: Gesetzmäßigkeiten der turbulenten Strömung in glatten Rohren. VDI-Forschungsheft **356**, 1–36 (1932)
25. Osiadacz, A.: Simulation of transient gas flows in networks. Int. J. Numer. Methods Fluids **4**, 13–24 (1984). https://doi.org/10.1002/fld.1650040103
26. Osiadacz, A.J., Chaczykowski, M.: Verification of transient gas flow simulation model. In: PSIG Annual Meeting, pp. 1–10 (2010). Available from: https://www.onepetro.org/conference-paper/PSIG-1010
27. Papay, J.: A Termelestechnologiai Parameterek Valtozasa a Gazlelepk Muvelese Soran, pp. 267–273. Tud. Kuzl., Budapest (1968)
28. Pfetsch, M.E., Fügenschuh, A., Geißler, B., Geißler, N., Gollmer, R., Hiller, B., Humpola, J., Koch, T., Lehmann, T., Martin, A., Morsi, A., Rövekamp, J., Schewe, L., Schmidt, M., Schultz, R., Schwarz, R., Schweiger, J., Stangl, C., Steinbach, M.C., Vigerske, S., Willert, B.M.: Validation of nominations in gas network optimization: models, methods, and solutions. Optim. Methods Softw. (2014). https://doi.org/10.1080/10556788.2014.888426
29. van der Hoeven, T.: Math in gas and the art of linearization. Ph.D. thesis, University of Groningen (2004). Available from: http://hdl.handle.net/11370/0bbb8138-6d96-4d79-aac3-e46983d1fd33
30. Zigrang, D.J., Sylvester, N.D.: A review of explicit friction factor equations. J. Energy Resour. Technol. **107**(2), 280–283 (1985). https://doi.org/10.1115/1.3231190

Topological Index Analysis Applied to Coupled Flow Networks

Ann-Kristin Baum, Michael Kolmbauer, and Günter Offner

Abstract This work is devoted to the analysis of multi-physics dynamical systems stemming from automated modeling processes in system simulation software. The multi-physical model consists of (simple connected) networks of different or the same physical type (liquid flow, electric, gas flow, heat flow) which are connected via interfaces or coupling conditions. Since the individual networks result in differential algebraic equations (DAEs), the combination of them gives rise to a system of DAEs. While for the individual networks existence and uniqueness results, including the formulation of index reduced systems, is available through the techniques of *modified nodal analysis* or *topological based index analysis*, topological results for coupled system are not available so far. We present an approach for the application of topological based index analysis for coupled systems of the same physical type and give the outline of this approach for coupled liquid flow networks. Exploring the network structure via graph theoretical approaches allows to develop topological criteria for the existence of solutions of the coupled systems. The conditions imposed on the coupled network are illustrated via various examples. Those results can be interpreted as a natural extensions of the topological existence and index criteria provided by the topological analysis for single connected circuits.

Keywords Coupled system · Differential-algebraic equation · Hydraulic network · Modified nodal analysis · Topological index criteria

PACS 02.30.Hq, 02.60.Lj, 02.10.Ox

Mathematics Subject Classification (2010) 65L80, 94C15, 34B45

A.-K. Baum (✉) · M. Kolmbauer
MathConsult GmbH, Linz, Austria
e-mail: ann-kristin.baum@mathconsult.co.at; michael.kolmbauer@mathconsult.co.at

G. Offner
AVL List GmbH, Graz, Austria
e-mail: guenter.offner@avl.com

S. Campbell et al. (eds.), *Applications of Differential-Algebraic Equations: Examples and Benchmarks*, Differential-Algebraic Equations Forum,
https://doi.org/10.1007/11221_2018_1

1 Introduction

Increasingly demanding emissions legislation specifies the performance requirements for the next generation of products from vehicle manufacturers. Conversely, the increasingly stringent emissions legislation is coupled with the trend in increased power, drivability and safety expectations from the consumer market. Promising approaches to meet these requirements are downsizing the internal combustion engines (ICE), the application of turbochargers, variable valve timing, advanced combustion systems or comprehensive exhaust after treatment but also different variants of combinations of the ICE with an electrical engine in terms of hybridization or even a purely electric propulsion. The challenges in the development of future powertrains do not only lie in the design of individual components but in the assessment of the power train as a whole. On a system engineering level it is required to optimize individual components globally and to balance the interaction of different sub-systems. A typical system engineering model comprises several sub-systems. For instance in case of a hybrid propulsion these can be the vehicle chassis, the drive line, the air path of the ICE including combustion and exhaust after treatment, the cooling and lubrication system of the ICE and battery packs, the electrical propulsion system including the engine and a battery pack, the air conditioning and passenger cabin models, waste heat recovery and finally according control systems.

State-of-the-art modeling and simulation packages such as Dymola,[1] OpenModelica,[2] Matlab/Simulink,[3] Flowmaster,[4] Amesim,[5] SimulationX,[6] or Cruise M[7] offer many concepts for the automatic generation of dynamic system models. Modeling is done in a modularized way, based on a network of subsystems which again consists of simple standardized sub-components. The automated modeling process allows the usage of various advanced libraries for different subcomponents of the system from possibly different physical domains. The connections between those subcomponents are typically based on physical coupling conditions or predefined controller interfaces. Furthermore the network structure (topology) carries the core information of the network properties and therefore is predestinated to be exploited for the analysis and numerical simulation of those. In the application of vehicle system simulation the equations of the subsystems are differential-algebraic equations (DAEs) of higher index. Hence, this type of modeling leads to systems of coupled large DAEs-systems. Consequently the analysis of existence

[1] http://www.dynasim.com.

[2] http://www.openmodelica.org.

[3] http://www.mathworks.com.

[4] http://www.mentor.com.

[5] http://www.plm.automation.siemens.com.

[6] http://www.iti.de.

[7] http://www.avl.com.

and uniqueness of solutions for both, the individual physical subsystems and the full coupled system of DAE-systems, is a delicate issue.

Topology based index analysis for networks connects the research fields of *Analysis for DAEs* [25] and *Graph Theory* [7] in order to provide the appropriate base to analyze DAEs stemming from automatic generated system models. So far it has been established for various types of networks, including electric circuits [28] (*Modified Nodal Analysis*), gas supply networks [9], thermal liquid flow networks [1, 2] and water supply networks [10, 11, 26]. Although all those networks share some similarities, an individual investigation is required due to their different physical nature. Recently, a unified modeling approach for different types of flow networks has been introduced in [12], aiming for a unified topology based index analysis for the different physical domains on an abstract level. In the mentioned approaches, the analysis of the different physical domains is always restricted to a simple connected network of one physical type. Anyhow, all the approaches have in common, that they provide an index reduced (differential-index 1 or strangeness-index 0, cf. [20]) formulation of the original DAE, which is suitable for numerical integration.

Due to the increasing complexity in vehicle system simulation the interchangeability of submodels is gaining increasing importance. Submodels are exchanged between different simulation environments in terms of white-box or black-box libraries describing a set of DAEs. The interconnection to the system of physical based DAEs is again established by predefined controller interface or physical coupling conditions. The individual subnetworks are assumed to be of index reduced form (d-index 1 or s-index 0). This can be achieved by the *Topological index analysis* or *Modified Nodal Analysis*. It is well known [24], that the combination of d-index 1 DAEs may not form a d-index 1 DAE.

The artificial coupling of circles of the same physical type via (defined) physical coupling conditions within one simulation package might appear superfluous, since the circuit could be modeled all at once. Due to increasing complexity also within one physical domain, the modeling of subcircuits is distributed among high specialized teams and finally combined to the complete circuit. Using physical coupling conditions allows to combine the subcircuits to a single circuit without modifying the developed submodels. For the case of DAEs of higher index, this is of special importance, since the set of feasible initial conditions is often defined by structural properties (e.g. chord sets or spanning trees) and they might change in a coupling process. Due to integrity of the overall modeling process, this type of change should be avoided. Typically the physical coupling conditions are defined to ensure that certain conservation laws are satisfied, e.g. conservation of mass in liquid flow networks or conservation of charge in electric systems. Consequently an appropriate treatment of those coupling conditions is a delicate issue.

In [10] a unified modeling approach for different types of flow networks (electric circuits, water and gas networks) has been stated. One specific part of this classifications are the boundary conditions, that prescribe a certain pressure or potential for node elements and flow sources. In the case of electric networks, those elements are voltage sources and current sources. In the case of gas and liquid flow networks, those are reservoirs and demand branches. Those boundary conditions provide the starting point for defining appropriate coupling and interface conditions. As an example we explore the coupling for the case of two liquid flow networks via reservoirs and demand branches. Providing pressure controlled flow sources and flow controlled pressure sources establishes a strong coupling of the individual liquid flow networks.

The structure of this work is the following. In Sect. 2 we state a simple model for an incompressible liquid flow network and summarize the existence and uniqueness results as well as DAE index results, that have been obtained in [2], in Sect. 3. Therein we especially focus on the methods, that are used to derive the index and existence results and provide a descriptive explanation in the context of linear algebra and graph theory. The red line of the analysis is accomplished by a set of suitable examples. In Sect. 4 we state a coupled model of incompressible flow networks. The challenges arising for these kind of models are described via a set of characteristic examples. An analysis for the coupled flow network is presented in Sect. 5. The analysis is specialized to some specific configurations, where topological conditions for the coupled flow networks can be obtained. Finally, Sect. 6 provides an overview of the addressed issues. Therein another major focus is put on the description of open topics and further research requirements.

2 A Network Model for Incompressible Flow Networks

We consider a liquid flow network

$$\mathcal{N} = \{\mathcal{PI}, \mathcal{PU}, \mathcal{DE}, \mathcal{JC}, \mathcal{RE}\}, \tag{2.1}$$

that is composed of pipes $\mathcal{PI}$, pumps $\mathcal{PU}$, demands $\mathcal{DE}$, junctions $\mathcal{JC}$ and reservoirs $\mathcal{RE}$ and that is filled with an incompressible fluid. The pipes and pumps are connected by the junctions, where the fluid is split or merged. The connection to the environment is modeled by the reservoirs and demands that impose a predefined pressure or mass flow to the network.

To set up a mathematical model describing the mass flows in the pipes and pumps and the pressures in the junctions, the network $\mathcal{N}$ is represented by a linear graph $\mathcal{G}_{\mathcal{N}}$. A linear graph $\mathcal{G}$ is a combination $\mathcal{G} = \{\mathcal{V}, \mathcal{E}\}$ of vertices $\mathcal{V} = \{v_1, \dots, v_{n_{\mathcal{V}}}\}$ and edges $\mathcal{E} = \{e_1, \dots, e_{n_{\mathcal{E}}}\}$, such that each edge $e_j \in \mathcal{E}$ corresponds to a pair of vertices $(v_{j_1}, v_{j_2}) \in \mathcal{V} \times \mathcal{V}$, i.e., $e_j = (v_{j_1}, v_{j_2})$, cp. [7, p. 2]. For a detailed introduction to graph theory, we refer the reader to, e.g., [3, 7, 27].

For the network $\mathcal{N}$, the pipes, pumps and demands correspond to the edges, i.e., $\mathcal{E} = \{\mathcal{PI}, \mathcal{PU}, \mathcal{DE}\}$, while the junctions and reservoirs correspond to the vertices, i.e., $\mathcal{V} = \{\mathcal{JC}, \mathcal{RE}\}$, cf. Fig. 1. To each junction Jc_i and reservoir Re_i we assign pressures p_{Jc_i} and p_{Re_i}, respectively, and to each pipe Pi_j, pump Pu_j and demand De_j mass flows q_{Pi_j}, q_{Pu_j} and q_{De_j}. To assign a direction to the mass flows, we

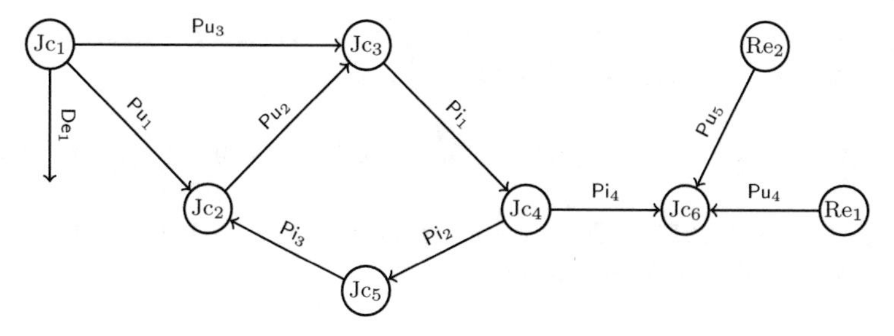

Fig. 1 Example of a graph $\mathcal{G}_{Ex1}$ of a network $\mathcal{N}_{Ex1}$

assign a direction to each edge element $e_j \in \mathcal{E}$, meaning that the pairs (v_{j_1}, v_{j_2}) with $e_j = (v_{j_1}, v_{j_2})$ are ordered. Then, the graph $\mathcal{G}_\mathcal{N}$ is oriented.

The connection structure of the network $\mathcal{N}$ is described by the incidence matrix $A_\mathcal{N} = (a_{ij})$, which is defined as, cp. e.g. [3, 7, 27],

$$a_{ij} = \begin{cases} 1, & \text{if the branch } j \text{ leaves the node } i, \\ -1, & \text{if the branch } j \text{ enters the node } i, \\ 0, & \text{else.} \end{cases}$$

Sorting the rows and columns of $A_\mathcal{N}$ according to the different element types, we obtain the incidence matrix as

$$A_\mathcal{N} = \begin{bmatrix} A_{\text{Jc,Pi}} & A_{\text{Jc,Pu}} & A_{\text{Jc,De}} \\ A_{\text{Re,Pi}} & A_{\text{Re,Pu}} & A_{\text{Re,De}} \end{bmatrix}.$$

Accordingly, the flows and pressures are summarized as

$$q = \begin{bmatrix} q_{\text{Pi}} \\ q_{\text{Pu}} \\ q_{\text{De}} \end{bmatrix}, \quad p = \begin{bmatrix} p_{\text{Jc}} \\ p_{\text{Re}} \end{bmatrix}.$$

Besides the connection structure, each network element is equipped with a characteristic equation describing the relation of the mass flow and pressure or pressure difference. In a pipe Pi_j, $j = 1, \ldots, n_{\text{Pi}}$, directed from node j_1 to node j_2, the mass flow $q_{\text{Pi},j}$ is specified by the transient momentum equation

$$\dot{q}_{\text{Pi},j} = c_{1,j} \Delta p_j + c_{2,j} |q_{\text{Pi},j}| q_{\text{Pi},j} + c_{3,j}, \qquad (\text{Pi}_j)$$

depending on the pressure difference $\Delta p_j = p_{j_1} - p_{j_2}$ between the adjacent nodes j_1, j_2 and constants $c_{i,j}$ depending, e.g., on the pipe diameter, length, inclination angle, and other physical properties, cp. Remark 2.1. Using the incidence matrix $A_\mathcal{N}$, the pressure drop $\Delta p_j = p_{j_1} - p_{j_2}$ along an edge $e_j = (v_{j_1}, v_{j_2})$ is given by

$e_j^T A_{\mathcal{N}}^T p = \Delta p_j$. Setting $C_I = \operatorname{diag}\left((c_{I,j})\right)_j$, $j = 1, \ldots n_{\text{Pi}}$, $I = 1, 2, 3$, we define the *pipe function* of the full network by

$$f_{\text{Pi}}(q_{\text{Pi}}, p_{\text{Jc}}, p_{\text{Re}}) := C_1(A_{\text{Jc,Pi}}^T p_{\text{Jc}} + A_{\text{Re,Pi}}^T p_{\text{Re}}) + C_2 \operatorname{diag}\left(|q_{\text{Pi},j}|\right)_j q_{\text{Pi}} + C_3.$$

with $f_{\text{Pi}} \in C^1(\Omega_{\text{Pi}}, \mathbb{R}^{n_{\text{Pi}}})$. Then, the transient momentum equations (Pi$_j$) can be summarized for the whole network as

$$\dot{q}_{\text{Pi}} = f_{\text{Pi}}(q_{\text{Pi}}, p_{\text{Jc}}, p_{\text{Re}}). \tag{Pi}$$

In a pump Pu$_j$, $j = 1, \ldots, n_{\text{Pu}}$, directed from node j_1 to node j_2, the mass flow $q_{\text{Pu},j}$ is specified algebraically by the pressure drop $\Delta p_j = p_{j_1} - p_{j_2}$, i.e.,

$$\Delta p_j = f_{\text{Pu}_j}(q_{\text{Pu},j}). \tag{Pu$_j$}$$

The function f_{Pu_j} is given by specialized pump models, cp. Remark 2.2. Like for the pipes, we use the incidence matrix $A_{\mathcal{N}}$ to summarize the pump functions in the function

$$f_{\text{Pu}} := [f_{\text{Pu},j}]_{j=1,\ldots,n_{\text{Pu}}},$$

where we assume that $f_{\text{Pu}} \in C^1(\Omega_{\text{Pu}}, \mathbb{R}^{n_{\text{Pu}}})$. Then, the pump equations (Pu$_j$) can be summarized for the whole network as

$$A_{\text{Jc,Pu}}^T p_{\text{Jc}} + A_{\text{Re,Pu}}^T p_{\text{Re}} = f_{\text{Pu}}(q_{\text{Pu}}). \tag{Pu}$$

In a junction Jc$_i$, $i = 1, \ldots, n_{\text{Jc}}$, the amount of mass entering and leaving Jc$_i$ is equal due to mass conservation. Summarizing the indices of pipes, pumps and demand branches that are incident to Jc$_i$ in the set $\hat{J}_i$, we thus get that

$$\sum_{j \in \hat{J}_i} q_j = 0. \tag{Jc$_i$}$$

Using the incidence matrix, the sum of all mass flows entering or leaving a junction Jc$_i$ is given by $e_i^T A q = \sum_{e_j \in \mathcal{E}_{inc}(\text{Jc}_i)} q_j$, such that the junction equations (Jc$_i$) can be summarized as

$$A_{\text{Jc,Pi}} q_{\text{Pi}} + A_{\text{Jc,Pu}} q_{\text{Pu}} + A_{\text{Jc,De}} q_{\text{De}} = 0. \tag{Jc}$$

In a demand branch De$_j$, $j = 1, \ldots, n_{\text{De}}$, the mass flow $q_{\text{De},j}$ is specified by a given function $\bar{q}_{\text{De},j}$, i.e.,

$$q_j = \bar{q}_{\text{De},j}. \tag{De$_j$}$$

Similarly, in a reservoir Re_i, $i = 1, \ldots, n_{\text{Re}}$, the pressure $p_{\text{Re},i}$ is specified by a given function $\bar{p}_{\text{Re},i}$, i.e.,

$$p_i = \bar{p}_{\text{Re},i}. \tag{Re$_i$}$$

Setting $\bar{q}_{\text{De}} := [q_{\text{De},j}]_{j=1,\ldots,n_{\text{De}}}$ and $\bar{p}_{\text{Re}} := [p_{\text{Re},i}]_{i=1,\ldots,n_{\text{Re}}}$, the boundary conditions (De_j) and (Re_i) are summarized as

$$q_{\text{De}} = \bar{q}_{\text{De}}, \tag{De}$$

$$p_{\text{Re}} = \bar{p}_{\text{Re}}. \tag{Re}$$

In conclusion, the dynamic of the network $\mathcal{N}$ is modeled by the DAE

$$\dot{q}_{\text{Pi}} = C_1(A_{\text{Jc,Pi}}^T p_{\text{Jc}} + A_{\text{Re,Pi}}^T p_{\text{Re}}) + C_2 \operatorname{diag}\left(|q_{\text{Pi},j}|\right) q_{\text{Pi}} + C_3 \tag{2.2a}$$

$$0 = A_{\text{Jc,Pu}}^T p_{\text{Jc}} + A_{\text{Re,Pu}}^T p_{\text{Re}} - f_{\text{Pu}}(q_{\text{Pu}}) \tag{2.2b}$$

$$0 = A_{\text{Jc,Pi}} q_{\text{Pi}} + A_{\text{Jc,Pu}} q_{\text{Pu}} + A_{\text{Jc,De}} q_{\text{De}} \tag{2.2c}$$

$$q_{\text{De}} = \bar{q}_{\text{De}} \tag{2.2d}$$

$$p_{\text{Re}} = \bar{p}_{\text{Re}}. \tag{2.2e}$$

The unknowns are given by $q(t)$ and $p(t)$. The system is square with size $n_{\text{Pi}} + n_{\text{Pu}} + n_{\text{De}} + n_{\text{Re}} + n_{\text{Jc}}$.

Remark 2.1 (Specific Pipe Model) For the transient momentum balance equation (Pi_j), we have used the reference [23] to obtain

$$\dot{q}_{\text{Pi}} = \frac{A}{L}\Delta p + \frac{1}{\rho}\frac{\phi\zeta}{2d_{hyd}A}|q_{\text{Pi}}|q_{\text{Pi}} + \rho A G \Delta h,$$

where A is the cross section, L is the length, d_{hyd} is the hydraulic diameter and ϕ is a shape factor of the pipe. The density of the liquid in the pipe is denoted by ρ. Δp and Δh are the pressure difference and the height difference across the pipe. g is the gravitational force. The friction coefficient ζ can be described, e.g., by Haaland formula [6]

$$\zeta = \begin{cases} \frac{64}{Re}, & \text{for laminar flow,} \\ \left(-1.81 \log_{10}\left(\frac{6.9}{Re} + \left(\frac{\varepsilon}{3.7d_{hyd}}\right)^{1.11}\right)\right)^{-0.5} & \text{for turbulent flow,} \end{cases}$$

where Re is the Reynolds number and ε is the surface roughness. From this, we define the constants directly from the geometry and the type of the liquid.

$$c_1 = \frac{A}{L}, \quad c_2 = \frac{1}{\rho}\frac{\phi\zeta}{2d_{hyd}A}, \quad c_3 = \rho A G \Delta h.$$

Remark 2.2 (Specific Pump Model) In system simulation, the pump equation (Pu_j) is typically characterized via pump curves. Pump curves provide information on the relationship between the total head (or pressure difference) and the mass flow rate. Examples for pump curves are given in Fig. 2. In our terminology the pump curves determine the pump function f_{Pu}. The actual pump curves differ from the theoretical pump curves, since they consider internal losses due to slip through the operating clearance (positive displacement pump) or friction and recirculation (centrifugal pump). In some cases the actual pump curves can be represented by polynomials, e.g. [8]. In practical applications it is quite difficult to determine the pump characteristics theoretically. Typically the pump characteristic is determined experimentally and provided (e.g. by the original pump manufacturer) in terms of a discrete data set of measurement points. Due to this proceeding, the actual system simulation can be kept to the real counterparts as close as possible. In many cases, see e.g. [5], the experimental curves differ from the theoretical models in terms of curve properties (e.g. invertibility, which is often determined in terms of strong monotonicity). Especially the dropping effect for mass flows tending to zero (starting phase of a pumping process) is a well observed phenomena and occurs frequently for experimental measured pump curves (see Fig. 3) or specific pump constellations, cf. [21, Figure 2.15].

In this constellation, there may exist more feasible mass flows q for a given pressure drop Δp, and consequently the solution may be not unique (serial pump constellations) or may oscillate between two feasible operation points (parallel pump constellations). In practical vehicle system simulations the pump characteristic is given by experimental data sets in almost all constellations. Hence (strong) monotonicity of the pump curve cannot be presupposed and has to be treated in an appropriate manner. An example of a pump characteristic used in system simulation software is displayed in Table 1.

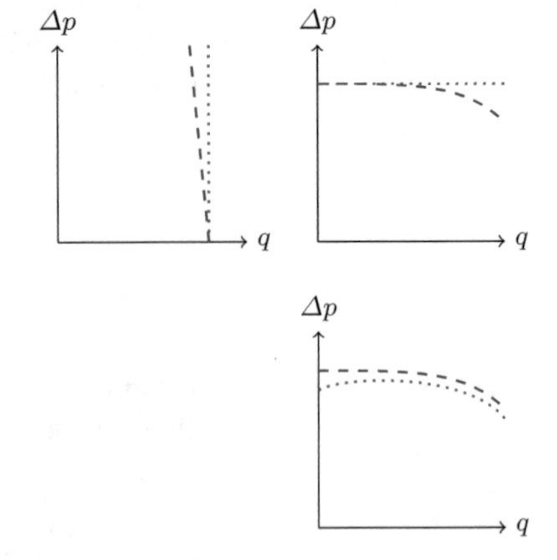

Fig. 2 Theoretical (dotted red) and actual (dashed blue) curves of positive displacement (right) and centrifugal (left) pumps (cf. [29])

Fig. 3 Comparison between theoretical (dashed blue) and experimental (dotted red) curves, where the experimental curves obtains a *dropping* structure (cf. [5])

Table 1 Example of a pump characteristic for a dynamic pump (pump speed = 5,000 rpm) used for an oil cooling and lubrication circuit in a vehicle system simulation in Cruise M

Mass flow q (kg/min)	Pressure rise Δp (bar)
0	3.7821
4.316	3.8123
8.632	3.8234
12.948	3.7756
17.264	3.7141
21.58	3.6407
25.896	3.5581
30.212	3.4611
34.528	3.3407
38.844	3.2662

Example 2.1 For the network $\mathcal{N}_{Ex1}$ given in Fig. 1, the incidence matrix is composed from the blocks

$$A_{\mathrm{Jc}^1,\mathrm{Pi}^1} = \begin{bmatrix} & Pi_1^1 & Pi_2^1 & Pi_3^1 & Pi_4^1 \\ Jc_1^1 & 0 & 0 & 0 & 0 \\ Jc_2^1 & 0 & 0 & -1 & 0 \\ Jc_3^1 & 1 & 0 & 0 & 0 \\ Jc_4^1 & -1 & 1 & 0 & 1 \\ Jc_5^1 & 0 & -1 & 1 & 0 \\ Jc_6^1 & 0 & 0 & 0 & -1 \end{bmatrix},$$

$$A_{\mathrm{Jc},\mathrm{Pu}} = \begin{bmatrix} & Pu_1^1 & Pu_2^1 & Pu_3^1 & Pu_4^1 & Pu_5^1 \\ Jc_1^1 & 1 & 0 & 1 & 0 & 0 \\ Jc_2^1 & -1 & 1 & 0 & 0 & 0 \\ Jc_3^1 & 0 & -1 & -1 & 0 & 0 \\ Jc_4^1 & 0 & 0 & 0 & 0 & 0 \\ Jc_5^1 & 0 & 0 & 0 & 0 & 0 \\ Jc_6^1 & 0 & 0 & 0 & -1 & -1 \end{bmatrix},$$

$$A_{\mathrm{Re}^1,\mathrm{Pi}^1} = \begin{bmatrix} & Pi_1^1 & Pi_2^1 & Pi_3^1 & Pi_4^1 \\ Re_1^1 & 0 & 0 & 0 & 0 \\ Re_2^1 & 0 & 0 & 0 & 0 \end{bmatrix},$$

$$A_{\mathrm{Re}^1,\mathrm{Pu}^1} = \begin{bmatrix} & Pu_1^1 & Pu_2^1 & Pu_3^1 & Pu_4^1 & Pu_5^1 \\ Re_1 & 0 & 0 & 0 & 1 & 0 \\ Re_2 & 0 & 0 & 0 & 0 & 1 \end{bmatrix},$$

$$A_{\mathrm{Re}^1,\mathrm{De}^1} = 0, \quad A_{\mathrm{Jc}^1,\mathrm{De}^1} = \begin{bmatrix} & Jc_1^1 & Jc_2^1 & Jc_3^1 & Jc_4^1 & Jc_5^1 & Jc_6^1 \\ De_1^1 & 1 & 0 & 0 & 0 & 0 & 0 \end{bmatrix}^T.$$

The DAE (2.2) modeling the dynamics of $\mathcal{N}_{Ex1}$ is given by

$$\dot{q}_{\mathrm{Pi}_1^1} = c_{11}(p_{\mathrm{Jc}_4^1} - p_{\mathrm{Jc}_3^1}) + c_{21}|q_{\mathrm{Pi}_1^1}|q_{\mathrm{Pi}_1^1} + c_{31} \tag{Pi_1^1}$$

$$\dot{q}_{\mathrm{Pi}_2^1} = c_{12}(p_{\mathrm{Jc}_4^1} - p_{\mathrm{Jc}_5^1}) + c_{22}|q_{\mathrm{Pi}_2^1}|q_{\mathrm{Pi}_2^1} + c_{32} \tag{Pi_2^1}$$

$$\dot{q}_{\mathrm{Pi}_3^1} = c_{13}(p_{\mathrm{Jc}_2^1} - p_{\mathrm{Jc}_5^1}) + c_{23}|q_{\mathrm{Pi}_3^1}|q_{\mathrm{Pi}_3^1} + c_{33} \tag{Pi_3^1}$$

$$\dot{q}_{\mathrm{Pi}_4^1} = c_{14}(p_{\mathrm{Jc}_6^1} - p_{\mathrm{Jc}_4^1}) + c_{24}|q_{\mathrm{Pi}_4^1}|q_{\mathrm{Pi}_4^1} + c_{34} \tag{Pi_4^1}$$

$$p_{\mathrm{Jc}_2^1} - p_{\mathrm{Jc}_1^1} = f_{\mathrm{Pu}_1^1}(q_{\mathrm{Pu}_1^1}) \tag{Pu_1^1}$$

$$p_{\mathrm{Jc}_3^1} - p_{\mathrm{Jc}_2^1} = f_{\mathrm{Pu}_2^1}(q_{\mathrm{Pu}_2^1}) \tag{Pu_2^1}$$

$$p_{\mathrm{Jc}_3^1} - p_{\mathrm{Jc}_1^1} = f_{\mathrm{Pu}_3^1}(q_{\mathrm{Pu}_3^1}) \tag{Pu_3^1}$$

$$p_{\mathrm{Jc}_6} - p_{\mathrm{Re}_1^1} = f_{\mathrm{Pu}_4^1}(q_{\mathrm{Pu}_4^1}) \tag{Pu_4^1}$$

$$p_{\mathrm{Jc}_6^1} - p_{\mathrm{Re}_2^1} = f_{\mathrm{Pu}_5^1}(q_{\mathrm{Pu}_5^1}) \tag{Pu_5^1}$$

$$0 = q_{\mathrm{De}_1^1} + q_{\mathrm{Pu}_1^1} + q_{\mathrm{Pu}_3^1} \tag{Jc_1^1}$$

$$0 = -q_{\mathrm{Pi}_3^1} - q_{\mathrm{Pu}_1^1} + q_{\mathrm{Pu}_2^1} \tag{Jc_2^1}$$

$$0 = q_{\mathrm{Pi}_1^1} - q_{\mathrm{Pu}_2^1} - q_{\mathrm{Pu}_3^1} \tag{Jc_3^1}$$

$$0 = -q_{\mathrm{Pi}_1^1} + q_{\mathrm{Pi}_2^1} + q_{\mathrm{Pi}_4^1} \tag{Jc_4^1}$$

$$0 = -q_{\mathrm{Pi}_2^1} + q_{\mathrm{Pi}_3^1} \tag{Jc_5^1}$$

$$0 = -q_{\mathrm{Pi}_4^1} - q_{\mathrm{Pu}_4^1} - q_{\mathrm{Pu}_5^1} \tag{Jc_6^1}$$

$$q_{\mathrm{De}^1} = \bar{q}_{\mathrm{De}^1}, \tag{De_1^1}$$

$$p_{\mathrm{Re}_1^1} = \bar{p}_{\mathrm{Re}_1^1} \tag{Re_1^1}$$

$$p_{\mathrm{Re}_2^1} = \bar{p}_{\mathrm{Re}_2^1}. \tag{Re_2^1}$$

Choosing the geometries of the pipes and a friction coefficients, the constants $c_{I,j}$, $j = 1, \ldots, 4$, $I = 1, 2, 3$ and hence the pipe functions $f_{\mathrm{Pi}_j^1}$ are fixed. Considering the pump characteristic given, e.g., in Table 1 and interpolate it, e.g., with cubic splines, the pump functions $f_{\mathrm{Pu}_j^1}$ are fixed. Choosing an input mass flow $\bar{q}_{\mathrm{De}_1^1}$ as well as input pressures $\bar{p}_{\mathrm{Re}_1^1}$, $\bar{p}_{\mathrm{Re}_2^1}$, the boundary conditions are fixed. In conclusion, the mass flows and pressures in the network given in Fig. 1 are determined by the DAE (Pi_1^1) to (Re_2^1).

3 Topology Based Index Analysis of a Single Network

To analyze the solvability of the DAE (2.2), we impose the following assumptions on the connection structure of the network $\mathcal{N}$.

Assumption 3.1 *Consider a network* $\mathcal{N}$ *as in* (2.1).

(N1) *Two junctions are connected at most by one pipe or one pump.*
(N2) *Each pipe, pump and demand has an assigned direction.*
(N3) *The network is connected, i.e., every pair of junctions and/or reservoirs can be reached by a sequence of pipes and pumps.*
(N4) *Every junction is adjacent to at most one demand branch. Every reservoir is connected at most to one pipe or pump.*

Under Assumption 3.1 the network graph is simple (N1), oriented (N2) and connected (N3). Assigning a direction to each pipe, pump and demand, allows to speak of a positive or negative mass flow. Note that this orientation of the pipes and pumps is arbitrary and only serves as a reference condition, it is not necessarily related with the true or expected direction of the fluid flow. As the reservoirs are end vertices and the demands are connected to junctions only, cf. (N4), implies that no reservoir is connected to a demand branch and hence the corresponding sub-matrix of the incidence matrix is zero, i.e. $A_{\mathrm{Re,De}} = 0$.

3.1 Graphtheoretical Prerequisites

In the following, we use graph theoretical concepts like paths, spanning trees, cycles, connected components, etc. A comprehensive introduction to this topic can be found, e.g., in [3, 7, 27].

For our purposes, we need these concepts for subsets describing the connection structure of two specific element types. Asking, e.g., for the solvability of the pump equations (2.2b), we are interested in the connection structure of the junction and pump subset $\{\mathcal{JC}, \mathcal{PU}\}$. This set is not necessarily a subgraph as it might contain isolated pumps (corresponding to a pump connecting two reservoirs), isolated junctions (corresponding to a junction connected to pipes and demands only) or loose edges (corresponding to a pump connecting a junction and a reservoir). Consequently, the connection matrix $A_{\mathrm{Jc,Pu}}$ does not have the usual entry pattern of an incidence matrix. Still, the ideas of trees, cycles, etc. and their correspondence to fundamental subspaces of the connection matrix can be easily extended, see [2].

Looking at the junction and pump subset $\mathcal{G}_{\mathrm{Jc,Pu}} := \{\mathcal{JC}, \mathcal{PU}\}$, we are interested in particular in the following substructures.

Substructure 3.1 *Substructure 1 of* $\mathcal{G}_{Jc,Pu}$.

a) Paths of pumps connecting two reservoirs.
b) Cycles of pumps.

On the substructures 1a) and 1b), the pressure difference is fixed. On a path of pumps between two reservoirs, the pressure drop across the path is fixed by the two reservoirs. On a cycle of pumps, the pressure difference vanishes as the path is closed. Regarding the solvability of the DAE, this means that on Substructure 3.1, the pumps have to work against their usual mode of operation. Instead of returning a pressure drop for a given mass flow, they have to adjust the mass flow to a given pressure. This means that the pump characteristic has to be invertible. Mathematically, this is reflected in the sense that the pumps of Substructure 3.1 correspond to the kernel of the connection matrix $A_{\mathrm{Jc,Pu}}$ of the set $\mathcal{G}_{\mathrm{Jc,Pu}}$. If $V_2 \in \mathbb{R}^{n_{\mathrm{Pu}} \times n_{V_2}}$ selects the paths of pumps between reservoirs as well as the cycles of pumps, then $\mathrm{span}(V_2) = \ker(A_{\mathrm{Jc,Pu}})$ [2]. An example of Substructure 3.1 and the matrix V_2 is given in Fig. 4 and Example 2.1.

Substructure 3.2 *Substructure 2 of* $\mathcal{G}_{Jc,Pu}$.

a) Connected components of junctions and pumps without loose pumps.

On a connected component of junctions and pumps *without* loose pumps, i.e., without a pump connection to a reservoir, the pumps only specify the pressure difference. Its absolute value has to be determined from the pipes connecting this connected component to a reservoir giving a reference value. Regarding the solvability of the DAE, this means that on junctions of Substructure 3.2, only

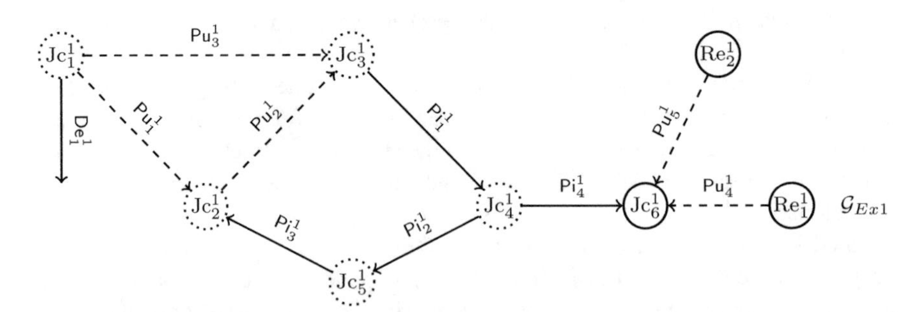

Fig. 4 Example of Substructure 3.1 (dashed pumps) and Substructure 3.2 (dotted junctions). The pumps $\mathrm{Pu}_4^1, \mathrm{Pu}_5^1$ form a path of pumps between the two reservoirs $\mathrm{Re}_1^1, \mathrm{Re}_2^1$, i.e., are of type 1a), while the pumps $\mathrm{Pu}_1^1, \mathrm{Pu}_2^1, \mathrm{Pu}_3^1$ form a cycle of pumps, i.e., are of type 1b). Together with the junctions $\mathrm{Jc}_1^1, \mathrm{Jc}_2^1, \mathrm{Jc}_3^1$, the pumps $\mathrm{Pu}_1^1, \mathrm{Pu}_2^1, \mathrm{Pu}_3^1$ form a connected component of $\mathcal{G}_{\mathrm{Jc}^1,\mathrm{Pu}^1}$ without loose pumps, i.e., belong to Substructure 3.2. The junctions $\mathrm{Jc}_4^1, \mathrm{Jc}_5^1$ are isolated in $\mathcal{G}_{\mathrm{Jc}^1,\mathrm{Pu}^1}$ as they are not incident to any and also belong to Substructure 3.2. Junction Jc_6 together with the pumps $\mathrm{Pu}_4^1, \mathrm{Pu}_5^1$ forms a connected component of $\mathcal{G}_{\mathrm{Jc}^1,\mathrm{Pu}^1}$, but as it is connected to reservoirs, it does not belong to Substructure 3.2

the pressure difference is specified by the pump equation. The absolute value is obtained from an additional equation, the *hidden constraint*, arising from the coupling of the pipe equation via the conservation of mass, see Theorem 3.3. Note that isolated junctions, i.e, junctions connected to pipes only, also are connected components in $\mathcal{G}_{\mathrm{Jc,Pu}}$ *without* loose pumps and hence belong to Substructure 3.2. Mathematically, this is reflected in the sense that the junctions of Substructure 3.2 correspond to the left kernel of the connection matrix $A_{\mathrm{Jc,Pu}}$ of the set $\mathcal{G}_{\mathrm{Jc,Pu}}$. If $U_2 \in \mathbb{R}^{n_{\mathrm{Jc}} \times n_{U_2}}$ is such that U_2 selects the junctions belonging to the connected components without loose pumps in $\mathcal{G}_{\mathrm{Jc,Pu}}$, then $\mathrm{coker}(A_{\mathrm{Jc,Pu}}) = \mathrm{span}(U_2)$ [2]. An example of Substructure 3.2 and the matrix U_2 are given in Fig. 4 and in Example 2.1.

Graphically, the action of U_2 on $A_{\mathrm{Jc,Pu}}$ corresponds to the *vertex identification* of the connected components $\mathcal{G}_{\mathrm{Jc,Pu};1}, \ldots, \mathcal{G}_{\mathrm{Jc,Pu};n_k}$ of $\mathcal{G}_{\mathrm{Jc,Pu}}$, i.e., we melt every connected component of pumps and junctions into a single junction

$$\overline{\mathrm{Jc}}_k := \bigcup_{i\,:\,\mathrm{Jc}_i \in \mathcal{G}_{\mathrm{Jc,Pu};k}} \mathrm{Jc}_i, \tag{3.1}$$

for $k = 1, \ldots, n_k$. We summarize the junctions $\overline{\mathrm{Jc}}_1, \ldots, \overline{\mathrm{Jc}}_{n_k}$ arising from the vertex identification (3.1) in the set

$$\overline{\mathcal{JC}} := \left\{\overline{\mathrm{Jc}}_1, \ldots, \overline{\mathrm{Jc}}_{n_k}\right\}$$

and consider the graph $\mathcal{G}_{\overline{\mathrm{Jc}},\mathrm{Pi}} := \{\overline{\mathcal{JC}}, \mathcal{PI}\}$. The connection matrix is given by

$$A_{\overline{\mathrm{Jc}},\mathrm{Pu}} = U_2^T A_{\mathrm{Jc,Pu}}.$$

An example of the graph $\mathcal{G}_{\overline{\mathrm{Jc}},\mathrm{Pi}}$ and the connection matrix $A_{\overline{\mathrm{Jc}},\mathrm{Pu}}$ is given in Fig. 5 and Example 2.1.

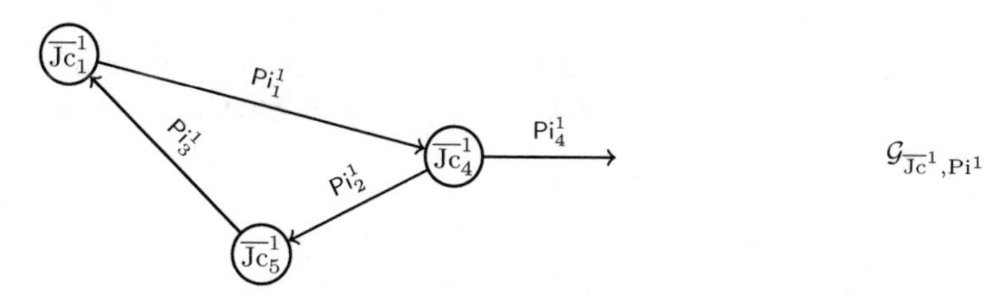

Fig. 5 Example of the vertex identification (3.1). For the graph $\mathcal{G}_{Ex1}$ of Fig. 4, the vertex identification of the connected components $\{\mathrm{Pu}_1^1, \mathrm{Pu}_2^1, \mathrm{Pu}_3^1; \mathrm{Jc}_1^1, \mathrm{Jc}_2^1, \mathrm{Jc}_3^1\}$, Jc_4^1, Jc_5^1 yields the graph $\mathcal{G}_{Ex1,\overline{\mathrm{Jc}}^1,\mathrm{Pi}^1}$ with junctions $\overline{\mathrm{Jc}}_1^1$ and $\overline{\mathrm{Jc}}_4^1$, $\overline{\mathrm{Jc}}_5^1$

For the graph $\mathcal{G}_{\overline{Jc},Pi}$, we consider the following substructures.

Substructure 3.3 *Substructure 3 of* $\mathcal{G}_{\overline{Jc},Pi}$.

a) A spanning tree, i.e., the largest subgraph without cycles.
b) The chord set belonging to the spanning tree, i.e., pipes that close a cycle.

On a spanning tree, the pressure difference across the edges is well-defined. The chord set refers to those edges that destroy this well-definiteness as they close a cycle. Regarding the solvability of the DAE, this means that on Substructure 3.3a), the pressure drop across the pipes is well-defined. Mathematically, this is reflected in the sense Substructure 3.3 corresponds to a permutation $[\Pi_1 \Pi_2]$, where Π_1 selects the edges on the spanning tree and Π_2 the edges on the chord set. Then, $\operatorname{corange}(A_{\overline{Jc},Pi}) = \operatorname{span}(\Pi_1)$ [2]. An example of Substructure 3.3 and the permutation Π are given in Fig. 6 and Example 2.1 *continued.*

Example 2.1 (continued) We consider again the network $\mathcal{N}_{Ex1}$ as given in Fig. 1. For this example, the Substructures 3.1 and 3.2 are illustrated in Fig. 4. The associated matrix $V_2^1 \in \mathbb{R}^{5\times 2}$ selecting the paths of pumps between reservoirs and the cycles of pumps as well the matrix $U_2^1 \in \mathbb{R}^{6\times 3}$ selecting the junctions belonging to the connected components without loose edges and the isolated junctions in $\mathcal{G}_{Jc^1,Pu^1}$ are given by

$$V_2^1 = \begin{bmatrix} Pu_1^1 & 1 & 0 \\ Pu_2^1 & 1 & 0 \\ Pu_3^1 & -1 & 0 \\ Pu_4^1 & 0 & 1 \\ Pu_5^1 & 0 & -1 \end{bmatrix}, \quad U_2^1 = \begin{bmatrix} Jc_1^1 & 1 & 0 & 0 \\ Jc_2^1 & 1 & 0 & 0 \\ Jc_3^1 & 1 & 0 & 0 \\ Jc_4^1 & 0 & 1 & 0 \\ Jc_5^1 & 0 & 0 & 1 \\ Jc_6^1 & 0 & 0 & 0 \end{bmatrix},$$

where the first column in V_2^1 selects the cycle of pumps Pu_1^1, Pu_2^1, Pu_3^1 and the second column in V_2^1 selects the path of pumps Pu_4^1 and Pu_5^1. Considering the connection matrix A_{Jc^1,Pu^1}, we verify that $\operatorname{span}(V_2^1) = \ker(A_{Jc^1,Pu^1})$ and $\operatorname{span}(U_2^1) = \operatorname{coker}(A_{Jc^1,Pu^1})$. The connection matrix $A_{\overline{Jc}^1,Pi^1}$ of the graph $\mathcal{G}_{\overline{Jc}^1,Pi^1}$ arising from the vertex identification is given by

$$A_{\overline{Jc}^1,Pi^1} = \begin{bmatrix} & Pi_1^1 & Pi_2^1 & Pi_3^1 & Pi_4^1 \\ \overline{Jc}_1^1 & 1 & 0 & -1 & 0 \\ \overline{Jc}_2^1 & 0 & 0 & 0 & -1 \\ Jc_4 & -1 & 1 & 0 & 1 \\ Jc_5 & 0 & -1 & 1 & 0 \end{bmatrix}.$$

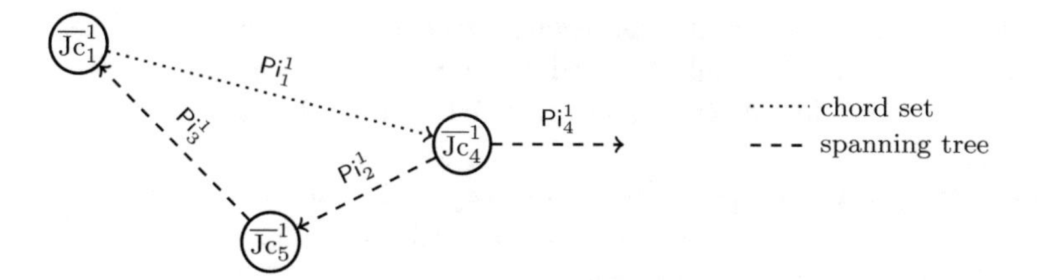

Fig. 6 Example of Substructure 3.3. For the graph $\mathcal{G}_{Ex1,\overline{\mathrm{Jc}}^1,\mathrm{Pi}^1}$ of Fig. 5, a spanning tree is given by the pipes Pi_2^1, Pi_3^1 and Pi_4^1. The associated chord set is given by Pi_1^1

A choice of a spanning tree and chord set of the graph $\mathcal{G}_{\overline{\mathrm{Jc}}^1,\mathrm{Pi}^1}$ is illustrated in Fig. 6. The associated permutation $[\Pi_1^1, \Pi_2^1]$ is given by

$$\Pi_1^1 = \begin{bmatrix} Pi_1^1 & 0 & 0 & 0 \\ Pi_2^1 & 1 & 0 & 0 \\ Pi_3^1 & 0 & 1 & 0 \\ Pi_4^1 & 0 & 0 & 1 \end{bmatrix}, \quad \Pi_2 = \begin{bmatrix} Pi_1^1 & 1 \\ Pi_2^1 & 0 \\ Pi_3^1 & 0 \\ Pi_4^1 & 0 \end{bmatrix}$$

and we verify that $\mathrm{span}(\Pi_1^1) = \mathrm{corange}(A_{\overline{\mathrm{Jc}}^1,\mathrm{Pi}^1})$.

3.2 Solvability Results

Now, we derive solvability conditions for the network DAE (2.2). We define the network function $F_{\mathcal{N}}\colon \mathbb{D} \to \mathbb{R}^n$, $\mathbb{D} \subset \mathbb{R} \times \mathbb{R}^n \times \mathbb{R}^n$ with

$$F_{\mathcal{N},1}(q_{\mathrm{Pi}}, p_{\mathrm{Jc}}, p_{\mathrm{Re}}) = \dot{q}_{\mathrm{Pi}} - f_{\mathrm{Pi}}(q_{\mathrm{Pi}}, p_{\mathrm{Jc}}, p_{\mathrm{Re}}) \tag{3.2a}$$

$$F_{\mathcal{N},2}(q_{\mathrm{Pu}}, p_{\mathrm{Jc}}, p_{\mathrm{Re}}) = A_{\mathrm{Jc,Pu}}^T p_{\mathrm{Jc}} + A_{\mathrm{Re,Pu}}^T p_{\mathrm{Re}} - f_{\mathrm{Pu}}(q_{\mathrm{Pu}}) \tag{3.2b}$$

$$F_{\mathcal{N},3}(q_{\mathrm{Pi}}, q_{\mathrm{Pu}}, q_{\mathrm{De}}) = A_{\mathrm{Jc,Pi}} q_{\mathrm{Pi}} + A_{\mathrm{Jc,Pu}} q_{\mathrm{Pu}} + A_{\mathrm{Jc,De}} q_{\mathrm{De}} \tag{3.2c}$$

$$F_{\mathcal{N},4}(q_{\mathrm{De}}) = q_{\mathrm{De}} - \bar{q}_{\mathrm{De}} \tag{3.2d}$$

$$F_{\mathcal{N},5}(p_{\mathrm{Re}}) = p_{\mathrm{Re}} - \bar{p}_{\mathrm{Re}}, \tag{3.2e}$$

where

$$f_{\mathrm{Pi}}(q_{\mathrm{Pi}}, p_{\mathrm{Jc}}, p_{\mathrm{Re}}) := C_1(A_{\mathrm{Jc,Pi}}^T p_{\mathrm{Jc}} + A_{\mathrm{Re,Pi}}^T p_{\mathrm{Re}}) + C_2 \,\mathrm{diag}\left(|q_{\mathrm{Pi},j}|\right) q_{\mathrm{Pi}} + C_3.$$

Furthermore, we define the set of consistent initial values

$$\mathcal{C}_{IV} := \{(t_0, q_0, p_0) \in \mathcal{I} \times \mathbb{R}^{n_\mathcal{E}} \times \mathbb{R}^{n_\mathcal{V}} \mid \exists \dot{q}_0, \dot{p}_0\colon F_{\mathcal{N}}(t_0, q_0, p_0, \dot{q}_0, \dot{p}_0) = 0\}.$$

Hence, the states q_0 and p_0 are consistent, if there exist vectors $\dot{q}_0$ and $\dot{p}_0$, such that the DAE (2.2) is algebraically satisfied. Usually, one needs more conditions on the set $\mathcal{C}_{IV}$, see [18]. In our setting, however, the DAE (3.2) is of s-index $\mu = 1$, see Theorem 3.1.

Combining the concept of derivative arrays [4] and the strangeness index as developed in [13, 15–17] with graph theoretical results, the unique solvability of the DAE model (2.2) can be characterized.

Theorem 3.1 ([2]) *Let $\mathcal{N}$ be a network given by* (2.1) *that satisfies Assumptions 3.1 and let $F_{\mathcal{N}} \in C^2(\mathbb{D}, \mathbb{R}^n)$. Let $n_{Re} > 0$ and let $V_2^T \, \mathrm{D}f_{Pu} V_2$ be pointwise nonsingular for* $\mathrm{span}(V_2) = \ker(A_{Jc,Pu})$. *Then,*

1. *The DAE* (2.2) *is regular and has strangeness index $\mu = 1$ (d-index 2).*
2. *The DAE* (2.2) *is uniquely solvable for every $(t_0, q_0, p_0) \in \mathcal{C}_{IV}$ and the solution is $(q, p) \in C^1(\mathcal{I}, \mathbb{R}^n)$.*

Translated as conditions on the network structure and its elements, the solvability conditions of Theorem 3.1 mean the following. As the transfer elements (the pipes and pumps) only specify the pressure difference, at least one reservoir is needed to specify a reference value for the pressure in the junctions. By construction, the matrix V_2 selects pumps lying on paths of pumps between reservoirs or cycles of pumps, i.e., structures on which the pressure difference is fixed, cp. Substructure 3.1. So instead of returning a pressure difference for given mass flow, pumps lying in $\mathrm{span}(V_2)$ must adjust their mass flow to a given pressure difference. Mathematically, this means that the corresponding pump function must be invertible, i.e., the matrix $V_2^T \, \mathrm{D}f_{\mathrm{Pu}} V_2$ must be pointwise nonsingular.

As the solvability conditions of Theorem 3.1 are formulated on the connection structure and the element functions, the plausibility of the network can be checked in a preprocessing step *before* the DAE is actually handed to a solver. If the solvability conditions are violated, the critical structures can be located in the network and advice can be given how to modify the model to obtain a physically reasonable system.

We can avoid the nonsingularity check of the matrix $V_2^T \, \mathrm{D}f_{\mathrm{Pu}} V_2$ by assuming that in every cycle of pumps and in every path of pumps between two reservoirs, there is at least one pipe.

Lemma 3.2 ([2]) *Let $\mathcal{N}$ be a network given by* (2.1) *that satisfies Assumptions 3.1. If on each path between two reservoirs and on each fundamental cycle there is at least one pipe, then* $\ker(A_{Jc,Pu}) = \{0\}$.

Lemma 3.2 gives a structural condition on the pumps in the network that is independent of the specific element functions. Stated as simple topological criteria, the assumption of Lemma 3.2 provides a very cheap and reliable preprocessing test for the solvability of the model under consideration.

So we can either impose a solvability condition on element level and check if $V_2^T \, \mathrm{D}f_{\mathrm{Pu}} V_2$ is pointwise nonsingular for a given pump specification or, to make sure that the model works for every kind of pump characteristics, impose the solvability condition on the structural level via the assumptions stated in Lemma 3.2.

The condition on element level, i.e., the non-singularity of $V_2^T \mathrm{D}f_{\mathrm{Pu}} V_2$, can be easily checked for not-to complicated pump constellations, allowing to use a broader class of pump functions. In some cases, the pump characteristic is a strictly monotone function and hence invertible.

The condition on structural level, i.e., the assumptions of Lemma 3.2, are useful for complex pump constellations and/or applications where the pump characteristics often change.

So depending on the topology of the network and the specific characteristic of the individual pumps, there are two options to ensure the global solvability.

Remark 3.1 Modeling single, smaller sized networks by hand, cycles of pumps or paths of pumps between reservoirs typically occur if serial or parallel pump constellations are considered. Furthermore, the characteristic pump equation (Pu_j) is also representative for the class of quasi-stationary pipes. Quasi-stationary pipes are used if the transient behavior is negligible and consequently (Pi_j) reduces to

$$c_{1,j} \Delta p_j = c_{2,j} |q_{\mathrm{Pi},j}| q_{\mathrm{Pi},j} + c_{3,j}.$$

Hence, considering networks consisting of transient pipes, quasi-stationary pipes, pumps, demand branches and reservoirs, the critical structures are paths of pumps and quasi-stationary pipes between reservoirs as well as cycles of pumps and quasi-stationary pipes. Indeed, this constellation occurs frequently in automatic modeling procedures.

3.3 Surrogate Model

Since the DAE (2.2) is of higher index, it is not suitable for a numerical simulation. Being assembled by simply gluing together the single elements, the DAE (2.2) contains hidden constraints, i.e., equations that every solution has to satisfy but which are not explicitly given in the representation (2.2). A simple example of such a hidden equation is given below. The hidden constraints might reduce the order of the method, might lead to drift of the numerical method and creates problems in the initialization, see e.g., [14, 18, 22]. Exploiting again the topology, we can locate these constraints in the network and assemble a surrogate model with better numerical performance.

Theorem 3.3 ([2]) *Let $\mathcal{N}$ be a network given by (2.1) that satisfies Assumptions 3.1 and let $F_{\mathcal{N}} \in C^2(\mathbb{D}, \mathbb{R}^n)$. Let $n_{Re} > 0$ and let $V_2^T \mathrm{D}f_{Pu} V_2$ be pointwise nonsingular for* $\mathrm{span}(V_2) = \ker(A_{Jc,Pu})$. *The* strangeness-free model *of $\mathcal{N}$ is* *given by*

$$\Pi_2^T \dot{q}_{Pi} = \Pi_2^T f_{Pi}(q_{Pi}, p_{Jc}, p_{Re}) \tag{3.3a}$$

$$0 = U_2^T A_{Jc,Pi} f_{Pi}(q_{Pi}, p_{Jc}, p_{Re}) + U_2^T A_{Jc,De} \dot{\bar{q}}_{De} \tag{3.3b}$$

$$0 = A_{Jc,Pu}^T p_{Jc} + A_{Re,Pu}^T p_{Re} - f_{Pu}(q_{Pu}) \tag{3.3c}$$

$$0 = A_{Jc,Pi} q_{Pi} + A_{Jc,Pu} q_{Pu} + A_{Jc,De} q_{De} \tag{3.3d}$$

$$q_{De} = \bar{q}_{De} \tag{3.3e}$$

$$p_{Re} = \bar{p}_{Re} \tag{3.3f}$$

where U_2 *is such that* $\operatorname{span}(U_2) = \operatorname{coker}(A_{Jc,Pu})$ *and* $[\Pi_1, \Pi_2]$ *is a permutation with* Π_1 *such that* $\operatorname{corange}(U_2^T A_{Jc,Pu}) = \operatorname{span}(\Pi_1)$.

1. *The* strangeness-free model *is regular and has strangeness index* $\mu = 0$ *(d-index 1).*
2. *A function* $(q, p) \in C^1(\mathcal{I}, \mathbb{R}^{n_\mathcal{E}} \times \mathbb{R}^{n_\mathcal{V}})$ *solves* (2.2) *if and only if it solves* (3.3).

The surrogate model (3.3) can be assembled based on network information only. The matrix U_2 selects the junctions of the connected components in $\mathcal{G}_{\mathrm{Jc,Pu}}$ and performs the vertex identification to construct the graph $\mathcal{G}_{\overline{\mathrm{Jc}},\mathrm{Pi}}$ of which Π_1 selects a spanning tree. Thus, the surrogate model (3.3) can be directly constructed from the network information, there is no need to compute (3.3) from (2.2) by symbolic or numerical manipulation, as it is necessary for example in a general modeling language like Modelica. In a simulation, this saves computational time as the system-to-solve (3.3) can be assembled directly from the network. Furthermore, the physical meaning of the equations and the states is preserved, i.e., in the DAE (3.3), each equation and each variable still has a physical counterpart. This is of special importance for the freely choosable initial conditions. Due to Theorem 3.3, the set of feasible initial conditions is determined by the chord set of $\mathcal{G}_{\overline{\mathrm{Jc}},\mathrm{Pi}}$. This means, that in a model assembled from a modular system simulation tool, only those elements are allowed to accept user defined initial conditions. The remaining ones are derived from the algebraic equations (3.3b)–(3.3f). At that point it is also clear that the set of feasible initial condition is not unique, since the choice of a spanning tree may not be unique. Thus, errors in the initialization or the simulation can be located in the network, allowing constructive error detection and handling.

Example We consider two pipes $\mathrm{Pi}_1, \mathrm{Pi}_2$ that are coupled by a junction Jc_1, cp. Fig. 7. For simplicity, we assume that the pipes are connected to two reservoirs Re_1 and Re_2. Then, we obtain the network DAE

$$\begin{aligned}
\dot{q}_{\mathrm{Pi},1} &= f_{\mathrm{Pi},1}(q_{\mathrm{Pi},1}, p_{\mathrm{Re},1} - p_{\mathrm{Jc},1}), \quad q_{\mathrm{Pi},1}(t_0) = q_{\mathrm{Pi},1,0}, \\
\dot{q}_{\mathrm{Pi},2} &= f_{\mathrm{Pi},2}(q_{\mathrm{Pi},2}, p_{\mathrm{Jc},1} - p_{\mathrm{Re},2}), \quad q_{\mathrm{Pi},2}(t_0) = q_{\mathrm{Pi},2,0}, \\
q_{\mathrm{Pi},1} &= q_{\mathrm{Pi},2}.
\end{aligned} \tag{3.4}$$

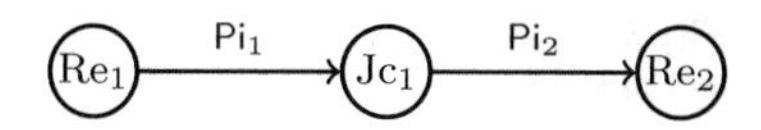

Fig. 7 Graph of the network associated with DAE (3.4)

The pipes specify the mass flows differentially, while the junction relates the flows algebraically. Consequently, only one mass flow evolves dynamically, the other one is fixed algebraically by the mass balance. In particular, only one initial value can be chosen. The pressure only occurs implicitly in the differential equations. Differentiating the algebraic equation and inserting the pipe equations for the derivatives of the mass flows, however, we discover the algebraic equation

$$f_{\mathrm{Pi},1}(q_{\mathrm{Pi},1}, \bar{p}_{\mathrm{Re},1} - p_{\mathrm{Jc},1}) = f_{\mathrm{Pi},2}(q_{\mathrm{Pi},2}, p_{\mathrm{Jc},1} - \bar{p}_{\mathrm{Re},2}). \tag{3.5}$$

As $\mathrm{D}_2(f_{\mathrm{Pi},2} - f_{\mathrm{Pi},1}) = c_{1,1} + c_{1,2}$ is nonsingular, (3.5) can be solved for the pressure $p_{\mathrm{Jc},1}$ and (3.4) is uniquely solvable. Hence, coupling two pipes by a junction, the network model (3.4) contains a hidden algebraic equation that is needed to specify the pressure in the coupling junction. Also, (3.4) does not correctly reflect the number of differential and algebraic variables as only one mass flow evolves dynamically. Thus, we consider the surrogate model

$$\begin{aligned}
\dot{q}_{\mathrm{Pi},1} &= f_{\mathrm{Pi},1}(q_{\mathrm{Pi},1}, p_{\mathrm{Re},1} - p_{\mathrm{Jc},1}), \qquad q_{\mathrm{Pi},1}(t_0) = q_{\mathrm{Pi},1,0}, \\
f_{\mathrm{Pi},1}(q_{\mathrm{Pi},1}, p_{\mathrm{Re},1} - p_{\mathrm{Jc},1}) &= f_{\mathrm{Pi},2}(q_{\mathrm{Pi},2}, p_{\mathrm{Jc},1} - p_{\mathrm{Re},2}), \\
q_{\mathrm{Pi},1} &= q_{\mathrm{Pi},2}.
\end{aligned}$$

which corresponds to the strangeness free representation of Eq. (3.3).

Example 2.1 (continued) We consider again the network $\mathcal{N}_{Ex1}$ as given in Fig. 1. The DAE modeling the dynamics of $\mathcal{N}_{Ex1}$ is given in Example 2.1. The matrices V_2^1, U_2^1, Π_2^1 selecting the relevant substructures are presented in Example 2.1. Assuming, that the pump functions $f_{\mathrm{Pu}_1^1}, \ldots, f_{\mathrm{Pu}_5^1}$ are invertible, the surrogate model of this DAE is given by

$$\dot{q}_{\mathrm{Pi}_1^1} = c_{11}(p_{\mathrm{Jc}_4^1} - p_{\mathrm{Jc}_3^1}) + c_{21}|q_{\mathrm{Pi}_1^1}|q_{\mathrm{Pi}_1^1} + c_{31}, \tag{Pi_1^1}$$

$$0 = \dot{q}_{\mathrm{De}_1^1} + f_{\mathrm{Pi}_1^1} - f_{\mathrm{Pi}_3^1} \tag{$\overline{\mathrm{Jc}_1^1}$}$$

$$0 = -f_{\mathrm{Pi}_1^1} + f_{\mathrm{Pi}_2^1} + f_{\mathrm{Pi}_4^1} \tag{$\overline{\mathrm{Jc}_4^1}$}$$

$$0 = -f_{\mathrm{Pi}_2^1} + f_{\mathrm{Pi}_3^1}, \tag{$\overline{\mathrm{Jc}_5^1}$}$$

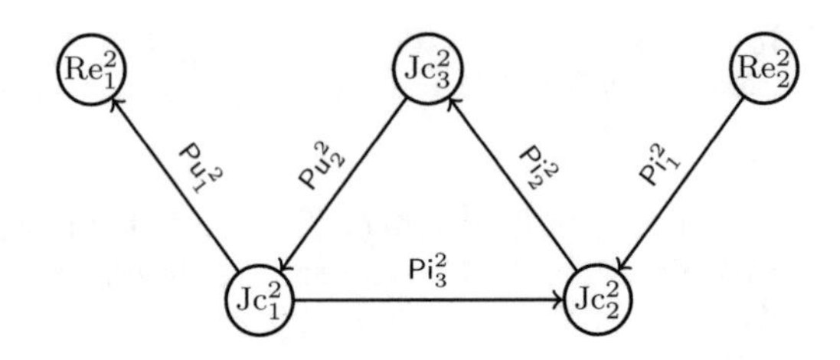

Fig. 8 Graph $\mathcal{G}_{Ex2}$ of the network $\mathcal{N}_{Ex2}$

as well as Eqs. (Pu$_1^1$)–(Pu$_5^1$), (Jc$_1^1$)–(Jc$_6^1$), (De$_1^1$), (Re$_1^1$), (Re$_2^1$). Hence, only the mass flow of Pi$_1^1$ is determined by a differential equation, the other mass flows are given by the equations (Jc$_1^1$)–(Jc$_6^1$) of the conservation of mass. Equations ($\overline{\text{Jc}}_1^1$), ($\overline{\text{Jc}}_4^1$), ($\overline{\text{Jc}}_5^1$) are the hidden constraints, which are needed to compute the pressures in these junctions.

Example 3.1 We consider the network $\mathcal{N}_{Ex2}$ as given in Fig. 8.

The incidence matrix of $\mathcal{N}_{Ex2}$ is given by

$$A_{\text{Jc,Pi}} = \begin{bmatrix} & Pi_1 & Pi_2 & Pi_3 \\ Jc_1 & 0 & 0 & 1 \\ Jc_2 & -1 & 1 & -1 \\ Jc_3 & 0 & -1 & 0 \end{bmatrix}, \qquad A_{\text{Jc,Pu}} = \begin{bmatrix} & Pu_1 & Pu_2 \\ Jc_1 & 1 & -1 \\ Jc_2 & 0 & 0 \\ Jc_3 & 0 & 1 \end{bmatrix},$$

$$A_{\text{Re,Pi}} = \begin{bmatrix} & Pi_1 & Pi_2 & Pi_3 \\ Re_1 & 0 & 0 & 0 \\ Re_2 & 1 & 0 & 0 \end{bmatrix}, \qquad A_{\text{Re,Pu}} = \begin{bmatrix} & Pu_1 & Pu_2 \\ Re_1 & -1 & 0 \\ Re_2 & 0 & 0 \end{bmatrix},$$

$$A_{\text{Jc,De}} = 0, \qquad A_{\text{Re,De}} = 0.$$

Specifying the pipe and pump functions f_{Pi}, f_{Pu} according to Remarks 2.1 and 2.2 as well as input pressures $\bar{p}_{\text{Re}_1^2}$, $\bar{p}_{\text{Re}_2^2}$, the DAE describing the dynamics of $\mathcal{N}_{Ex2}$ is given by

$$\dot{q}_{\text{Pi}_1^2} = c_{11}(p_{\text{Jc}_2^2} - p_{\text{Re}_2^2}) + c_{21}|q_{\text{Pi}_1^2}|q_{\text{Pi}_1^2} + c_{31} \tag{Pi$_1^2$}$$

$$\dot{q}_{\text{Pi}_2^2} = c_{12}(p_{\text{Jc}_3^2} - p_{\text{Jc}_2^2}) + c_{22}|q_{\text{Pi}_2^2}|q_{\text{Pi}_2^2} + c_{32} \tag{Pi$_2^2$}$$

$$\dot{q}_{\text{Pi}_3^2} = c_{13}(p_{\text{Jc}_2^2} - p_{\text{Jc}_1^2}) + c_{23}|q_{\text{Pi}_3^2}|q_{\text{Pi}_3^2} + c_{33} \tag{Pi$_3^2$}$$

$$p_{\text{Jc}_2^2} - p_{\text{Jc}_1^2} = f_{\text{Pu}_1^2}(q_{\text{Pu}_1^2}) \tag{Pu$_1^2$}$$

$$p_{\text{Jc}_3^2} - p_{\text{Jc}_2^2} = f_{\text{Pu}_2^2}(q_{\text{Pu}_2^2}) \tag{Pu$_2^2$}$$

$$0 = q_{\text{Pi}_3^2} + q_{\text{Pu}_1^2} - q_{\text{Pu}_2^2} \tag{Jc$_1^2$}$$

$$0 = -q_{\text{Pi}_1^2} + q_{\text{Pi}_2^2} - q_{\text{Pi}_3^2} \tag{Jc$_2^2$}$$

$$0 = -q_{\mathrm{Pi}_2^2} - q_{\mathrm{Pu}_2^2} \qquad (\mathrm{Jc}_3^2)$$

$$p_{\mathrm{Re}_1^2} = \bar{p}_{\mathrm{Re}_1^2} \qquad (\mathrm{Re}_1^2)$$

$$p_{\mathrm{Re}_2^2} = \bar{p}_{\mathrm{Re}_2^2}. \qquad (\mathrm{Re}_2^2)$$

The pumps in network $\mathcal{N}_{Ex2}$ neither form a cycle nor connect two reservoirs, meaning that $\mathcal{N}_{Ex2}$ does not contains elements belonging to Substructure 3.1. Hence, $\ker(A_{\mathrm{Jc,Pu}}) = \{0\}$ and the associated matrix V_2 is of rank 0.

The connected components in $\mathcal{G}_{Ex2;\mathrm{Jc,Pu}}$ are given by $\{\mathrm{Pu}_1^2, \mathrm{Pu}_2^2, \mathrm{Jc}_1^2, \mathrm{Jc}_2^2\}$ and Jc_2^2. Since Pu_1^2 is connected to a reservoir, the component $\{\mathrm{Pu}_1^2, \mathrm{Pu}_2^2, \mathrm{Jc}_1^2, \mathrm{Jc}_2^2\}$ does not belong to Substructure 3.2, whereas the isolated junction Jc_2^2 does, see Fig. 9. Hence, the associated matrix U_2 satisfying $\mathrm{span}(U_2) = \mathrm{coker}(A_{\mathrm{Jc,Pu}})$ is given by

$$U_2 = \begin{bmatrix} Jc_1 & 0 \\ Jc_2 & 1 \\ Jc_3 & 0 \end{bmatrix}.$$

The graph $\mathcal{G}_{Ex2,\overline{\mathrm{Jc}},\mathrm{Pi}}$ arising from the vertex identification is illustrated in Fig. 10. The associated connection matrix is given by

$$A_{\overline{\mathrm{Jc}},\mathrm{Pi}} = \begin{bmatrix} & Pi_1 & Pi_2 & Pi_3 \\ Jc_2 & -1 & 1 & -1 \end{bmatrix}.$$

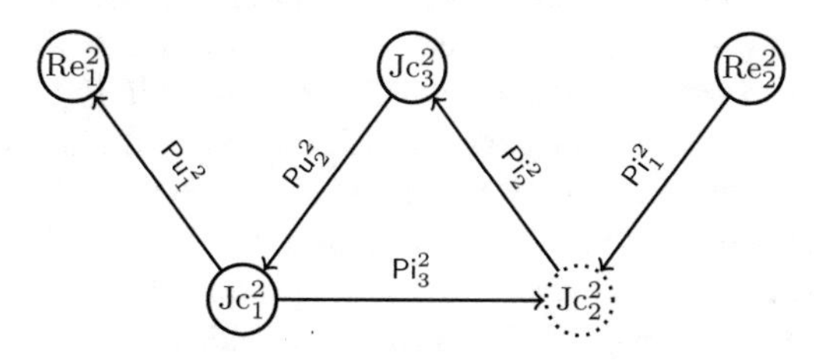

Fig. 9 Substructures 3.2 (dotted junctions) of the network $\mathcal{N}_{Ex2}$. The junction Jc_2^2 is isolated in $\mathcal{G}_{\mathrm{Jc}^2,\mathrm{Pu}^2}$ and thus belongs to Substructure 3.2

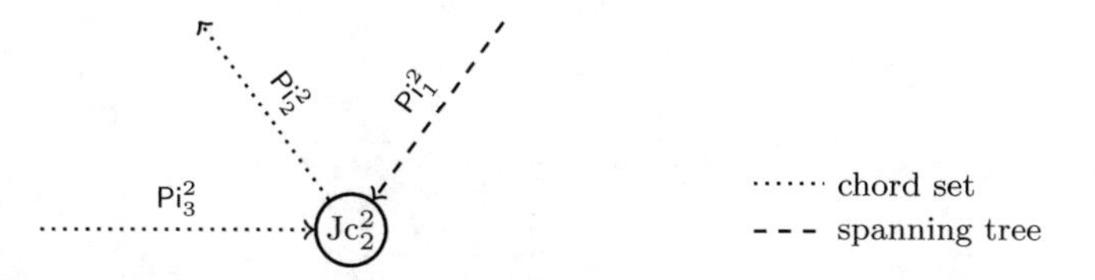

Fig. 10 Example of Substructure 3.3. For the graph $\mathcal{G}_{Ex2,\overline{\mathrm{Jc}},\mathrm{Pi}}$ of Fig. 9, a spanning tree is given by the pipes Pi_2^2, Pi_3^2 and Pi_4. The associated chord set is given by Pi_1^2

A choice of a spanning tree and chord set of $\mathcal{G}_{Ex2,\overline{\mathrm{Jc}},\mathrm{Pi}}$, i.e., of Substructure 3.3, is given in Fig. 10. The associated permutation Π is given by

$$\Pi_1 = \begin{bmatrix} Pi_1 & 1 \\ Pi_2 & 0 \\ Pi_3 & 0 \end{bmatrix}, \quad \Pi_2 = \begin{bmatrix} Pi_1 & 0 & 0 \\ Pi_2 & 1 & 0 \\ Pi_3 & 0 & 1 \end{bmatrix}.$$

As $n_{\mathrm{Re}} = 2$ and $\mathrm{rank}(V_2) = 0$, the DAE modeling the dynamics of $\mathcal{N}_{Ex2}$ is strangeness-free and uniquely solvable for every consistent initial value. With the matrices U_2, Π_2, we obtain the surrogate model

$$\dot{q}_{\mathrm{Pi}_2^2} = c_{12}(p_{\mathrm{Jc}_3^2} - p_{\mathrm{Jc}_2^2}) + c_{22}|q_{\mathrm{Pi}_2^2}|q_{\mathrm{Pi}_2^2} + c_{32} \tag{Pi_2^2}$$

$$\dot{q}_{\mathrm{Pi}_3^2} = c_{13}(p_{\mathrm{Jc}_2^2} - p_{\mathrm{Jc}_1^2}) + c_{23}|q_{\mathrm{Pi}_3^2}|q_{\mathrm{Pi}_3^2} + c_{33} \tag{Pi_3^2}$$

$$0 = -f_{\mathrm{Pi}_1^2} + f_{\mathrm{Pi}_2^2} - f_{\mathrm{Pi}_3^2} \tag{$\overline{\mathrm{Jc}_2^2}$}$$

as well as Eqs. (Pu_1^2), (Pu_2^2), (Jc_1^2)–(Jc_3^2) and (Re_1^2), (Re_2^2).

4 A Model for Coupled Flow Networks

In this section we consider multiple networks as defined in Sect. 2 and analyzed in Sect. 3 and couple them via defined coupling conditions. All individual networks are assumed to fulfill Assumption 3.1 and that Theorem 3.1 as well as Theorem 3.3 are applicable. An example of a coupled network is given in Fig. 11.

We start by presentation some examples of coupled liquid flow network in order to point out the difficulties, that arise when dealing with such kind of problems. In all the shown cases one of the assumptions imposed in Theorem 3.1 or Theorem 3.3 is not satisfied for the coupled system. Therein the coupling is represented based on the network structure, cf. Fig. 12. The boundary condition imposed on the state (■)

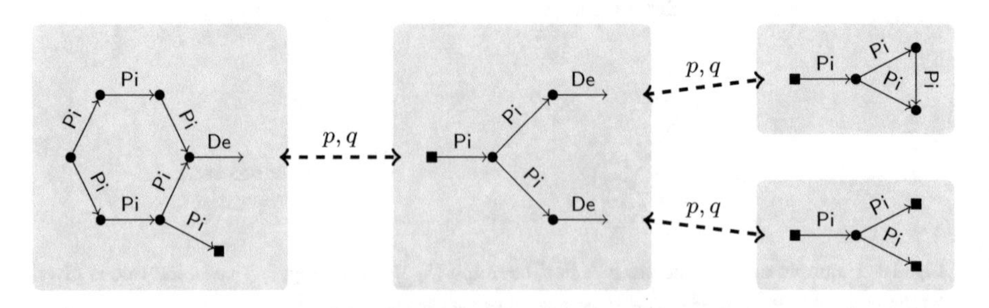

Fig. 11 Example of a coupled network consisting of four liquid flow networks

and the boundary condition imposed on the flow (De) are melt together to a junction (•) via a cycling coupling of the flow q and the state p.

In practical applications the coupling as defined in Fig. 12 is realized via directed information databusses, see Fig. 13. Eliminating the trivial relations leads to the equivalent representation of Fig. 12. Hence for the analysis, the representation of Fig. 12 is sufficient. At that point we also mention, that one important part of the coupling in Fig. 13 is the availability of the derivatives of the coupling variables p and q. This means, that not only p and q are communicated via databusses, but also their derivatives with respect to time $\dot{p}$ and $\dot{q}$. This requirement is automatically fulfilled via the representation in Fig. 12.

Example 4.1 (Missing Reference Pressure) Consider the network of coupled liquid flow networks as displayed in Fig. 14. Clearly, both subnetworks are unique solvable. But the coupled network is not uniquely solvable, since the reference pressure is lost through the coupling procedure.

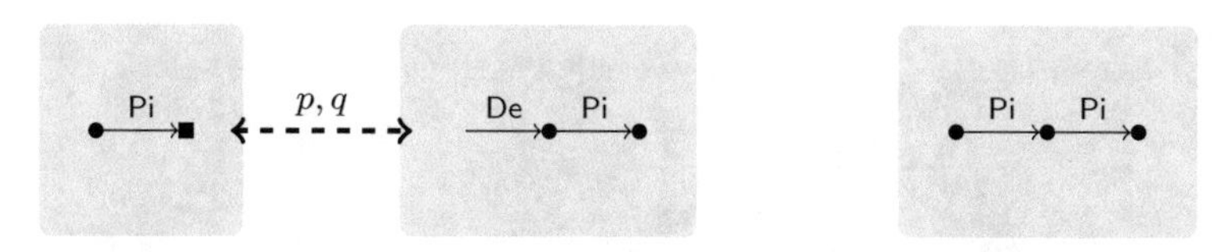

Fig. 12 Definition of the coupling of two networks (left) and the coupled equivalent network (right)

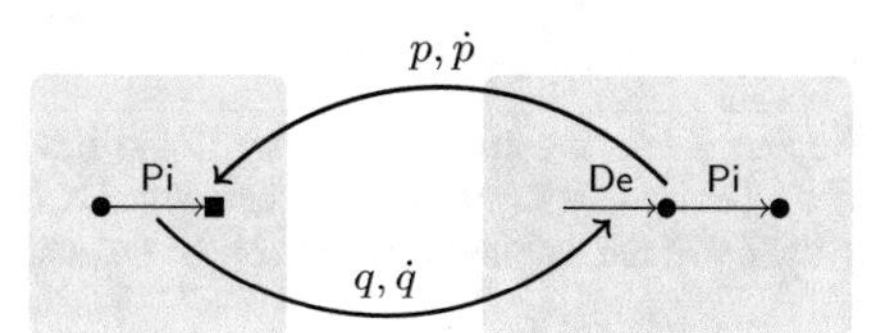

Fig. 13 Definition of the coupling of two networks via a directed databus connection

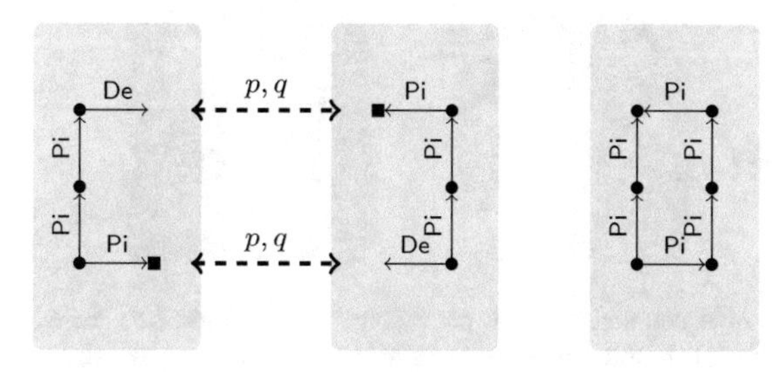

Fig. 14 Example of a coupled network consisting of two liquid flow networks (left) and the equivalent network (right). The coupled network is not uniquely solvable, since there remains no reservoir in the coupled network

Example 4.2 (Cycle of Pumps) Consider the network of coupled liquid flow networks as displayed in Fig. 15. In contrast to Example 4.1 we replace some pipes by pumps and add an additional reservoir in one of the subnetworks. Clearly, both subnetworks are uniquely solvable. But the coupled network may not be solvable at all, since a cycle consisting solely of pumps is obtained through the coupling procedure.

Example 4.3 (Spanning Tree) Consider the network of coupled liquid flow networks as displayed in Fig. 16. In contrast to Example 4.1 we add an additional reservoir in one of the subnetworks. Clearly, both subnetworks are uniquely solvable and also the coupled network is uniquely solvable. Determining the spanning trees of the subnetworks and the combined networks, we observe, that the spanning tree of the combined network does not form a proper spanning tree of the new network (since it is not a tree). It can easily be seen, that another choice of the spanning tree in the subnetworks leads to a valid combined result.

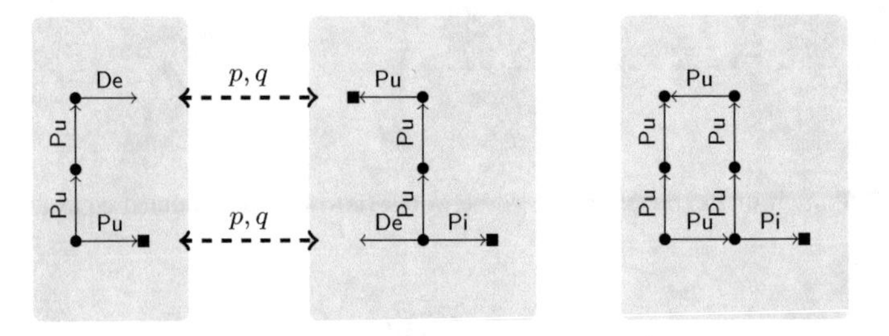

Fig. 15 Example of a coupled network consisting of two liquid flow networks (left) and the equivalent network (right). In contrast to Example 4.1, there remains a reservoir in the coupled network. Anyhow, the solvability of the coupled network cannot be guaranteed, since there arises a cycle of pumps

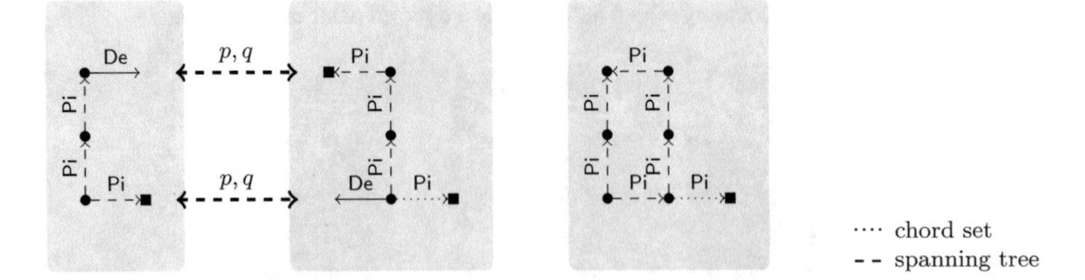

Fig. 16 Example of a coupled network consisting of two liquid flow networks (left) and the equivalent network (right). Both subnetworks as well as the coupled network are uniquely solvable. The surrogate model for the coupled network cannot be derived straight forward by combining the surrogate models of the subnetworks. Indeed, the combination of the spanning trees of the subnetworks does not form a proper spanning tree for the coupled network

In the next section, the coupling addressed in Fig. 12 is defined algebraically. Based on this definition an analysis is established, that gives answers to the issues raised in Examples 4.1–4.3.

5 Topology Based Index Analysis for Coupled Flow Networks

We consider a set of networks $\mathcal{N}^1, \ldots, \mathcal{N}^K$ with graphs $\mathcal{G}^1, \ldots, \mathcal{G}^K$ and network functions $F_{\mathcal{N}}^1, \ldots, F_{\mathcal{N}}^K$. For $k = 1, \ldots, K$, we assume that $\mathcal{N}^k$ satisfies the Assumptions 3.1 as well as the solvability assumptions of Theorem 3.1. Then, the DAE (2.2) modeling the dynamics of $\mathcal{N}^k$ is regular, has strangeness index $\mu = 1$ and is uniquely solvable for every consistent initial value.

These networks $\mathcal{N}^1, \ldots, \mathcal{N}^K$ are now connected to one large network $\mathcal{N}$ with graph $\mathcal{G}$ and network function $F_{\mathcal{N}}$. The coupling of the networks is performed via the boundary conditions, by the reservoirs and demands. Before we specify the coupling procedure, we point out the issues we are interested in.

1. Under which conditions does the coupled network $\mathcal{N}$ satisfy the Assumptions 3.1?
2. Can we assemble the network function $F_{\mathcal{N}}$ of the coupled network $\mathcal{N}$ from the individual network functions $F_{\mathcal{N}}^k$?
3. Under which conditions does the coupled network $\mathcal{N}$ satisfy the solvability assumptions of Theorem 3.1?
4. Can we construct the surrogate model (3.3) of the coupled system from the surrogate models of the subsystems?
5. How do we specify the consistent initial values of the coupled system from the consistent initial values of the subsystems?

For a network $\mathcal{N}^k$, we denote the boundary conditions that serve as coupling points by Re_c^k and De_c^k and summarize them in the sets $\mathcal{RE}_c^k$ and $\mathcal{DE}_c^k$, respectively. The boundary conditions that are not coupled are denoted by $\mathrm{Re}_{\not c}^k$ and $\mathrm{De}_{\not c}^k$ and summarized in the sets $\mathcal{RE}_{\not c}^k$ and $\mathcal{DE}_{\not c}^k$, respectively. Then, $\mathcal{RE}^k = \mathcal{RE}_c^k \cup \mathcal{RE}_{\not c}^k$ and $\mathcal{DE}^k = \mathcal{DE}_c^k \cup \mathcal{DE}_{\not c}^k$. We call the elements of $\mathcal{RE}_c^k$ and $\mathcal{DE}_c^k$ coupling reservoirs and coupling demands. Accordingly, we partition the junctions and edges incident to a coupling demand or reservoir by Jc_c^k and Pi_c^k, Pu_c^k and summarize them in the sets $\mathcal{JC}_c^k$ and $\mathcal{PI}_c^k$, $\mathcal{PU}_c^k$, respectively. The junctions and edges that are not incident to a coupling boundary condition are denoted by $\mathrm{Jc}_{\not c}^k$ and $\mathrm{Pi}_{\not c}^k$, $\mathrm{Pu}_{\not c}^k$ and summarized in the sets $\mathcal{JC}_{\not c}^k$ and $\mathcal{PI}_{\not c}^k$, $\mathcal{PU}_{\not c}^k$. Then, $\mathcal{JC}^k = \mathcal{JC}_c^k \cup \mathcal{JC}_{\not c}^k$ and $\mathcal{PI}^k = \mathcal{PI}_c^k \cup \mathcal{PI}_{\not c}^k$, $\mathcal{PU}^k = \mathcal{PU}_c^k \cup \mathcal{PU}_{\not c}^k$. We call the elements of $\mathcal{JC}_c^k$ and $\mathcal{PI}_c^k$, $\mathcal{PU}_c^k$ coupling junctions and coupling edges. In the following, we frequently summarize the set of pipes and pumps as $\mathcal{P} := \mathcal{PI} \cup \mathcal{PU}$ and denote its elements by P. The partitioning into coupling and non-coupling elements straightforward extends to $\mathcal{P}$ and its elements. For the coupling edges, we indicate the incident nodes where necessary by $\mathrm{P}(\mathrm{Jc}_i, \mathrm{Jc}_j)$ if P is an edge incident to Jc_i and Jc_j.

As the considered networks satisfy Assumption 3.1, every reservoir is incident to exactly one pipe or pump, and every junction is incident to at most one demand. This one-to-one correspondence allows to number the coupling elements such that the coupling reservoir $\mathrm{Re}^k_{c,l}$ is incident to the coupling edge $\mathrm{P}^k_{c,l}$ and the coupling demand $\mathrm{De}^k_{c,m}$ is incident to the coupling junction $\mathrm{Jc}^k_{c,m}$.

With this notation, we define the coupling procedure of two networks.

Definition 5.1 Consider two networks $\mathcal{N}^1$, $\mathcal{N}^2$ as in (2.1). Let $\mathrm{Re}^1_c \in \mathcal{RE}^1$ be a coupling reservoir with coupling edge $\mathrm{P}^1_c(\mathrm{Jc}^1, \mathrm{Re}^1_c) \in \mathcal{P}^1_c$ for $\mathrm{Jc}^1 \in \mathcal{JC}^1$. Let $\mathrm{De}^2_c \in \mathcal{DE}^2_c$ be a coupling demand with coupling junction $\mathrm{Jc}^2_c \in \mathcal{JC}^2$. The coupling of $\mathcal{N}^1, \mathcal{N}^2$ via $(\mathrm{Re}^1_c, \mathrm{De}^2_c)$ yields the network

$$\mathcal{N} = \left(\mathcal{N}^1 \setminus \{\mathrm{Re}^1_c, \mathrm{P}^1_c(\mathrm{Jc}^1, \mathrm{Re}^1_c)\}\right) \cup \left\{\mathrm{P}^1_c(\mathrm{Jc}^1, \mathrm{Jc}^2_c)\right\} \cup \left(\mathcal{N}^2 \setminus \{\mathrm{De}^2_c\}\right).$$

Hence, the pair $(\mathrm{Re}^1_c, \mathrm{De}^2_c)$ thus means that the coupling boundary conditions $\mathrm{Re}^1_c, \mathrm{De}^2_c$ are removed, while the coupling edge P^1_c is connected to the coupling junction Jc^2_c. An example of the coupling procedure is given in Fig. 17.

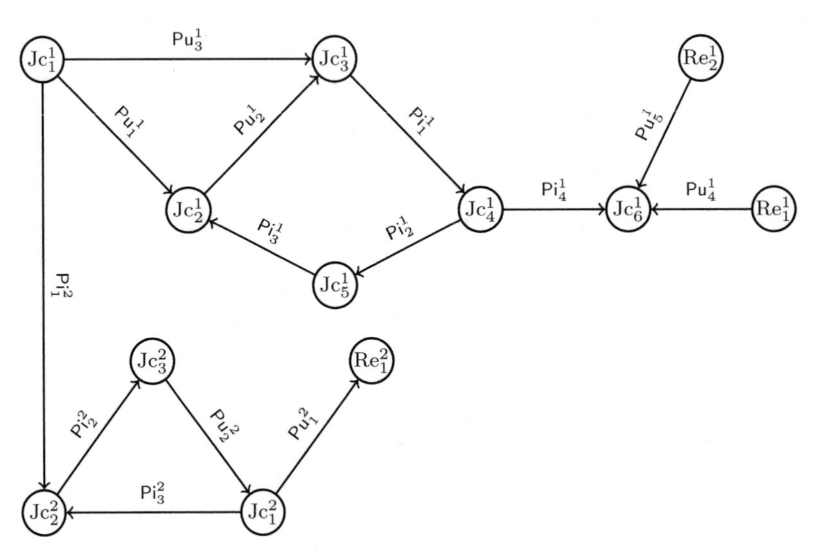

Fig. 17 Example of a coupled network. Coupling $\mathcal{N}_{Ex1}$, $\mathcal{N}_{Ex2}$ from Examples 2.1, 3.1 via the boundary conditions De^1_1, Re^2_2, we obtain the network $\mathcal{N}_{Ex3}$ with graph $\mathcal{G}_{Ex3}$

The incidence matrix $A_{\mathcal{N}}$ of the coupled network $\mathcal{N}$ reflects this coupling procedure as follows. With $A_{\mathcal{N}}$ given by

$$A_{\mathcal{N}} = \begin{bmatrix} A_{\mathrm{Jc}^1,\mathrm{P}^1_e} & A_{\mathrm{Jc}^1,\mathrm{P}^1_c} & 0 & 0 & A_{\mathrm{Jc}^1,\mathrm{De}^1_e} & 0 \\ 0 & 0 & A_{\mathrm{Jc}^2_e,\mathrm{P}^2} & 0 & 0 & A_{\mathrm{Jc}^2_e,\mathrm{De}^2_c} \\ 0 & A_{\mathrm{Re}^1_c,\mathrm{P}^1_c} & 0 & A_{\mathrm{Jc}^2_c,\mathrm{P}^2} & 0 & 0 \\ A_{\mathrm{Re}^1_e,\mathrm{P}^1_e} & 0 & 0 & 0 & 0 & 0 \\ 0 & 0 & 0 & A_{\mathrm{Re}^2_e,\mathrm{P}^2} & 0 & 0 \end{bmatrix}, \tag{5.1}$$

we see that the coupling boundary conditions $\mathrm{Re}^1_c, \mathrm{De}^2_c$ are removed, while the connection information of the coupling reservoir, i.e., the block $A_{\mathrm{Re}^1_c,\mathrm{P}^1_c}$, moves to the row of the coupling junction Jc^2_c. If $\mathrm{sgn}(\mathrm{P}^1_c) = \mathrm{sgn}(\mathrm{De}^2_c)$, then $A_{\mathrm{Re}^1_c,\mathrm{P}^1_c} = A_{\mathrm{Jc}^2_c,\mathrm{De}^2_c}$ and we can equivalently move the connection information of the coupling demand, i.e., the block $A_{\mathrm{Jc}^2_c,\mathrm{De}^2_c}$, to the column of the coupling edge P^1_c.

Considering several networks $\mathcal{N}^1, \ldots, \mathcal{N}^K$, the coupling procedure of Definition 5.1 is successively applied to couple $\mathcal{N}^1, \ldots, \mathcal{N}^K$ into a single network. The information how the subnetworks are connected is stored in the *adjacency matrix* $B \in \mathbb{R}^{K \times K}$ defined by

$$B_{kl} = \begin{cases} 1, & k = l, \\ 1, & k \neq l \text{ and } \mathcal{N}^k, \mathcal{N}^l \text{ are connected according to the coupling} \\ & \text{procedure of Definition 5.1 via the coupling pair } (\mathrm{Re}^k_c, \mathrm{De}^l_c), \\ 0, & \text{else.} \end{cases}$$

The graph $\mathcal{G}_{\mathrm{coup}}$ associated with B is called the *coupling graph*.

In the following, we assume that two networks $\mathcal{N}^k$, $\mathcal{N}^l$ are coupled at most by one pair of boundary conditions. Coupling two networks via several boundary conditions corresponds to coupling a network with itself, which would result in a change of its internal structure. Hence, in the following, we assume that the coupling graph $\mathcal{G}_{\mathrm{coup}}$ is simple. Coupling two networks $\mathcal{N}^k, \mathcal{N}^l$ by a coupling reservoir from $\mathcal{N}^k$ and a demand from $\mathcal{N}^l$, we can thus number the elements such that the coupling is performed by the pair $(\mathrm{Re}^k_c, \mathrm{De}^l_c)$.

With the adjacency matrix B, we specify the structure of the coupled network $\mathcal{N}$.

Lemma 5.1 *Consider networks $\mathcal{N}^1, \ldots, \mathcal{N}^K$ as in* (2.1). *Let $B \in \mathbb{R}^{K \times K}$ be the adjacency matrix of a simple coupling graph $\mathcal{G}_{coup}$. Coupling of $\mathcal{N}^1, \ldots, \mathcal{N}^K$ according to the adjacency matrix B yields the network $\mathcal{N}$*

$$\mathcal{N} = \bigcup_{\substack{k,l \in \{1,\ldots,K\} \\ s.t.\ B_{kl}=1}} \left(\mathcal{N}^k \setminus \{Re^k_c, P^k_c(Jc^k, Re^k_c)\}\right) \cup \left\{P^k_c(Jc^k, Jc^l_c)\right\} \cup \left(\mathcal{N}^l \setminus \{De^l_c\}\right).$$

The incidence matrix of $\mathcal{N}$ is given by

$$A_{\mathcal{N}} = \begin{bmatrix} A_{Jc,P} & A_{Jc,De} \\ A_{Re,P} & 0 \end{bmatrix},$$

with

$$A_{Jc,De} = \begin{bmatrix} A_{Jc^1,De_e^1} & & \\ & \ddots & \\ & & A_{Jc^K,De_e^K} \end{bmatrix}, \quad A_{Re,P} = \begin{bmatrix} A_{Re_c^1,P^1} & & \\ & \ddots & \\ & & A_{Re_c^K,P^K} \end{bmatrix},$$

$$A_{Jc,P} = \begin{bmatrix} A_{Jc^1,P^1} & & A_{coup,kl} \\ & \ddots & \\ A_{coup,kl} & & A_{Jc^K,P^K} \end{bmatrix},$$

with

$$A_{Jc,P} = \begin{bmatrix} A_{Jc_e^k,P_e^k} & A_{Jc_e^k,P_c^k} \\ A_{Jc_c^k,P_e^k} & A_{Jc_c^k,P_c^k} \end{bmatrix}, \quad A_{coup,kl} = \begin{cases} \begin{bmatrix} 0 & 0 \\ 0 & A_{Re_{c,l}^k,P_{c,l}^k} \end{bmatrix}, & \text{if } B_{kl} = 1, \\ 0, & \text{if } B_{kl} = 0. \end{cases}$$

Proof The assertion follows from Definition 5.1 and the structure of the incidence matrix (5.1). □

If the coupling graph $\mathcal{G}_{\text{coup}}$ is simple, then $\mathcal{N}$ satisfies Assumption 3.1 if the subnetworks $\mathcal{N}^1, \ldots, \mathcal{N}^K$ do.

Lemma 5.2 *Consider networks $\mathcal{N}^1, \ldots, \mathcal{N}^K$ as in* (2.1). *Let $B \in \mathbb{R}^{K \times K}$ be the adjacency matrix of a simple coupling graph $\mathcal{G}_{coup}$ and let $\mathcal{N}$ be the coupling of $\mathcal{N}^1, \ldots, \mathcal{N}^K$ according to B. If $\mathcal{N}^1, \ldots, \mathcal{N}^K$ satisfy Assumption 3.1, then the coupled network $\mathcal{N}$ satisfies Assumption 3.1.*

Proof As the coupling graph $\mathcal{G}_{\text{coup}}$ is simple, two networks $\mathcal{N}^k, \mathcal{N}^l$ are connected at most by one coupling edge. By assumption, $\mathcal{N}^1, \ldots, \mathcal{N}^K$ are simple and it follows that also $\mathcal{N}$ satisfies Assumption 3.1 (N1). As the coupling edges keep their orientation and the networks $\mathcal{N}^1, \ldots, \mathcal{N}^K$ are oriented, also $\mathcal{N}$ is oriented, hence (N2) is satisfied. As $\mathcal{N}^1, \ldots, \mathcal{N}^K$ are connected, connecting them to a new graph $\mathcal{N}$, also $\mathcal{N}$ is connected, hence (N3) is satisfied. The coupling procedure of Definition 5.1 only removes boundary conditions, it does not add any reservoirs or demands. As $\mathcal{N}^1, \ldots, \mathcal{N}^K$ satisfy (N4), also $\mathcal{N}$ satisfies (N4). □

Example 5.1 To illustrate the coupling procedure in Definition 5.1, we consider the networks $\mathcal{N}_{Ex1}$ and $\mathcal{N}_{Ex2}$ from Examples 2.1 and 3.1. Coupling these networks via the boundary conditions De_1^1, Re_2^2, we obtain the network $\mathcal{N}_{Ex3}$ with graph $\mathcal{G}_{Ex3}$, see Fig. 17. The incidence matrix of the coupled network $\mathcal{N}_{Ex3}$ is given by

$$A_{\mathrm{Jc,Pi}} = \left[\begin{array}{cccccccc} & Pi_1^1 & Pi_2^1 & Pi_3^1 & Pi_4^1 & Pi_1^2 & Pi_2^2 & Pi_3^2 \\ Jc_1^1 & 0 & 0 & 0 & 0 & 1 & 0 & 0 \\ Jc_2^1 & 0 & 0 & -1 & 0 & 0 & 0 & 0 \\ Jc_3^1 & 1 & 0 & 0 & 0 & 0 & 0 & 0 \\ Jc_4^1 & -1 & 1 & 0 & 1 & 0 & 0 & 0 \\ Jc_5^1 & 0 & -1 & 1 & 0 & 0 & 0 & 0 \\ Jc_6^1 & 0 & 0 & 0 & -1 & 0 & 0 & 0 \\ Jc_1^2 & 0 & 0 & 0 & 0 & 0 & 0 & 1 \\ Jc_2^2 & 0 & 0 & 0 & 0 & -1 & 1 & -1 \\ Jc_3^2 & 0 & 0 & 0 & 0 & 0 & -1 & 0 \end{array}\right],$$

$$A_{\mathrm{Jc,Pu}} = \left[\begin{array}{cccccccc} & Pu_1^1 & Pu_2^1 & Pu_3^1 & Pu_4^1 & Pu_5^1 & Pu_1^2 & Pu_2^2 \\ Jc_1^1 & 1 & 0 & 1 & 0 & 0 & 0 & 0 \\ Jc_2^1 & -1 & 1 & 0 & 0 & 0 & 0 & 0 \\ Jc_3^1 & 0 & -1 & -1 & 0 & 0 & 0 & 0 \\ Jc_4^1 & 0 & 0 & 0 & 0 & 0 & 0 & 0 \\ Jc_5^1 & 0 & 0 & 0 & 0 & 0 & 0 & 0 \\ Jc_6^1 & 0 & 0 & 0 & -1 & -1 & 0 & 0 \\ Jc_1^2 & 0 & 0 & 0 & 0 & 0 & 1 & -1 \\ Jc_2^2 & 0 & 0 & 0 & 0 & 0 & 0 & 0 \\ Jc_3^2 & 0 & 0 & 0 & 0 & 0 & 0 & 1 \end{array}\right],$$

$$A_{\mathrm{Re,Pi}} = \left[\begin{array}{cccccccc} & Pi_1^1 & Pi_2^1 & Pi_3^1 & Pi_4^1 & Pi_1^2 & Pi_2^2 & Pi_3^2 \\ Re_1^1 & 0 & 0 & 0 & 0 & 0 & 0 & 0 \\ Re_2^1 & 0 & 0 & 0 & 0 & 0 & 0 & 0 \\ Re_1^2 & 0 & 0 & 0 & 0 & 0 & 0 & 0 \end{array}\right],$$

$$A_{\mathrm{Re,Pu}} = \left[\begin{array}{cccccccc} & Pu_1^1 & Pu_2^1 & Pu_3^1 & Pu_4^1 & Pu_5^1 & Pu_1^2 & Pu_2^2 \\ Re_1^1 & 0 & 0 & 0 & 1 & 0 & 0 & 0 \\ Re_2^1 & 0 & 0 & 0 & 0 & 1 & 0 & 0 \\ Re_1^2 & 0 & 0 & 0 & 0 & 0 & -1 & 0 \end{array}\right].$$

To assemble the network function of the coupled network $\mathcal{N}$, we partition the network function $F_{\mathcal{N}^k}$ of the individual networks $\mathcal{N}^k$ according to the coupling elements, i.e.,

$$F_{\mathcal{N}^k,1_e}(q_{\mathrm{Pi}_e^k}, p_{\mathrm{Jc}^k}, p_{\mathrm{Re}_e^k}) = \dot{q}_{\mathrm{Pi}_e^k} - f_{\mathrm{Pi}_e^k}(q_{\mathrm{Pi}_e^k}, p_{\mathrm{Jc}^k}, p_{\mathrm{Re}_e^k}) \tag{5.2a}$$

$$F_{\mathcal{N}^k,1_c}(q_{\mathrm{Pi}_c^k}, p_{\mathrm{Jc}^k}, p_{\mathrm{Re}_c^k}) = \dot{q}_{\mathrm{Pi}_c^k} - f_{\mathrm{Pi}_c^k}(q_{\mathrm{Pi}_c^k}, p_{\mathrm{Jc}^k}, p_{\mathrm{Re}_c^k}) \tag{5.2b}$$

$$F_{\mathcal{N}^k,2_e}(q_{\mathrm{Pu}_e^k}, p_{\mathrm{Jc}^k}, p_{\mathrm{Re}_e^k}) = A^T_{\mathrm{Jc}^k,\mathrm{Pu}_e^k} p_{\mathrm{Jc}^k} + A^T_{\mathrm{Re}_e^k,\mathrm{Pu}_e} p_{\mathrm{Re}_e^k} - f_{\mathrm{Pu}_e^k}(q_{\mathrm{Pu}_e^k}) \tag{5.2c}$$

$$F_{\mathcal{N}^k,2_c}(q_{\mathrm{Pu}_c^k}, p_{\mathrm{Jc}^k}, p_{\mathrm{Re}_c^k}) = A^T_{\mathrm{Jc}^k,\mathrm{Pu}_c^k} p_{\mathrm{Jc}^k} + A^T_{\mathrm{Re}_c^k,\mathrm{Pu}_c} p_{\mathrm{Re}_c^k} - f_{\mathrm{Pu}_c^k}(q_{\mathrm{Pu}_c^k}) \tag{5.2d}$$

$$F_{\mathcal{N}^k,3_e}(q_{\mathrm{Pi}^k}, q_{\mathrm{Pu}^k}, q_{\mathrm{De}_e^k}) = A_{\mathrm{Jc}_e^k,\mathrm{Pi}^k} q_{\mathrm{Pi}^k} + A_{\mathrm{Jc}_e^k,\mathrm{Pu}^k} q_{\mathrm{Pu}^k} + A_{\mathrm{Jc}_e^k,\mathrm{De}^k} q_{\mathrm{De}_e^k} \tag{5.2e}$$

$$F_{\mathcal{N}^k,3_c}(q_{\mathrm{Pi}^k}, q_{\mathrm{Pu}^k}, q_{\mathrm{De}_c^k}) = A_{\mathrm{Jc}_c^k,\mathrm{Pi}^k} q_{\mathrm{Pi}^k} + A_{\mathrm{Jc}_c^k,\mathrm{Pu}^k} q_{\mathrm{Pu}^k} + A_{\mathrm{Jc}_c^k,\mathrm{De}_c^k} q_{\mathrm{De}_e^k} \tag{5.2f}$$

$$F_{\mathcal{N}^k,4_e}(q_{\mathrm{De}_e^k}) = q_{\mathrm{De}_e^k} - \bar{q}_{\mathrm{De}_e^k} \tag{5.2g}$$

$$F_{\mathcal{N}^k,4_c}(q_{\mathrm{De}_c^k}) = q_{\mathrm{De}_c^k} - \bar{q}_{\mathrm{De}_c^k} \tag{5.2h}$$

$$F_{\mathcal{N}^k,5_e}(p_{\mathrm{Re}_e^k}) = p_{\mathrm{Re}_e^k} - \bar{p}_{\mathrm{Re}_e^k} \tag{5.2i}$$

$$F_{\mathcal{N}^k,5_c}(p_{\mathrm{Re}_c^k}) = p_{\mathrm{Re}_c^k} - \bar{p}_{\mathrm{Re}_c^k}, \tag{5.2j}$$

where $f_{\mathrm{Pi}_e^k}(q_{\mathrm{Pi}_e^k}, p_{\mathrm{Jc}^k}, p_{\mathrm{Re}_e^k}) = [f_{\mathrm{Pi}_{e,l}^k}(q_{\mathrm{Pi}_{e,l}^k}, p_{\mathrm{Jc}^k}, p_{\mathrm{Re}_e^k})]_{l=1,\ldots,|\mathcal{PI}_e|}$ and $f_{\mathrm{Pi}_c^k}(q_{\mathrm{Pi}_c^k}, p_{\mathrm{Jc}^k}, p_{\mathrm{Re}_c^k}) = [f_{\mathrm{Pi}_{c,l}^k}(q_{\mathrm{Pi}_{c,j}^k}, p_{\mathrm{Jc}^k}, p_{\mathrm{Re}_c^k})]_{l=1,\ldots,|\mathcal{PI}_e|}$ as well as $f_{\mathrm{Pu}_e^k}(q_{\mathrm{Pu}_e^k}) = [f_{\mathrm{Pu}_{e,l}^k}(q_{\mathrm{Pu}_{e,l}^k})]_{l=1,\ldots,|\mathcal{PU}_e|}]$ and $f_{\mathrm{Pu}_c^k}(q_{\mathrm{Pu}_c^k}) = [f_{\mathrm{Pu}_{c,l}^k}(q_{\mathrm{Pu}_{c,l}^k})]_{l=1,\ldots,|\mathcal{PU}_c|}]$. Furthermore, we set $q_{\mathrm{Pi}} = [q_{\mathrm{Pi}^k}]_{k=1,\ldots,K}$, $q_{\mathrm{Pu}} = [q_{\mathrm{Pu}^k}]_{k=1,\ldots,K}$, $q_{\mathrm{De}} = [q_{\mathrm{De}^k}]_{k=1,\ldots,K}$ and $p_{\mathrm{Jc}} = [p_{\mathrm{Jc}^k}]_{k=1,\ldots,K}$, $p_{\mathrm{Re}} = [p_{\mathrm{Re}^k}]_{k=1,\ldots,K}$.

Then, the DAE of the coupled system is given as follows.

Lemma 5.3 *Consider networks $\mathcal{N}^1, \ldots, \mathcal{N}^K$ as in* (2.1) *that satisfy Assumption 3.1. Let $B \in \mathbb{R}^{K\times K}$ be the adjacency matrix of a simple coupling graph $\mathcal{G}_{coup}$ and let $\mathcal{N}$ be the coupling of $\mathcal{N}^1, \ldots, \mathcal{N}^K$ according to B. The network function $F_\mathcal{N}$ of $\mathcal{N}$ is given by*

$$F_{\mathcal{N},1}(q_{Pi}, p_{Jc}, p_{Re_e}) = \begin{bmatrix} F_{\mathcal{N}^k,1_e}(q_{Pi_e^k}, p_{Jc^k}, p_{Re_e^k}) \\ F_{\mathcal{N}^k,1_c}(q_{Pi_c^k}, p_{Jc^k}, p_{Jc_c^l}) \end{bmatrix}_{k=1,\ldots,K,\, l\neq k\in\{1,\ldots,K\}} \tag{5.3a}$$

$$F_{\mathcal{N},2}(q_{Pu_e}, p_{Jc}, p_{Re_e}) = \begin{bmatrix} F_{\mathcal{N}^k,2_e}(q_{Pu_e^k}, p_{Jc^k}, p_{Re_e^k}) \\ F_{\mathcal{N}^k,2_c}(q_{Pu_c^k}, p_{Jc^k}, p_{Jc_c^l}) \end{bmatrix}_{k=1,\ldots,K,\, l\neq k\in\{1,\ldots,K\}} \tag{5.3b}$$

$$F_{\mathcal{N},3}(q_{Pi}, q_{Pu}, q_{De_e}) = \begin{bmatrix} F_{\mathcal{N}^k,3_e}(q_{Pi^k}, q_{Pu^k}, q_{De_e^k}) \\ F_{\mathcal{N}^k,3_c}(q_{Pi^k}, q_{Pu^k}, q_{P_c^l}) \end{bmatrix}_{k=1,\ldots,K,\, l\neq k\in\{1,\ldots,K\}} \tag{5.3c}$$

$$F_{\mathcal{N},4}(q_{De_e}) = \begin{bmatrix} F_{\mathcal{N}^k,4_e}(q_{De_e^k}) \end{bmatrix}_{k=1,\ldots,K} \tag{5.3d}$$

$$F_{\mathcal{N},5}(p_{Re_e}) = \begin{bmatrix} F_{\mathcal{N}^k,5_e}(p_{Re_e^k}) \end{bmatrix}_{k=1,\ldots,K}, \tag{5.3e}$$

If $F_{\mathcal{N}^k} \in C^2(\mathbb{D}^k, \mathbb{R}^n)$ for $k = 1, \ldots, K$, then $F_\mathcal{N} \in C^2(\mathbb{D}, \mathbb{R}^n)$.

Proof The coupling procedure of Definition 5.1 does not change the internal structure of the networks $\mathcal{N}^1, \ldots, \mathcal{N}^K$, so the solution q, p of the coupled network

$\mathcal{N}$ must naturally satisfy the DAEs

$$F_{\mathcal{N}^k}(q^k, p^k) = 0, \quad k = 1, \dots, K \tag{5.4}$$

with $F_{\mathcal{N}^k}$ given by (5.2). By the coupling procedure, we obtain the additional conditions

$$q_{\mathrm{De}^k_{c,l}} = q_{\mathrm{P}^l_{c,k}}, \qquad p_{\mathrm{Re}^k_{c,m}} = p_{\mathrm{Jc}^m_{c,k}} \tag{5.5}$$

for $l = 1, \dots, |\mathcal{DE}_c|$, $m = 1, \dots, |\mathcal{RE}_c|$ and $k = 1, \dots, K$. With (5.5), we can eliminate the coupling boundary conditions in (5.4) to obtain (5.3). The smoothness of $F_{\mathcal{N}}$ then follows directly from the smoothness of the subnetwork functions. Alternatively, we can construct (5.3) in the same manner as (3.2) using the incidence matrix $A_{\mathcal{N}}$. □

Hence, the dynamics of the coupled network $\mathcal{N}$ are described by the DAE

$$F_{\mathcal{N}}(q_{\mathrm{Pi}}, q_{\mathrm{Pu}}, p_{\mathrm{Jc}}, p_{\mathrm{Re}_e}, q_{\mathrm{De}_e}) = 0 \tag{5.6}$$

with $F_{\mathcal{N}}$ given by (5.3). The set of initial values is defined as

$$\mathcal{C}_{IV} := \{(t_0, q_0, p_0) \in \mathcal{I} \times \mathbb{R}^{n_\mathcal{E}} \times \mathbb{R}^{n_\mathcal{V}} \mid \exists \dot{q}_0, \dot{p}_0 \colon F_{\mathcal{N}}(t_0, q_0, p_0, \dot{q}_0, \dot{p}_0) = 0\}.$$

Example 5.2 For the network $\mathcal{N}_{Ex3}$ as given in Fig. 17, the DAE modeling the dynamics of $\mathcal{N}$ is given by

$$0 = q_{\mathrm{Pi}^2_1} + q_{\mathrm{Pu}^1_1} + q_{\mathrm{Pu}^1_3} \tag{Jc^3_1}$$

$$\dot{q}_{\mathrm{Pi}^2_1} = c_{11}(p_{\mathrm{Jc}^2_2} - p_{\mathrm{Jc}^1_1}) + c_{21}|q_{\mathrm{Pi}^2_1}|q_{\mathrm{Pi}^2_1} + c_{31} \tag{Pi^3_1}$$

as well as Eqs. (Pi^1_1)-(Pi^1_4), (Pi^1_2), (Pi^1_3), (Pu^1_1)-(Pu^1_5), (Pu^2_1),(Pu^2_2), (Jc^1_1)-(Jc^1_6), (Jc^2_1)-(Jc^2_3), (Re^1_1), (Re^1_2), (Re^2_1).

We characterize the solvability of the DAE (5.6).

Theorem 5.4 *Consider networks $\mathcal{N}^1, \dots, \mathcal{N}^K$ as in* (2.1). *Let $B \in \mathbb{R}^{K \times K}$ be the adjacency matrix of a simple coupling graph $\mathcal{G}_{coup}$ and let $\mathcal{N}$ be the coupling of $\mathcal{N}^1, \dots, \mathcal{N}^K$ according to B. Let $n_{Re^k_e} > 0$ for at least one $k \in \{1, \dots, K\}$ and let $V_2^T \,\mathrm{D}f_{Pu} V_2$ be pointwise nonsingular for* $\mathrm{span}(V_2) = \ker(A_{Jc,Pu})$. *The following assertions hold.*

1. *The DAE* (5.6) *is regular and has strangeness index $\mu = 1$ (d-index 2).*
2. *The DAE* (5.6) *is uniquely solvable for every $(t_0, q_0, p_0) \in \mathcal{C}_{IV}$ and the solution is $(q, p) \in C^1(\mathcal{I}, \mathbb{R}^n)$.*

Proof Considering the reservoir part (5.3e) of the coupled network function, we find that $n_{\text{Re}} = \sum_{k=1}^{K} n_{\text{Re}_e^k}$. Hence, if $n_{\text{Re}_e^k} > 0$ for at least one $k \in \{1, \ldots, K\}$, then $n_{\text{Re}} > 0$. If, in addition, $V_2^T \, \mathrm{D}f_{\text{Pu}} V_2$ is pointwise nonsingular for $\operatorname{span}(V_2) = \ker(A_{\text{Jc,Pu}})$, then the network function (5.3) satisfies the assumptions of Theorem 3.1 and the assertions (1) and (2) follow. As the DAE is strangeness-free, the set of initial values corresponds to the consistent initial values. □

Hence, coupling the networks $\mathcal{N}^1, \ldots, \mathcal{N}^K$ with a simple coupling graph, the solvability and index result of Theorem 3.1 straightforward extend to the coupled network $\mathcal{N}$.

The structural part of the solvability condition, i.e., $n_{\text{Re}} > 0$, can be easily verified. If there is at least one non-coupling reservoir in one of the subnetworks, there is at least one reservoir in the coupled network. Under certain conditions, the element part of the solvability condition, i.e., that $V_2^T \, \mathrm{D}f_{\text{Pu}} V_2$ is pointwise nonsingular for $\operatorname{span}(V_2) = \ker(A_{\text{Jc,Pu}})$, can be also deduced from the subnetworks, cp. Lemma 5.6.

Regarding the simulation of the coupled network $\mathcal{N}$, we consider the strangeness-free surrogate model of $\mathcal{N}$.

Theorem 5.5 *Consider networks $\mathcal{N}^1, \ldots, \mathcal{N}^K$ as in (2.1). Let $B \in \mathbb{R}^{K \times K}$ be the adjacency matrix of a simple coupling graph $\mathcal{G}_{coup}$ and let $\mathcal{N}$ be the coupling of $\mathcal{N}^1, \ldots, \mathcal{N}^K$ according to B. Let $n_{\text{Re}_e^k} > 0$ for at least one $k \in \{1, \ldots, K\}$ and let $V_2^T \, \mathrm{D}f_{Pu} V_2$ be pointwise nonsingular for* $\operatorname{span}(V_2) = \ker(A_{Jc,Pu})$. *The* strangeness-free model *of $\mathcal{N}$ is given by*

$$0 = \Pi_2^T F_{\mathcal{N},1}(q_{Pi}, p_{Jc}, p_{Re_e}) \tag{5.7a}$$

$$\begin{aligned} 0 = {} & U_{2,e}^T A_{Jc_e,Pi} f_{Pi_e}(q_{Pi}, p_{Jc}, p_{Re_e}) + U_{2,e}^T A_{Jc_e,De_e} \dot{q}_{De_e^k} \\ & + U_{2,c}^T A_{Jc_c,Pi} f_{Pi_c}(q_{Pi}, p_{Jc}, p_{Re_c}) + U_{2,c}^T A_{Jc_c,De_c} f_{Pi_c}(q_{Pi_c}, p_{Jc}, p_{Jc_c}) \end{aligned} \tag{5.7b}$$

$$0 = F_{\mathcal{N},2}(q_{Pu_e}, p_{Jc}, p_{Re_e}) \tag{5.7c}$$

$$0 = F_{\mathcal{N},3}(q_{Pi}, q_{Pu}, q_{De_e}) \tag{5.7d}$$

$$0 = F_{\mathcal{N},4}(q_{De_e}) \tag{5.7e}$$

$$0 = F_{\mathcal{N},5}(p_{Re_e}), \tag{5.7f}$$

where $f_{Pi} = [f_{Pi_e}^T, f_{Pi_c}^T]^T$, $U_2 = [U_{2,e}^T, U_{2,c}^T]^T$ is such that $\operatorname{span}(U_2) = \operatorname{coker}(A_{Jc,Pu})$ *and $[\Pi_1, \Pi_2]$ is a permutation with Π_1 such that* $\operatorname{corange}(U_2^T A_{Jc,Pu}) = \operatorname{span}(\Pi_1)$.

1. *The* strangeness-free model *is regular and has strangeness index $\mu = 0$ (d-index 1).*
2. *A function $(q, p) \in C^1(\mathcal{I}, \mathbb{R}^{n_\mathcal{E}} \times \mathbb{R}^{n_\mathcal{V}})$ solves (5.6) if and only if it solves (5.7).*

Proof The surrogate model (5.7) as well the assertions 1. and 2. follow straightforward from Theorem 3.3. Note that in the hidden constraints (5.7b), the coupling edges P_c play the role of the demands De_c as the coupling implies that $q_{\mathrm{P}_c} = q_{\mathrm{De}_c}$. With $\dot{q}_{\mathrm{P}_c} = f_{\mathrm{Pi}_c}(q_{\mathrm{Pi}_c}, p_{\mathrm{Jc}}, p_{\mathrm{Jc}_c})$, Eq. (5.7b) follows. □

Having specified the solvability conditions as well as the surrogate model of the coupled network, we ask how the knowledge about the Substructures 3.1–3.3 can be exploited to assemble the corresponding substructures of the coupled network $\mathcal{N}$.

Lemma 5.6 *Consider networks* $\mathcal{N}^1, \ldots, \mathcal{N}^K$ *as in* (2.1). *Let* $B \in \mathbb{R}^{K \times K}$ *be the adjacency matrix of a simple coupling graph* $\mathcal{G}_{coup}$ *and let* $\mathcal{N}$ *be the coupling of* $\mathcal{N}^1, \ldots, \mathcal{N}^K$ *according to* B. *For* $k = 1, \ldots, K$, *let* V_2^k, U_2^k *be such that* $\mathrm{span}(V_2^k) = \ker(A_{Jc^k,Pu^k})$, $\mathrm{span}(U_2^k) = \mathrm{coker}(A_{Jc^k,Pu^k})$ *and let* $[\Pi_1^k, \Pi_2^k]$ *be a permutation such that* $\mathrm{span}(\Pi_1^k) = \mathrm{corange}(A_{\overline{Jc}^k,Pu^k})$. *For the coupled network* $\mathcal{N}$, *let* V_2, U_2 *be such that* $\mathrm{span}(V_2) = \ker(A_{Jc,Pu})$, $\mathrm{span}(U_2) = \mathrm{coker}(A_{Jc,Pu})$ *and let* $[\Pi_1, \Pi_2]$ *be a permutation with* $\mathrm{span}(\Pi_1) = \mathrm{corange}(A_{\overline{Jc},Pu})$.

1. *For* $k = 1, \ldots, K$, *if* $\mathcal{PU}_c^k = \emptyset$, *then* $V_2 = diag(V_2^k)_k$, $U_2 = diag(U_2^k)_k$ *and* $\Pi_1 = diag(\Pi_1^k)_k$.
2. *For* $k = 1, \ldots, K$, *if* $\ker(A_{Jc^k,Pu^k}) = \{0\}$, *then* $\ker(A_{Jc,Pu}) = \{0\}$.
3. *For* $k = 1, \ldots, K$, *if the network* $\mathcal{N}^k$ *does not contain paths of pumps connecting elements of* $\mathcal{RE}_e^k$ *and* $\mathcal{RE}_c^k$, *then* $V_2 = diag(V_2^k)_k$.

Proof The connection matrix $A_{\mathrm{Jc,Pu}}$ of the junction and pump set $\mathcal{G}_{\mathrm{Jc,Pu}}$ is given by, cp. Lemma 5.1,

$$A_{\mathrm{Jc,Pu}} = \begin{bmatrix} A_{\mathrm{Jc}^1,\mathrm{Pu}^1} & & A_{\mathrm{coup},kl} \\ & \ddots & \\ A_{\mathrm{coup},kl} & & A_{\mathrm{Jc}^K,\mathrm{Pu}^K} \end{bmatrix},$$

with

$$A_{\mathrm{Jc}^k,\mathrm{Pu}^k} = \begin{bmatrix} A_{\mathrm{Jc}_e^k,\mathrm{Pu}_e^k} & A_{\mathrm{Jc}_e^k,\mathrm{Pu}_c^k} \\ A_{\mathrm{Jc}_c^k,\mathrm{Pu}_e^k} & A_{\mathrm{Jc}_c^k,\mathrm{Pu}_c^k} \end{bmatrix}, \quad A_{\mathrm{coup},kl} = \begin{cases} \begin{bmatrix} 0 & 0 \\ 0 & A_{\mathrm{Re}_{c,k}^l,\mathrm{Pu}_{c,k}^l} \end{bmatrix}, & \text{if } B_{kl} = 1, \\ 0, & \text{if } B_{kl} = 0. \end{cases}$$

1. If the coupling is performed by pipes only, then $A_{\mathrm{coup},kl} = 0$ for $k, l = 1, \ldots, K$, and $A_{\mathrm{Jc,Pu}}$ is block diagonal. It follows that $\ker(A_{\mathrm{Jc,Pu}}) = \mathrm{span}(\mathrm{diag}(V_2^k)_k)$ with $\mathrm{span}(V_2^k) = \ker(A_{\mathrm{Jc}^k,\mathrm{Pu}^k})$ and $U_2 = \mathrm{diag}(U_2^k)_k$ with $\mathrm{span}(U_2^k) = \mathrm{coker}(A_{\mathrm{Jc}^k,\mathrm{Pu}^k})$. From the latter, in particular, it follows that $A_{\overline{\mathrm{Jc}},\mathrm{Pu}} = \mathrm{diag}(A_{\overline{\mathrm{Jc}}^k,\mathrm{Pu}^k})$. Hence, $\mathrm{corange}(U_2^T A_{\mathrm{Jc,Pu}}) = \mathrm{span}(\Pi_1)$

for $\Pi_1 = \mathrm{diag}(\Pi_1^k)_k$ with $\mathrm{span}(\Pi_1^k) = \mathrm{corange}(U_2^{k,T} A_{\mathrm{Jc}^k,\mathrm{Pu}^k})$ for $k = 1,\dots,K$.

2. If the coupling graph $\mathcal{G}_{\mathrm{coup}}$ is a tree, then $\mathcal{G}_{\mathrm{coup}}$ is not strongly connected and its adjacency matrix B is reducible cp. [19]. Hence, there exists a permutation, such that $A_{\mathrm{Jc,Pu}}$ is block triangular. If $\ker(A_{\mathrm{Jc}^k,\mathrm{Pu}^k}) = \{0\}$ for $k = 1,\dots,K$, then the triangular block structure of $A_{\mathrm{Jc,Pu}}$ implies that $\ker(A_{\mathrm{Jc,Pu}}) = \{0\}$. In the same manner, we find that $\mathrm{coker}(A_{\mathrm{Jc,Pu}}) = \{0\}$ if $\mathrm{coker}(A_{\mathrm{Jc}^k,\mathrm{Pu}^k}) = \{0\}$ for $k = 1,\dots,K$. From the latter, in particular, we get that $U_2^T A_{\mathrm{Jc,Pu}} = \mathrm{diag}(A_{\overline{\mathrm{Jc}}^k,\mathrm{Pu}^k})$. Hence, $\mathrm{corange}(U_2^T A_{\mathrm{Jc,Pu}}) = \mathrm{span}(\Pi_1)$ for $\Pi_1 = \mathrm{diag}(\Pi_1^k)_k$ with $\mathrm{span}(\Pi_1^k) = \mathrm{corange}(A_{\overline{\mathrm{Jc}}^k,\mathrm{Pu}^k})$ for $k = 1,\dots,K$.
3. WLOG, we assume that the matrix V_2^k with $\mathrm{span}(V_2^k) = \ker(A_{\mathrm{Jc}^k,\mathrm{Pu}^k})$ is sorted such that $V_2^k = [V_{2,1}^{k,T}, V_{2,2}^{k,T}]^T$, where $V_{2,1}^k$ selects paths of pumps connecting elements of $\mathcal{RE}_{\mathfrak{e}}^k$ and $\mathcal{RE}_{\mathfrak{e}}^k$ and cycles of pumps and $V_{2,2}^k$ selects paths of pumps connecting elements of $\mathcal{RE}_{\mathfrak{e}}^k$ and $\mathcal{RE}_c^k$. If $\mathcal{N}^k$ is free of paths of pumps connecting elements of $\mathcal{RE}_{\mathfrak{e}}^k$ and $\mathcal{RE}_c^k$, then $V_{2,2}^k = 0$, implying that $A_{\mathrm{coup},kl} V_2^k = 0$. As the coupling graph is simple, there is only one non-zero off-diagonal entry in each column associated with a coupling pump. Hence, $\ker(A_{\mathrm{Jc,Pu}}) = \mathrm{span}(\mathrm{diag}(V_2^k)_k)$ with $\mathrm{span}(V_2^k) = \ker(A_{\mathrm{Jc}^k,\mathrm{Pu}^k})$. □

If the coupling of $\mathcal{N}^1,\dots,\mathcal{N}^K$ is performed by pipes only, the Substructures 3.1–3.3 of the coupled network $\mathcal{N}$ are simply the union of the Substructures 3.1–3.3 of the individual networks $\mathcal{N}^1,\dots,\mathcal{N}^K$. Consequently, the selecting matrices V_2, U_2, Π_2 have block diagonal form.

If the coupling by pumps is allowed as well, we can still make the following observations. If the coupling graph $\mathcal{G}_{\mathrm{coup}}$ is a tree and the individual networks $\mathcal{N}^k$ do not contain cycles of pumps or paths of pumps connecting two reservoirs for $k = 1,\dots,K$, then the coupled network $\mathcal{N}$ neither contains elements of Substructure 3.1. If, for $k = 1,\dots,K$, the network $\mathcal{N}^k$ does not contain paths of pumps connecting elements of $\mathcal{RE}_{\mathfrak{e}}^k$ and $\mathcal{RE}_c^k$, then $V_2 = \mathrm{diag}(V_2^k)_k$.

As the Jacobian of the pump function is given by $\mathrm{D}f_{\mathrm{Pu}} = \mathrm{diag}(\mathrm{D}f_{\mathrm{Pu}^k})_k$, it follows that $V_2^T \mathrm{D}f_{\mathrm{Pu}} V_2 = \mathrm{diag}(V_2^{k,T} \mathrm{D}f_{\mathrm{Pu}^k} V_2^k)_k$ if $V_2 = \mathrm{diag}(V_2^k)_k$. Hence, if one of the assertions of Lemma 5.6 is satisfied, then $V_2^T \mathrm{D}f_{\mathrm{Pu}} V_2$ is pointwise nonsingular if $V_2^{k,T} \mathrm{D}f_{\mathrm{Pu}^k} V_2^k$ are pointwise nonsingular.

Corollary 5.7 *Consider networks $\mathcal{N}^1,\dots,\mathcal{N}^K$ as in (2.1). Let $B \in \mathbb{R}^{K\times K}$ be the adjacency matrix of a simple coupling graph $\mathcal{G}_{coup}$ and let $\mathcal{N}$ be the coupling of $\mathcal{N}^1,\dots,\mathcal{N}^K$ according to B. Let $V_2^{k,T} \mathrm{D}f_{Pu^k} V_2^k$ be pointwise nonsingular for $k = 1,\dots,K$. If one of the assertions of Lemma 5.6 is satisfied, then $V_2^T \mathrm{D}f_{Pu} V_2$ is pointwise nonsingular.*

Hence, under one of the assertions of Lemma 5.6, the solvability of the coupled network $\mathcal{N}$ is characterized by the subnetworks $\mathcal{N}^1, \ldots, \mathcal{N}^K$. There is no need for an extra solvability analysis of the coupled system.

If the coupling is performed by pipes only, then not only the solvability of the coupled network $\mathcal{N}$ is characterized by the subnetworks $\mathcal{N}^1, \ldots, \mathcal{N}^K$, but also the surrogate model of the coupled network can be assembled directly from the subnetworks. The hidden constraints of the coupled network correspond to the hidden constraints of the subnetworks and the set of differential equations of the coupled system corresponds to the differential equations of the subnetworks. In particular, this implies that the set of consistent initial values $\mathcal{C}_{IV}$ of $\mathcal{N}$ corresponds to the union of the sets of consistent initial values $\mathcal{C}_{IV}^k$ of $\mathcal{N}^k$.

Example 4.1 (continued) We consider the coupled network $\mathcal{N}_{Ex3}$ from Example 5.1. As the coupling is performed by a coupling pipe, Lemma 5.6, 1. applies. Hence, the Substructures 3.1–3.3 of the coupled network $\mathcal{N}$ are the union of the Substructures 3.1–3.3 of the networks $\mathcal{N}^1, \mathcal{N}^2$ and the selecting matrices V_2, U_2, Π_2 have block diagonal form. Indeed, in $\mathcal{N}$ the elements belonging to Substructure 3.1 are the cycle and the path of pumps between the reservoirs in $\mathcal{N}_{Ex1}$, see Fig. 18. $\mathcal{N}_{Ex2}$ does not bring in elements of Substructure 3.1. The junctions belonging to Substructure 3.2 in $\mathcal{N}$ are the isolated junctions of $\mathcal{N}_{Ex1;\mathrm{Jc,Pu}}$, $\mathcal{N}_{Ex2;\mathrm{Jc,Pu}}$ as well as junctions of the connected component without loose pumps of $\mathcal{N}_{Ex1}$, see Fig. 18. Consequently, with $\mathrm{rank}(V_2^2) = 0$, the associated matrices are

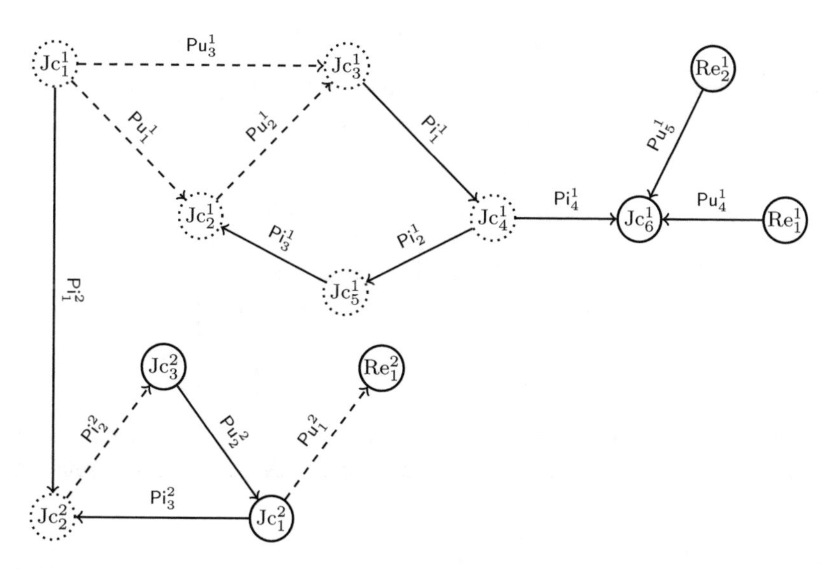

Fig. 18 Substructure 3.1 (dashed pumps) and Substructure 3.2 (dotted junctions) of the network $\mathcal{N}_{Ex3}$. The elements of Substructures 3.1 and 3.2 of the coupled graph $\mathcal{N}_{Ex3}$ are the union of the Substructures 3.1 and 3.2 of $\mathcal{N}_{Ex1}$ and $\mathcal{N}_{Ex2}$, see Figs. 4 and 9

given by

$$V_2 = \operatorname{diag}\left(V_2^1, V_2^2\right) = \begin{bmatrix} Pu_1^1 & 1 & 0 \\ Pu_2^1 & 1 & 0 \\ Pu_3^1 & -1 & 0 \\ Pu_4^1 & 0 & 1 \\ Pu_5^1 & 0 & -1 \\ Pu_1^2 & 0 & 0 \\ Pu_2^2 & 0 & 0 \end{bmatrix}, \quad U_2 = \operatorname{diag}\left(U_2^1, U_2^2\right) = \begin{bmatrix} Jc_1^1 & 1 & 0 & 0 & 0 \\ Jc_2^1 & 1 & 0 & 0 & 0 \\ Jc_3^1 & 1 & 0 & 0 & 0 \\ Jc_4^1 & 0 & 1 & 0 & 0 \\ Jc_5^1 & 0 & 0 & 1 & 0 \\ Jc_6^1 & 0 & 0 & 0 & 0 \\ Jc_1^2 & 0 & 0 & 0 & 0 \\ Jc_2^2 & 0 & 0 & 0 & 1 \\ Jc_3^2 & 0 & 0 & 0 & 0 \end{bmatrix}$$

Hence, as $V_2^{k,T} \, \mathrm{D}f_{\mathrm{Pu}^k} V_2^k$ are pointwise nonsingular for $k = 1, 2$, this implies that the DAE modeling the dynamics of $\mathcal{N}_{Ex3}$ is strangeness-free and uniquely solvable for every consistent initial value.

To construct the surrogate model for $\mathcal{N}_{Ex3}$, we consider the graph $\mathcal{G}_{\overline{\mathrm{Jc}},\mathrm{Pi}}$ arising from the vertex identification, see Fig. 19, whose connection matrix $A_{\overline{\mathrm{Jc}},\mathrm{Pi}}$ is given by

$$A_{\overline{\mathrm{Jc}},\mathrm{Pi}} = \begin{bmatrix} & Pi_1^1 & Pi_2^1 & Pi_3^1 & Pi_4^1 & Pi_1^2 & Pi_2^2 & Pi_3^2 \\ \hline \overline{Jc_1} & 1 & 0 & -1 & 0 & 0 & -1 & 1 \\ Jc_4^1 & -1 & 1 & 0 & 1 & 0 & 0 & 0 \\ Jc_5^1 & 0 & -1 & 1 & 0 & 0 & 0 & 0 \\ Jc_2^2 & 0 & 0 & 0 & 0 & -1 & 1 & -1 \end{bmatrix}.$$

A spanning tree of $\mathcal{G}_{\overline{\mathrm{Jc}},\mathrm{Pi}}$ is given by the composition of the spanning trees of $\mathcal{N}_{Ex1}$ and $\mathcal{N}_{Ex2}$, see Fig. 19, and we obtain the permutation

$$\Pi_1 = \begin{bmatrix} Pi_1^1 & 0 & 0 & 0 & 0 & 0 \\ Pi_2^1 & 1 & 0 & 0 & 0 & 0 \\ Pi_3^1 & 0 & 1 & 0 & 0 & 0 \\ Pi_4^1 & 0 & 0 & 1 & 0 & 0 \\ Pi_1^2 & 0 & 0 & 0 & 1 & 0 \\ Pi_2^2 & 0 & 0 & 0 & 0 & 1 \\ Pi_3^2 & 0 & 0 & 0 & 0 & 0 \end{bmatrix}, \quad \Pi_2 = \begin{bmatrix} Pi_1^1 & 1 & 0 \\ Pi_2^1 & 0 & 0 \\ Pi_3^1 & 0 & 0 \\ Pi_4^1 & 0 & 0 \\ Pi_1^2 & 0 & 0 \\ Pi_2^2 & 0 & 0 \\ Pi_3^2 & 0 & 1 \end{bmatrix}.$$

Then, the surrogate model (5.7) of the coupled network is obtained from the surrogate models of $\mathcal{N}_{Ex1}$ and $\mathcal{E}_{Ex2}$ by removing the coupling boundary conditions

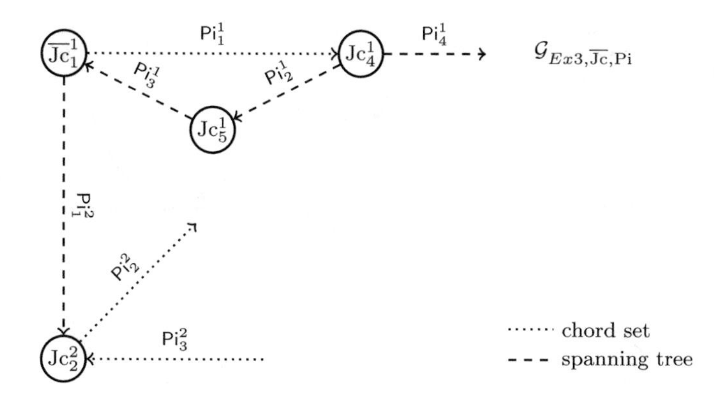

Fig. 19 The graph $\mathcal{G}_{Ex3,\overline{\mathrm{Jc}},\mathrm{Pu}}$ arising from the vertex identification of $\mathcal{G}_{Ex3,\mathrm{Jc}^3,\mathrm{Pu}^3}$. A spanning tree of $\mathcal{G}_{Ex3,\overline{\mathrm{Jc}},\mathrm{Pu}}$ is given by the union of the spanning trees of the individual networks $\mathcal{G}_{Ex1}$ and $\mathcal{G}_{Ex2}$

Re_2^1, De_1^1 and adding the coupling condition $p_{\mathrm{Jc}_1^1} = p_{\mathrm{Re}_2^1}$ and $q_{\mathrm{Pi}_1^2} = q_{\mathrm{De}_1^1}$. From this, we get the system

$$\dot{q}_{\mathrm{Pi}_1^1} = c_{11}(p_{\mathrm{Jc}_4^1} - p_{\mathrm{Jc}_3^1}) + c_{21}|q_{\mathrm{Pi}_1^1}|q_{\mathrm{Pi}_1^1} + c_{31}, \tag{Pi_1^3}$$

$$\dot{q}_{\mathrm{Pi}_2^2} = c_{12}(p_{\mathrm{Jc}_3^1} - p_{\mathrm{Jc}_2^1}) + c_{22}|q_{\mathrm{Pi}_2^2}|q_{\mathrm{Pi}_2^2} + c_{32} \tag{Pi_2^3}$$

$$\dot{q}_{\mathrm{Pi}_3^2} = c_{13}(p_{\mathrm{Jc}_2^1} - p_{\mathrm{Jc}_1^1}) + c_{23}|q_{\mathrm{Pi}_3^2}|q_{\mathrm{Pi}_3^2} + c_{33} \tag{Pi_3^3}$$

$$0 = f_{\mathrm{Pi}_1^2} + f_{\mathrm{Pi}_1^1} - f_{\mathrm{Pi}_3^1} \tag{$\overline{\mathrm{Jc}_1^3}$}$$

$$0 = -f_{\mathrm{Pi}_1^1} + f_{\mathrm{Pi}_2^1} + f_{\mathrm{Pi}_4^1} \tag{$\overline{\mathrm{Jc}_4^3}$}$$

$$0 = -f_{\mathrm{Pi}_2^1} + f_{\mathrm{Pi}_3^1} \tag{$\overline{\mathrm{Jc}_5^3}$}$$

$$0 = -f_{\mathrm{Pi}_1^2} + f_{\mathrm{Pi}_2^2} - f_{\mathrm{Pi}_3^2} \tag{$\overline{\mathrm{Jc}_2^3}$}$$

together with (Pu_1^1)-(Pu_5^1), (Pu_1^2),(Pu_2^2), (Jc_1^1)-(Jc_6^1), (Jc_1^2)–(Jc_3^2), (Re_1^1), (Re_2^1), (Re_1^2).

Example 5.3 As a second example, we couple the networks $\mathcal{N}_{Ex1}$ and $\mathcal{N}_{Ex2}$ from Examples 2.1 and 3.1 via a coupling pump, i.e., via the boundary conditions De_1^1, Re_1^2. The graph $\mathcal{G}_{Ex4}$ of the resulting network $\mathcal{N}_{Ex4}$ is illustrated in Fig. 20.

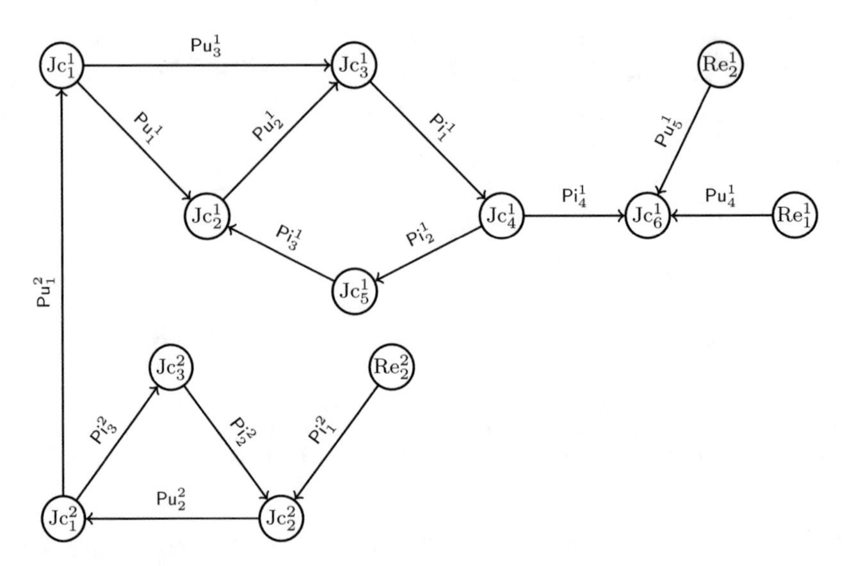

Fig. 20 Graph $\mathcal{G}_{Ex4}$ of the network $\mathcal{N}_{Ex4}$

The incidence matrix of $\mathcal{N}_{Ex4}$ is given by

$$
A_{\mathrm{Jc,Pi}} = \left[\begin{array}{cccccccc}
 & Pi_1^1 & Pi_2^1 & Pi_3^1 & Pi_4^1 & Pi_1^2 & Pi_2^2 & Pi_3^2 \\
Jc_1^1 & 0 & 0 & 0 & 0 & 0 & 0 & 0 \\
Jc_2^1 & 0 & 0 & -1 & 0 & 0 & 0 & 0 \\
Jc_3^1 & 1 & 0 & 0 & 0 & 0 & 0 & 0 \\
Jc_4^1 & -1 & 1 & 0 & 1 & 0 & 0 & 0 \\
Jc_5^1 & 0 & -1 & 1 & 0 & 0 & 0 & 0 \\
Jc_6^1 & 0 & 0 & 0 & -1 & 0 & 0 & 0 \\
Jc_1^2 & 0 & 0 & 0 & 0 & 0 & 0 & 1 \\
Jc_2^2 & 0 & 0 & 0 & 0 & -1 & 1 & -1 \\
Jc_3^2 & 0 & 0 & 0 & 0 & 0 & -1 & 0
\end{array}\right],
$$

$$
A_{\mathrm{Jc,Pu}} = \left[\begin{array}{cccccccc}
 & Pu_1^1 & Pu_2^1 & Pu_3^1 & Pu_4^1 & Pu_5^1 & Pu_1^2 & Pu_2^2 \\
Jc_1^1 & 1 & 0 & 1 & 0 & 0 & -1 & 0 \\
Jc_2^1 & -1 & 1 & 0 & 0 & 0 & 0 & 0 \\
Jc_3^1 & 0 & -1 & -1 & 0 & 0 & 0 & 0 \\
Jc_4^1 & 0 & 0 & 0 & 0 & 0 & 0 & 0 \\
Jc_5^1 & 0 & 0 & 0 & 0 & 0 & 0 & 0 \\
Jc_6^1 & 0 & 0 & 0 & -1 & -1 & 0 & 0 \\
Jc_1^2 & 0 & 0 & 0 & 0 & 0 & 1 & -1 \\
Jc_2^2 & 0 & 0 & 0 & 0 & 0 & 0 & 0 \\
Jc_3^2 & 0 & 0 & 0 & 0 & 0 & 0 & 1
\end{array}\right],
$$

$$A_{\mathrm{Re,Pi}} = \begin{bmatrix} & Pi_1^1 & Pi_2^1 & Pi_3^1 & Pi_4^1 & Pi_1^2 & Pi_2^2 & Pi_3^2 \\ Re_1^1 & 0 & 0 & 0 & 0 & 0 & 0 & 0 \\ Re_2^1 & 0 & 0 & 0 & 0 & 0 & 0 & 0 \\ Re_2^2 & 0 & 0 & 0 & 0 & 01 & 0 & 0 \end{bmatrix},$$

$$A_{\mathrm{Re,Pu}} = \begin{bmatrix} & Pu_1^1 & Pu_2^1 & Pu_3^1 & Pu_4^1 & Pu_5^1 & Pu_1^2 & Pu_2^2 \\ Re_1^1 & 0 & 0 & 0 & 1 & 0 & 0 & 0 \\ Re_2^1 & 0 & 0 & 0 & 0 & 1 & 0 & 0 \\ Re_2^2 & 0 & 0 & 0 & 0 & 0 & 0 & 0 \end{bmatrix}.$$

The network function $F_{\mathcal{N}_{Ex4}}$ of $\mathcal{N}_{Ex4}$ is again composed from the network functions $F_{\mathcal{N}_{Ex1}}$ and $F_{\mathcal{N}_{Ex2}}$. In detail, the DAE modeling the dynamics of $\mathcal{N}$ is given by

$$0 = q_{\mathrm{Pu}_1^2} + q_{\mathrm{Pu}_1^1} + q_{\mathrm{Pu}_3^1} \qquad (\mathrm{Jc}_1^1)$$

$$p_{\mathrm{Jc}_1^1} - p_{\mathrm{Jc}_1^2} = f_{\mathrm{Pu}_1^2}(q_{\mathrm{Pu}_1^2}) \qquad (\mathrm{Pu}_1^2)$$

as well as Eqs. (Pi_1^1)–(Pi_4^1), (Pi_1^1)–(Pi_3^1), (Pu_1^1)–(Pu_5^1), (Pu_2^2), (Jc_1^1)–(Jc_6^1), (Jc_1^2)–(Jc_3^2), (Re_1^1), (Re_2^1), (Re_1^2).

To analyze the solvability of this DAE, we consider the Substructures 3.1–3.3 of $\mathcal{N}_{Ex4}$. As the coupling is performed by a coupling pump and $\mathcal{N}_{Ex1}$ contains elements of Substructure 3.1, Lemma 5.6 (1) and (2) do not apply. However, as neither $\mathcal{N}_{Ex1}$ nor $\mathcal{N}_{Ex2}$ contains a path of pumps connecting a coupling and a non-coupling reservoir, Lemma 5.6 (3) applies. The Substructure 3.1 of the coupled network $\mathcal{N}$ is the union of the Substructure 3.1 of the networks $\mathcal{N}^1, \mathcal{N}^2$, see Fig. 21, and the selecting matrix V_2 has block diagonal form. For the matrix U_2, however, we note that the components belonging to Substructure 3.2 are given by the isolated junctions $\mathrm{Jc}_4^1, \mathrm{Jc}_5^1, \mathrm{Jc}_3^2$ as well as by $\{\mathrm{Jc}_1^1, \mathrm{Jc}_2^1, \mathrm{Jc}_3^1, \mathrm{Jc}_1^2, \mathrm{Jc}_2^2, \mathrm{Pu}_1^1, \mathrm{Pu}_2^1, \mathrm{Pu}_3^1, \mathrm{Pu}_1^2\mathrm{Pu}_2^1, \mathrm{Pu}_3^1\}$. Hence, for $\mathcal{N}_{Ex3}$, Substructure 3.2 is not the union of the Substructure 3.2 of $\mathcal{N}_{Ex1}$ and $\mathcal{N}_{Ex2}$. In conclusion, the associated matrices are given by

$$V_2 = \begin{bmatrix} Pu_1^1 & 1 & 0 \\ Pu_2^1 & 1 & 0 \\ Pu_3^1 & -1 & 0 \\ Pu_4^1 & 0 & 1 \\ Pu_5^1 & 0 & -1 \\ Pu_1^2 & 0 & 0 \\ Pu_2^2 & 0 & 0 \end{bmatrix}, \qquad U_2 = \begin{bmatrix} Jc_1^1 & 1 & 0 & 0 & 0 \\ Jc_2^1 & 1 & 0 & 0 & 0 \\ Jc_3^1 & 1 & 0 & 0 & 0 \\ Jc_4^1 & 0 & 1 & 0 & 0 \\ Jc_5^1 & 0 & 0 & 1 & 0 \\ Jc_6^1 & 0 & 0 & 0 & 0 \\ Jc_1^2 & 1 & 0 & 0 & 0 \\ Jc_2^2 & 0 & 0 & 0 & 1 \\ Jc_3^2 & 1 & 0 & 0 & 0 \end{bmatrix}.$$

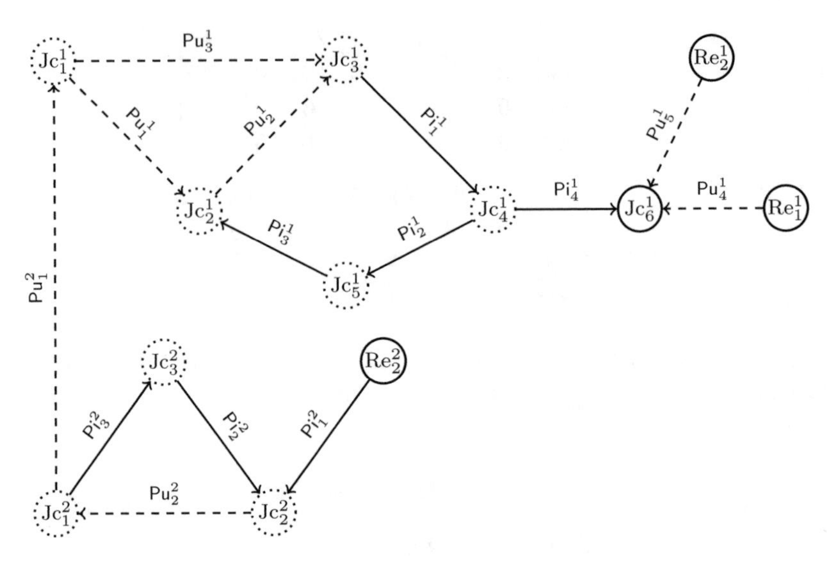

Fig. 21 For $\mathcal{N}_{Ex4}$, the elements belonging to Substructure 3.1 (dashed pumps) are the cycle and the path of pumps between the reservoirs in $\mathcal{N}_{Ex1}$, see Fig. 4, while $\mathcal{N}_{Ex2}$ does not bring in elements of Substructure 3.1. As the coupling edge is a pump, the elements of Substructure 3.2 (dotted junctions), however, are the isolated junctions Jc_4^1, Jc_5^1, Jc_3^2 and the connected component $\{\mathrm{Jc}_1^1, \dots, \mathrm{Jc}_3^1, \mathrm{Jc}_1^2, \mathrm{Jc}_2^2, \mathrm{Pu}_1^1, \dots, \mathrm{Pu}_3^1, \mathrm{Pu}_1^2, \mathrm{Pu}_2^1\}$

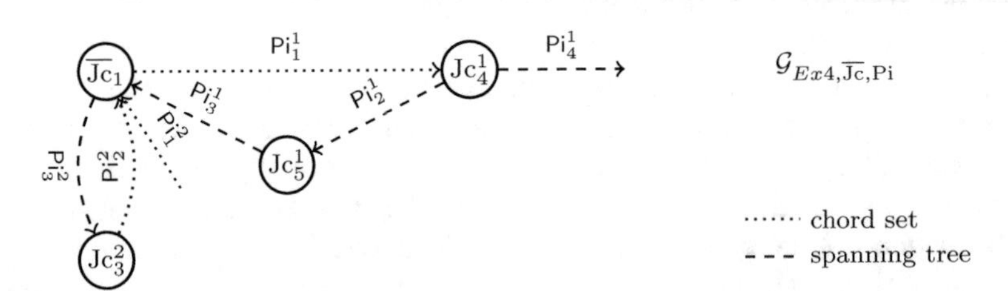

Fig. 22 The graph $\mathcal{G}_{Ex3,\overline{\mathrm{Jc}},\mathrm{Pu}}$ arising from the vertex identification of $\mathcal{G}_{Ex3,\mathrm{Jc}^3,\mathrm{Pu}^3}$. A spanning tree of $\mathcal{G}_{Ex3,\overline{\mathrm{Jc}},\mathrm{Pu}}$ is given by the spanning tree of $\mathcal{G}_{Ex1,\overline{\mathrm{Jc}},\mathrm{Pu}}$ plus Pi_2^2 or Pi_3^2

The graph $\mathcal{G}_{\overline{\mathrm{Jc}},\mathrm{Pi}}$ arising from the vertex identification is illustrated in Fig. 22. The associated connection matrix is given by

$$A_{\overline{\mathrm{Jc}},\mathrm{Pi}} = \begin{bmatrix} & Pi_1^1 & Pi_2^1 & Pi_3^1 & Pi_4^1 & Pi_1^2 & Pi_2^2 & Pi_3^2 \\ \overline{Jc_1} & 1 & 0 & -1 & 0 & 0 & -1 & 1 \\ Jc_4^1 & -1 & 1 & 0 & 1 & 0 & 0 & 0 \\ Jc_5^1 & 0 & -1 & 1 & 0 & 0 & 0 & 0 \\ Jc_2^2 & 0 & 0 & 0 & 0 & -1 & 1 & -1 \end{bmatrix},$$

A spanning tree of $\mathcal{G}_{\overline{\mathrm{Jc}},\mathrm{Pi}}$ is selected, e.g., by

$$\Pi_1 = \begin{bmatrix} Pi_1^1 & 0&0&0&0&0 \\ Pi_2^1 & 1&0&0&0&0 \\ Pi_3^1 & 0&1&0&0&0 \\ Pi_4^1 & 0&0&1&0&0 \\ Pi_1^2 & 0&0&0&1&0 \\ Pi_2^2 & 0&0&0&0&0 \\ Pi_3^2 & 0&0&0&0&1 \end{bmatrix}, \quad \Pi_2 = \begin{bmatrix} Pi_1^1 & 1&0&0 \\ Pi_2^1 & 0&0&0 \\ Pi_3^1 & 0&0&0 \\ Pi_4^1 & 0&0&0 \\ Pi_1^2 & 0&1&0 \\ Pi_2^2 & 0&0&1 \\ Pi_3^2 & 0&0&0 \end{bmatrix}$$

see Fig. 22. Hence, coupling two networks with a pump, the spanning tree of the coupled network is not necessarily the union of the spanning trees of the individual networks.

The surrogate model (5.7) of $\mathcal{N}_{Ex4}$ is given by

$$\dot{q}_{\mathrm{Pi}_1^1} = c_{11}(p_{\mathrm{Jc}_4^1} - p_{\mathrm{Jc}_3^1}) + c_{21}|q_{\mathrm{Pi}_1^1}|q_{\mathrm{Pi}_1^1} + c_{31}, \qquad (\mathrm{Pi}_1^3)$$

$$\dot{q}_{\mathrm{Pi}_1^2} = c_{13}(p_{\mathrm{Jc}_2^1} - p_{\mathrm{Jc}_1^1}) + c_{23}|q_{\mathrm{Pi}_1^2}|q_{\mathrm{Pi}_1^2} + c_{33} \qquad (\mathrm{Pi}_1^3)$$

$$\dot{q}_{\mathrm{Pi}_2^2} = c_{12}(p_{\mathrm{Jc}_3^1} - p_{\mathrm{Jc}_2^1}) + c_{22}|q_{\mathrm{Pi}_2^2}|q_{\mathrm{Pi}_2^2} + c_{32} \qquad (\mathrm{Pi}_2^3)$$

$$0 = f_{\mathrm{Pi}_1^2} + f_{\mathrm{Pi}_1^1} - f_{\mathrm{Pi}_3^1} \qquad (\overline{\mathrm{Jc}_1^3})$$

$$0 = -f_{\mathrm{Pi}_1^1} + f_{\mathrm{Pi}_2^1} + f_{\mathrm{Pi}_4^1} \qquad (\overline{\mathrm{Jc}_4^3})$$

$$0 = -f_{\mathrm{Pi}_2^1} + f_{\mathrm{Pi}_3^1} \qquad (\overline{\mathrm{Jc}_5^3})$$

$$0 = -f_{\mathrm{Pi}_1^2} + f_{\mathrm{Pi}_2^2} - f_{\mathrm{Pi}_3^2} \qquad (\overline{\mathrm{Jc}_2^3})$$

together with (Pu_1^1)–(Pu_5^1), (Pu_2^2), (Jc_1^1)–(Jc_6^1), (Jc_1^2)–(Jc_3^2), (Re_1^1), (Re_2^1), (Re_2^2).

6 Conclusion and Discussion

So far physical networks have been considered mainly as isolated and stand alone, but in many application they are not. The derivation of physical based topological conditions is also required for coupled systems of physical DAEs. We have shown, that it is very promising to derive additional topology based rules for coupled system. Nevertheless they seem to be not sufficient for all type of constellations. It is quite remarkable, that already the analysis of the uniform network has to be done in an appropriate way to provide the basic framework for a constructive analysis of the coupled networks. As a specific graph theoretical problem we have identified the *unique* and *good* choice of the spanning trees of the graphs of the underlying network. Consequently, the existing tools for single networks (Modified nodal analysis, Topological based index analysis) have to be re-evaluated against the possibility to be used for the topological analysis in multi-network structures.

The main driving feature for the presented approach is the possibility to combine assembled models without changing the models itself (or changing initial conditions). Indeed this feature can not be guaranteed by using purely graph theoretical approaches like Pantelides or the Σ-Method. Anyhow, those algorithms have their right to exists in all applications, where a tight connection to the underlying physics is not relevant or not available. For modular system simulation software the strong connection to the physics increases the applicability in engineering approaches and therefore has to preferred.

Furthermore, the analysis of multi-network structures provides the basic tools for the treatment of black-box elements (which physical coupling conditions) within physical networks. At that point it is required to marriage purely graph theoretical approaches with physical based topological methods in order to extract the advantages of both worlds. One very recent example in automotive applications is the incorporation of Functional Mock-up Units (FMUs)[8] in physical networks. Therein FMUs with appropriate coupling conditions provide internal dependency graphs, that can be re-interpreted as additional class of components. E.g. in liquid flow networks those components can form an additional class next to pipes, pumps, demands, junctions and reservoirs. In this case the topological criteria may be extended to this new classes.

In this work we have considered liquid flow network as a representative example. Indeed, those topics are also relevant for other physical domains like gas–dynamics and electric networks. To the authors best knowledge, the coupling of electrical network via defined interface and coupling conditions has not been considered so far. For the *Modified nodal analysis* applied to networks of electric networks, the challenge definitely is hidden in the identification and correct treatment of CV-loops and IL-cutsets, that arise through the coupling procedure. At least the case of CV-loops in electrical networks may be equivalent to the case of pump circles in liquid flow networks.

Acknowledgements Part of this work has been supported by the government of Upper Austria within the programme "Innovatives Oberösterreich 2020".

References

1. Baum, A.-K., Kolmbauer, M.: Solvability and topological index criteria for thermal management in liquid flow networks. Technical Report RICAM-Report 2015–21, Johann Radon Institute for Computational and Applied Mathematics, Austrian Academy of Sciences, (2015)

[8]http://fmi-standard.org/.

2. Baum, A.-K., Kolmbauer, M., Offner, G.: Topological solvability and DAE-index conditions for mass flow controlled pumps in liquid flow networks. Electron. Trans. Numer. Anal. **46**, 395–423 (2017)
3. Biggs, N.: Algebraic Graph Theory. Cambridge Mathematical Library. Cambridge University Press, Cambridge (1974)
4. Campbell, S.L.: A general form for solvable linear time varying singular systems of differential equations. SIAM J. Math. Anal. **18**, 1101–1115 (1987)
5. Chalghoum, I., Elaoud, S., Akrout, M., Taieb, E.H.: Transient behavior of a centrifugal pump during starting period. Appl. Acoust. **109**, 82–89 (2016)
6. Clamond, D.: Efficient resolution of the colebrook equation. Ind. Eng. Chem. Res. **48**(7), 3665–3671 (2009)
7. Diestel, R.: Graduate Texts in Mathematics: Graph Theory. Springer, Heidelberg (2000)
8. EPANET: Software that models the hydraulic and water quality behaviour of water distribution piping systems. (online) (2014). http://www.epa.gov/nrmrl/wswrd/dw/epanet.html
9. Grundel, S., Jansen, L., Hornung, N., Clees, T., Tischendorf, C., Benner, P.: Model order reduction of differential algebraic equations arising from the simulation of gas transport networks. In: Progress in Differential-Algebraic Equations, pp. 183–205. Springer, Berlin (2014)
10. Huck, C., Jansen, L., Tischendorf, C.: A topology based discretization of PDAEs describing water transportation networks. Proc. Appl. Math. Mech. **14**(1), 923–924 (2014)
11. Jansen, L., Pade, J.: Global unique solvability for a quasi-stationary water network model. Technical Report P-13-11, Institut für Mathematik, Humboldt-Universität zu Berlin (2013)
12. Jansen, L., Tischendorf, C.: A unified (P)DAE modeling approach for flow networks. In: Progress in Differential-Algebraic Equations, pp. 127–151. Springer, Berlin (2014)
13. Kunkel, P., Mehrmann, V.: Canonical forms for linear differential-algebraic equations with variable coefficients. J. Comput. Appl. Math. **56**, 225–251 (1994)
14. Kunkel, P., Mehrmann, V.: A new class of discretization methods for the solution of linear differential-algebraic equations. SIAM J. Numer. Anal. **33**, 1941–1961 (1996)
15. Kunkel, O., Mehrmann, V.: Local and global invariants of linear differential algebraic equations and their relation. Electron. Trans. Numer. Anal. **4**, 138–157 (1996)
16. Kunkel, P., Mehrmann, V.: Regular solutions of nonlinear differential-algebraic equations and their numerical determination. Numer. Math. **79**, 581–600 (1998)
17. Kunkel, P., Mehrmann, V.: Differential-Algebraic Equations. Analysis and Numerical Solution. EMS Publishing House, Zürich (2006)
18. Kunkel, P., Mehrmann, V.: Stability properties of differential-algebraic equations and spin-stabilized discretizations. Electron. Trans. Numer. Anal. **26**, 385–420 (2007)
19. Lancaster, P., Tismenetsky, M.: The Theory of Matrices. Academic, New York (1985)
20. Mehrmann, V.: Index Concepts for Differential-Algebraic Equations, pp. 676–681. Springer, Berlin, Heidelberg (2015)
21. Neumaier, R.: Hermetic Pumps - The Latest Innovations and Industrial Applications of Sealess Pumps. Gulf Publishing Company, Houston (1997)
22. Petzold, L.R.: Differential/algebraic equations are not ODEs. SIAM J. Sci. Statist. Comput. **3**, 367–384 (1982)
23. Rogalla, B.-U., Wolters, A.: Slow transients in closed conduit flow — Part I - Numerical Methods. In: NATO ASI Series Applied Sciences, vol. 274, pp. 613–642. Springer, Dordrecht (1994)
24. Scholz, L., Steinbrecher, A.: Structural-algebraic regularization for coupled systems of DAEs. BIT Numer. Math. **56**(2), 777–804 (2016)
25. Spijker, M.N.: Contractivity in the numerical solution of initial-value-problems. Numer. Math. **42**, 271–290 (1983)
26. Steinbach, M.C.: Topological Index Criteria in DAE for Water Networks. Konrad-Zuse-Zentrum für Informationstechnik, Berlin (2005)
27. Thulasiraman, K., Swamy, M.N.S.: Graphs: Theory and Algorithms. Wiley, New York (2011)

28. Tischendorf, C.: Topological index calculation of differential-algebraic equations in circuit simulation. Surv. Math. Ind. **8**, 187–199 (1999)
29. Urquhart, C.D., Wordley, M.: A comparison of the application of centrifugal and positive displacement pumps. In: Proceedings of the International Pump Users Symposium (17th Pump), pp. 145–152 (2000)

Nonsmooth DAEs with Applications in Modeling Phase Changes

Peter Stechlinski, Michael Patrascu, and Paul I. Barton

Abstract A variety of engineering problems involve dynamic simulation and optimization, but exhibit a mixture of continuous and discrete behavior. Such hybrid continuous/discrete behavior can cause failure in traditional methods; theoretical and numerical treatments designed for smooth models may break down. Recently it has been observed that, for a number of operational problems, such hybrid continuous/discrete behavior can be accurately modeled using a nonsmooth differential-algebraic equations (DAEs) framework, now possessing a foundational well-posedness theory and a computationally relevant sensitivity theory. Numerical implementations that scale efficiently for large-scale problems are possible for nonsmooth DAEs. Moreover, this modeling approach avoids undesirable properties typical in other frameworks (e.g., hybrid automata); in this modeling paradigm, extraneous (unphysical) variables are often avoided, unphysical behaviors (e.g., Zeno phenomena) from modeling abstractions are not prevalent, and *a priori* knowledge of the evolution of the physical system (e.g., phase changes experienced in a flash process execution) is not needed. To illustrate this nonsmooth modeling paradigm, thermodynamic phase changes in a simple, but widely applicable flash process are modeled using nonsmooth DAEs.

Keywords Dynamic simulation · Dynamic optimization · Nonsmooth analysis · Generalized derivatives · Process systems engineering

Mathematics Subject Classification (2010) 34A09, 49J52, 34A12

P. Stechlinski (✉)
Process Systems Engineering Laboratory, Massachusetts Institute of Technology, Cambridge, MA, USA

Department of Mathematics and Statistics, University of Maine, Orono, ME, USA
e-mail: pstechli@mit.edu; peter.stechlinski@maine.edu

M. Patrascu · P. I. Barton
Process Systems Engineering Laboratory, Massachusetts Institute of Technology, Cambridge, MA, USA
e-mail: mikesp@mit.edu; pib@mit.edu

S. Campbell et al. (eds.), *Applications of Differential-Algebraic Equations: Examples and Benchmarks*, Differential-Algebraic Equations Forum,
https://doi.org/10.1007/11221_2018_7

1 Introduction

The significant presence of discrete phenomena during a process operation is a major challenge in dynamic simulation and optimization. "Discrete" phenomena broadly refers to behavior that can be characterized by event-like transitions, such as discontinuities in state variables or changes in functional form of the governing equations at discrete events. Such discrete phenomena can be caused by (planned or unplanned) discrete control actions or physiochemical switches, such as the transition between flow regimes in a pipe or thermodynamic phase changes. The mixture of such discrete behavior with "continuous" behavior (i.e., continuous evolution of the dynamic system) is referred to as hybrid behavior. Prevalent hybrid frameworks for modeling operational problems include hybrid automata [5, 18, 38, 39, 64] and complementarity systems [23, 54, 65], though a variety of formalisms for hybrid dynamic systems [19] exist.

Studies of physical problems modeled using hybrid automata are found in [4, 5, 26, 39]. Of present interest, hybrid automata frameworks have been used for handling thermodynamic phase changes in equation-oriented simulation and optimization [3]; in this approach, the model is split into an invariant set of equations, which remains unchanged regardless of the thermodynamic phase regime, and a variant set of equations specific to each phase regime. At each point in time, the dynamic model is constructed as the union of the invariant set and the active equations in the variant set. The mole fraction vector of a phase that is not present is undefined, necessitating assignment of arbitrary constant values to such mole fractions. This leads to discontinuous behavior when that phase appears since the mole fractions jump to their physical values, and makes the model unsuited for simulation, sensitivity analysis, and optimization. A detailed discussion of theoretical limitations to the hybrid automata modeling framework, as well as pathological (unphysical) behaviors they can introduce, is given in [60].

An alternate approach for modeling thermodynamic phase changes is determining the phase regime by solving a nonlinear optimization problem [10]. Here, the vapor-liquid equilibrium (VLE) conditions appear in the form of constraints, and are automatically "relaxed" when there is only one thermodynamic phase present instead of a VLE. Such a relaxation allows for a continuous extension of the mole fractions into the thermodynamic phase regimes where they are undefined, preventing discontinuities at phase regime transitions. Moreover, this approach can be modified to an equation-solving problem for the case of the two-phase equilibrium by reformulating the optimization problem [22]. The resulting system of equations contains complementarity conditions (i.e., a complementarity system) that represent changes in the thermodynamic phase regime. Numerical issues associated with the presence of complementarity conditions are circumvented by requiring additional user-defined (unphysical) parameters, artificial variables and binary variables [1, 7] (e.g., in smoothing methods and mixed-integer reformulations), which may not be exact and may increase the running time of algorithms dramatically. For examples of relevant operational problems modeled as complementarity systems, the reader is referred to [7, 48].

Recently, a nonsmooth formulation for modeling the appearance and disappearance of phases has been suggested [52]. It was shown that a formulation with complementarity conditions is equivalent to a nonsmooth model. The proposed approach leads to a system of nonsmooth differential-algebraic equations (DAEs), which can be solved using any DAE integrator with proper event-detection capabilities. This approach was recently applied to plant-wide dynamic simulations and optimization [43, 44]. It models phase changes in a compact way, and requires no optimization nor does it introduce discontinuities in the model equations. In fact, in general this nonsmooth framework naturally avoids many of the issues outlined above, such as the need to add unphysical parameters and have *a priori* knowledge of the system dynamics. Mirroring the classical theory for smooth DAEs (see, e.g., [2, 9, 34]), nonsmooth semi-explicit DAEs now possess a notion of generalized differentiation index one, which implies regular mathematical properties such as well-posedness [59] and sensitivity theory [57] that is computationally relevant for existing nonsmooth methods (e.g., equation-solving [16, 46] and optimization [33, 37]).

In this article, we demonstrate a nonsmooth DAEs model of thermodynamic phase changes and analyze its mathematical properties; a model for a simple constant volume flash process is introduced in Sect. 2 that includes a nonsmooth algebraic equation that determines the thermodynamic phase regime. The mathematical foundation of the nonsmooth DAEs framework is presented in Sect. 3, which includes a discussion of consistency, initial consistency and regularity of nonsmooth DAEs (Sect. 3.1) suitable for generalizing the notion of differentiation index one, followed by nonsmooth DAE well-posedness theory (Sect. 3.2). Next, we analyze and simulate different formulations of the flash process model, given common forms to express the phase thermodynamic properties. Specifically, we point out the difficulties associated with the attempt to simulate all phase regimes using simplified expressions for the thermodynamic properties. Conclusions and future research directions are given in Sect. 5.

2 Modeling Flash Processes with Nonsmooth DAEs

A very simple and common unit operation in chemical engineering is modeled by nonsmooth DAEs in this paper: in a flash process, a stream is split into two phases, a liquid and a vapor phase, depending on the thermodynamic conditions. Usually the purpose is to increase the concentration of a desired component (chemical species) in one stream compared to the other (or decrease the concentration of an unwanted species). To simplify the analysis we focus on a single component system, which represents a simple evaporation or condensation process. Similar models have been proposed to simulate multiple component systems [52].

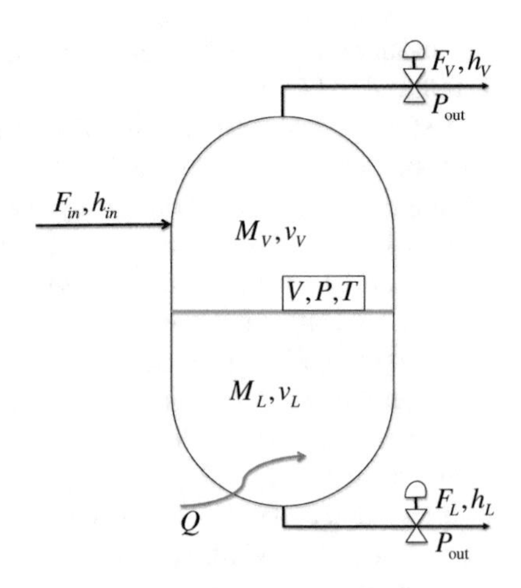

Fig. 1 A schematic of a flash vessel, showing the dynamic variables of interest. Q can be positive (heating) or negative (cooling)

The mass and energy balances of a constant volume flash process with a pure component, as schematically illustrated in Fig. 1, are given by Eq. (2.1). It is assumed that the temperature, T, and pressure, P, are uniform in space (point variables), and that the phase properties are homogeneous.

$$\frac{dM}{dt}(t) = F_{\text{in}}(t) - F_V(t) - F_L(t), \tag{2.1a}$$

$$\frac{dU}{dt}(t) = F_{\text{in}}(t)h_{\text{in}}(t) - F_L(t)h_L(t) - F_V(t)h_V(t) + Q(t), \tag{2.1b}$$

$$M(t) = M_L(t) + M_V(t), \tag{2.1c}$$

$$U(t) = M_L(t)\left(h_L(t) - P(t)v_L(t)\right) + M_V(t)\left(h_V(t) - P(t)v_V(t)\right), \tag{2.1d}$$

$$V = M_L(t)v_L(t) + M_V(t)v_V(t), \tag{2.1e}$$

where M, U and V are the total number of moles, total internal energy and (constant) volume of the vessel, respectively. F is the flow rate in mol/s, h is the specific enthalpy in kJ/mol and v is the specific volume in L/mol. Q is the heat duty in kW, which can be positive (heating) or negative (cooling). Subscripts L and V denote properties related to the liquid and vapor phase, respectively. The existence of the vapor and/or liquid phase is determined from the nonsmooth algebraic equation:

$$0 = \text{mid}\left(M_V(t), P(t) - P^{\text{sat}}(T(t)), -M_L(t)\right). \tag{2.2}$$

The mid function selects the median value of its three arguments, such that in case that the pressure in the vessel is higher than the saturation pressure, $P - P^{\text{sat}} > 0$, only a liquid phase exists, thus $0 = M_V$. Alternatively, if the pressure is lower than the saturation pressure only the vapor phase exists, thus $0 = -M_L$ is enforced.

This formulation and its derivation from thermodynamic principles are extensively discussed in [52]. Here, we will illustrate the use of this expression and elaborate further in Sect. 4.2.3.

P^{sat} depends on the temperature, T, and can be calculated from an equation of state using the equilibrium condition (see Sect. 4.1) or from the empirical Antoine expression:

$$P^{\text{sat}}(T) = 10^{\left(A - \frac{B}{T+C}\right)}, \tag{2.3}$$

where A, B, C are constants associated with a pure species, and the temperature, T, is in Kelvin. Before proceeding with analyzing and simulating this model, nonsmooth DAEs are formally introduced and their mathematical properties are highlighted.

3 Nonsmooth Generalized Differentiation Index-One DAEs

The following initial value problem (IVP) in semi-explicit DAEs is the main focus of this work:

$$\dot{\mathbf{x}}(t, \mathbf{p}) = \mathbf{f}(t, \mathbf{p}, \mathbf{x}(t, \mathbf{p}), \mathbf{y}(t, \mathbf{p})), \tag{3.1a}$$

$$\mathbf{0}_{n_y} = \mathbf{g}(t, \mathbf{p}, \mathbf{x}(t, \mathbf{p}), \mathbf{y}(t, \mathbf{p})), \tag{3.1b}$$

$$\mathbf{x}(t_0, \mathbf{p}) = \mathbf{f}_0(\mathbf{p}), \tag{3.1c}$$

where t is the independent variable, $\mathbf{p}$ is a vector of the problem parameters, $\mathbf{x}$ is the differential state variable, $\mathbf{y}$ is the algebraic state variable, and the functions

$$\mathbf{f} : D_t \times D_p \times D_x \times D_y \to \mathbb{R}^{n_x},$$

$$\mathbf{g} : D_t \times D_p \times D_x \times D_y \to \mathbb{R}^{n_y},$$

$$\mathbf{f}_0 : D_p \to D_x,$$

are not necessarily differentiable at all points on their respective domains ($D_t \subset \mathbb{R}$, $D_p \subset \mathbb{R}^{n_p}$, $D_y \subset \mathbb{R}^{n_y}$, and $D_x \subset \mathbb{R}^{n_x}$ are open and connected sets). Note that Eqs. (2.1) and (2.2) fall into this framework.

The notational conventions chosen here are consistent with those laid out in [29, 30, 57]: a set is denoted by an uppercase letter (e.g., H), vector-valued functions and vectors in $\mathbb{R}^n$ are denoted by lowercase boldface letters (e.g., $\mathbf{h}$) and matrix-valued functions and matrices in $\mathbb{R}^{m\times n}$ are denoted by uppercase boldface letters (e.g., $\mathbf{H}$). A well-defined vertical block matrix (or vector)

$$\begin{bmatrix} \mathbf{H}_1 \\ \mathbf{H}_2 \end{bmatrix}$$

is often written as $(\mathbf{H}_1, \mathbf{H}_2)$. The kth column vector of $\mathbf{H} \in \mathbb{R}^{m\times n}$ is denoted by $\mathbf{h}_{(k)} \in \mathbb{R}^m$ and the kth row of $\mathbf{H}$ is denoted by $\mathbf{H}_k \in \mathbb{R}^{1\times n}$. The ith component of a vector $\mathbf{h}$ is denoted by h_i. The Jacobian matrix of a differentiable function $\mathbf{h} : \mathbb{R}^n \to \mathbb{R}^m$ at $\mathbf{x} \in \mathbb{R}^n$ is denoted by $\mathbf{Jh}(\mathbf{x}) \in \mathbb{R}^{m\times n}$. The $n \times n$ identity matrix, n-dimensional zero vector and $m \times n$ zero matrix are denoted by $\mathbf{I}_n$, $\mathbf{0}_n$ and $\mathbf{0}_{m\times n}$, respectively. A neighborhood of $\mathbf{h} \in \mathbb{R}^n$ is a set of points $B_\delta(\mathbf{h})$ (i.e., the open ball of radius δ centered at $\mathbf{h}$) for some $\delta > 0$. A neighborhood of a set $H \subset \mathbb{R}^n$ is given by $B_\delta(H) := \cup_{\mathbf{h}\in H} B_\delta(\mathbf{h})$ for some $\delta > 0$.

3.1 Regularity and Consistency of Nonsmooth DAEs

Following the exposition in [59], the following assumptions are considered in analyzing the IVP in DAEs (3.1).

Assumption 3.1 *Suppose that the following conditions hold:*

(i) $\mathbf{f}(\cdot, \mathbf{p}, \boldsymbol{\eta}_x, \boldsymbol{\eta}_y)$ *is measurable on* D_t *for each* $(\mathbf{p}, \boldsymbol{\eta}_x, \boldsymbol{\eta}_y) \in D_p \times D_x \times D_y$*;*
(ii) $\mathbf{f}(t, \cdot, \cdot, \cdot)$ *is continuous on* $D_p \times D_x \times D_y$ *for each* $t \in D_t \setminus Z_{\mathbf{f}}$*, where* $Z_{\mathbf{f}}$ *is a zero-measure subset;*
(iii) there exists a Lebesgue integrable function $m_{\mathbf{f}} : D_t \to \mathbb{R}_+ \cup \{+\infty\}$ *such that*

$$\|\mathbf{f}(t, \mathbf{p}, \boldsymbol{\eta}_x, \boldsymbol{\eta}_y)\| \leq m_{\mathbf{f}}(t), \quad \forall t \in D_t, \quad \forall (\mathbf{p}, \boldsymbol{\eta}_x, \boldsymbol{\eta}_y) \in D_p \times D_x \times D_y;$$

(iv) $\mathbf{g}$ *is locally Lipschitz continuous on* $D := D_t \times D_p \times D_x \times D_y$*; and*
(v) $\mathbf{f}_0$ *is locally Lipschitz continuous on* D_p*.*

Assumption 3.2 *Suppose that there exists a Lebesgue integrable function* $k_{\mathbf{f}} : D_t \to \mathbb{R}_+ \cup \{+\infty\}$ *such that, for any* $t \in D_t$ *and any* $(\mathbf{p}_1, \boldsymbol{\eta}_{x_1}, \boldsymbol{\eta}_{y_1}), (\mathbf{p}_2, \boldsymbol{\eta}_{x_2}, \boldsymbol{\eta}_{y_2}) \in D_p \times D_x \times D_y$*,*

$$\|\mathbf{f}(t, \mathbf{p}_1, \boldsymbol{\eta}_{x_1}, \boldsymbol{\eta}_{y_1}) - \mathbf{f}(t, \mathbf{p}_2, \boldsymbol{\eta}_{x_2}, \boldsymbol{\eta}_{y_2})\| \leq k_{\mathbf{f}}(t)\|(\mathbf{p}_1, \boldsymbol{\eta}_{x_1}, \boldsymbol{\eta}_{y_1}) - (\mathbf{p}_2, \boldsymbol{\eta}_{x_2}, \boldsymbol{\eta}_{y_2})\|.$$

The conditions in Assumptions 3.1(i)–(iii) and 3.2 need only hold on an open subset $N \subset D_t \times D_p \times D_x \times D_y$ for the theory outlined in this section to hold (see [59]). Such a relaxation matches the classical Carathéodory ODE existence and uniqueness conditions [14, 17] (which are implied by local Lipschitz continuity of $\mathbf{f}$). The choice of presentation made here is for clarity of exposition.

Given a connected set $T \subset D_t$ containing t_0 and $P \subset D_p$, a mapping $\mathbf{z} \equiv (\mathbf{x}, \mathbf{y}) : T \times P \to D_x \times D_y$ is called a *solution of* (3.1) *on* $T \times P$ if, for each $\mathbf{p} \in P$, $\mathbf{z}(\cdot, \mathbf{p})$ is an absolutely continuous function satisfying (3.1a) for almost every $t \in T$, (3.1b) for every $t \in T$, and (3.1c) at $t = t_0$. Given a solution $\widetilde{\mathbf{z}} \equiv (\widetilde{\mathbf{x}}, \widetilde{\mathbf{y}})$ of (3.1) on $T \times \{\mathbf{p}_0\}$, for some reference parameter $\mathbf{p}_0 \in D_p$, the classical characterization of differentiation index one is not viable in the setting of Assumption 3.1; since $\mathbf{g}$ may not be differentiable on its domain, the requirement that

$$\frac{\partial \mathbf{g}}{\partial \mathbf{y}}(t, \mathbf{p}_0, \widetilde{\mathbf{x}}(t, \mathbf{p}_0), \widetilde{\mathbf{y}}(t, \mathbf{p}_0)) \tag{3.2}$$

be nonsingular for all $t \in T$ must be generalized in a suitable way. This is accomplished as follows: define the *regularity* set of (3.1) by

$$G_\mathrm{R} := \{(t, \mathbf{p}, \boldsymbol{\eta}_x, \boldsymbol{\eta}_y) \in D : (t^*, \mathbf{p}^*, \boldsymbol{\eta}_x^*, \boldsymbol{\eta}_y^*) \mapsto (t^*, \mathbf{p}^*, \boldsymbol{\eta}_x^*, \mathbf{g}(t^*, \mathbf{p}^*, \boldsymbol{\eta}_x^*, \boldsymbol{\eta}_y^*))$$
$$\text{is a Lipschitz homeomorphism on some neighborhood of } (t, \mathbf{p}, \boldsymbol{\eta}_x, \boldsymbol{\eta}_y)\}.$$

Then $\widetilde{\mathbf{z}}$ is called a *regular* solution of (3.1) on $T \times \{\mathbf{p}_0\}$ if

$$\{(t, \mathbf{p}_0, \widetilde{\mathbf{x}}(t, \mathbf{p}_0), \widetilde{\mathbf{y}}(t, \mathbf{p}_0)) : t \in T\} \subset G_\mathrm{R},$$

and (3.1) is said to have *generalized differentiation index one* (a local characteristic along a solution trajectory). If $\mathbf{g}$ is C^1, then nonsingularity of the partial Jacobian matrix in (3.2) for all $t \in T$ implies

$$(t^*, \mathbf{p}^*, \boldsymbol{\eta}_x^*, \boldsymbol{\eta}_y^*) \mapsto (t^*, \mathbf{p}^*, \boldsymbol{\eta}_x^*, \mathbf{g}(t^*, \mathbf{p}^*, \boldsymbol{\eta}_x^*, \boldsymbol{\eta}_y^*))$$

is a C^1-diffeomorphism, and the classical notion of differentiation index one is recovered in this setting.

To verify regularity *a priori* based on the functional form of $\mathbf{g}$ in a global manner, a sufficient condition for regularity is derived using generalized derivatives. The *B-subdifferential* (sometimes called the *limiting Jacobian*) of $\mathbf{g}$ at $(t, \mathbf{p}, \boldsymbol{\eta}_x, \boldsymbol{\eta}_y) \in D_t \times D_p \times D_x \times D_y$ is given by

$$\partial_\mathrm{B}\mathbf{g}(t, \mathbf{p}, \boldsymbol{\eta}_x, \boldsymbol{\eta}_y) := \left\{\lim_{i\to\infty} \mathbf{J}\mathbf{g}(\boldsymbol{\omega}_{(i)}) : \lim_{i\to\infty} \boldsymbol{\omega}_{(i)} = (t, \mathbf{p}, \boldsymbol{\eta}_x, \boldsymbol{\eta}_y),\ \boldsymbol{\omega}_{(i)} \in D \setminus Z_\mathbf{g}, \forall i \in \mathbb{N}\right\},$$

where $Z_\mathbf{g} \subset D$ is the zero (Lebesgue) measure subset for which $\mathbf{g}$ is not differentiable (by Rademacher's theorem). The Clarke (generalized) *Jacobian* [13] of $\mathbf{g}$ at $(t, \mathbf{p}, \boldsymbol{\eta}_x, \boldsymbol{\eta}_y)$ is the convex hull of the B-subdifferential;

$$\partial\mathbf{g}(t, \mathbf{p}, \boldsymbol{\eta}_x, \boldsymbol{\eta}_y) := \operatorname{conv} \partial_\mathrm{B}\mathbf{g}(t, \mathbf{p}, \boldsymbol{\eta}_x, \boldsymbol{\eta}_y).$$

If $\mathbf{g}$ is C^1 at $(t, \mathbf{p}, \boldsymbol{\eta}_x, \boldsymbol{\eta}_y)$, then

$$\partial\mathbf{g}(t, \mathbf{p}, \boldsymbol{\eta}_x, \boldsymbol{\eta}_y) = \partial_\mathrm{B}\mathbf{g}(t, \mathbf{p}, \boldsymbol{\eta}_x, \boldsymbol{\eta}_y) = \{\mathbf{J}\mathbf{g}(t, \mathbf{p}, \boldsymbol{\eta}_x, \boldsymbol{\eta}_y)\}.$$

A nonsmooth extension of the partial Jacobian matrix in (3.2) is constructed as follows: the *projection of the* Clarke *Jacobian* [13] with respect to $\mathbf{y}$ at $(t, \mathbf{p}, \boldsymbol{\eta}_x, \boldsymbol{\eta}_y)$

is defined as

$$\pi_{\mathbf{y}}\partial \mathbf{g}(t,\mathbf{p},\boldsymbol{\eta}_x,\boldsymbol{\eta}_y) := \{\mathbf{Y}\in\mathbb{R}^{n_y\times n_y} : \exists[\mathbf{T}\quad \mathbf{P}\quad \mathbf{X}\quad \mathbf{Y}]\in\partial\mathbf{g}(t,\mathbf{p},\boldsymbol{\eta}_x,\boldsymbol{\eta}_y)\},\tag{3.3}$$

and satisfies

$$\pi_{\mathbf{y}}\partial \mathbf{g}(t,\mathbf{p},\boldsymbol{\eta}_x,\boldsymbol{\eta}_y) = \left\{\frac{\partial \mathbf{g}}{\partial \mathbf{y}}(t,\mathbf{p},\boldsymbol{\eta}_x,\boldsymbol{\eta}_y)\right\}$$

if $\mathbf{g}$ is C^1 at $(t,\mathbf{p},\boldsymbol{\eta}_x,\boldsymbol{\eta}_y)$.

Said projection can be used to provide a sufficient condition for regularity of solutions of the IVP in DAEs (3.1) (and thus generalized differentiation index one): let

$$G_{\mathrm{R}}^{\pi} :=\{(t,\mathbf{p},\boldsymbol{\eta}_x,\boldsymbol{\eta}_y)\in D : \pi_{\mathbf{y}}\partial\mathbf{g}(t,\mathbf{p},\boldsymbol{\eta}_x,\boldsymbol{\eta}_y) \text{ contains no singular matrices}\}.$$

Then $G_{\mathrm{R}}^{\pi}\subset G_{\mathrm{R}}$ since nonsingularity of every matrix in $\pi_{\mathbf{y}}\partial\mathbf{g}(t,\mathbf{p},\boldsymbol{\eta}_x,\boldsymbol{\eta}_y)$ implies the mapping $(t^*,\mathbf{p}^*,\boldsymbol{\eta}_x^*,\boldsymbol{\eta}_y^*)\mapsto(t^*,\mathbf{p}^*,\boldsymbol{\eta}_x^*,\mathbf{g}(t^*,\mathbf{p}^*,\boldsymbol{\eta}_x^*,\boldsymbol{\eta}_y^*))$ is a homeomorphism on a neighborhood of $(t,\mathbf{p},\boldsymbol{\eta}_x,\boldsymbol{\eta}_y)$ [13]. Hence, the condition

$$\{(t,\mathbf{p}_0,\widetilde{\mathbf{x}}(t,\mathbf{p}_0),\widetilde{\mathbf{y}}(t,\mathbf{p}_0)) : t\in T\}\subset G_{\mathrm{R}}^{\pi},$$

implies $\widetilde{\mathbf{z}}$ is a regular solution of (3.1) on $T\times\{\mathbf{p}_0\}$ (and thus (3.1) is generalized differentiation index one). Again, this condition corresponds to the classical notion of differentiation index one when $\mathbf{g}$ is continuously differentiable.

If $\mathbf{g}$ is piecewise differentiable (PC^1) in the sense of Scholtes [53], which is a broad class of functions including all C^1 functions, the absolute-value, min, max and mid functions, and compositions thereof, then

$$\partial_{\mathrm{B}}\mathbf{g}(t,\mathbf{p},\boldsymbol{\eta}_x,\boldsymbol{\eta}_y) = \left\{\mathbf{J}\mathbf{g}_{(i)}(t,\mathbf{p},\boldsymbol{\eta}_x,\boldsymbol{\eta}_y) : i=1,\ldots,n_{\mathrm{ess}}\right\},$$

where $\mathbf{J}\mathbf{g}_{(i)}(t,\mathbf{p},\boldsymbol{\eta}_x,\boldsymbol{\eta}_y)$ is the Jacobian matrix of $\mathbf{g}_{(i)}$ evaluated at $(t,\mathbf{p},\boldsymbol{\eta}_x,\boldsymbol{\eta}_y)$ and n_{ess} is a positive integer corresponding to the number of (essentially active) C^1 selection functions $\mathbf{g}_{(i)}$. In this case, the PC^1-*regularity* set of (3.1) is defined as

$$G_{\mathrm{R}}^{PC} := \{(t,\mathbf{p},\boldsymbol{\eta}_x,\boldsymbol{\eta}_y)\in D : (t^*,\mathbf{p}^*,\boldsymbol{\eta}_x^*,\boldsymbol{\eta}_y^*)\mapsto(t^*,\mathbf{p}^*,\boldsymbol{\eta}_x^*,\mathbf{g}(t^*,\mathbf{p}^*,\boldsymbol{\eta}_x^*,\boldsymbol{\eta}_y^*))$$
$$\text{is a } PC^1\text{-homeomorphism on some neighborhood of } (t,\mathbf{p},\boldsymbol{\eta}_x,\boldsymbol{\eta}_y)\}.$$

Clearly $G_{\mathrm{R}}^{PC}\subset G_{\mathrm{R}}$, implying regularity of a solution contained in this set (and therefore the generalized differentiation index one property of (3.1)). Moreover, a verifiable condition can be developed in the PC^1 setting, using the theory in [49]: let

$$G_{\mathrm{R}}^{CCO} :=\{(t,\mathbf{p},\boldsymbol{\eta}_x,\boldsymbol{\eta}_y)\in D : \mathrm{sign}(\det(\mathbf{Y})) = a\ \forall\mathbf{Y}\in\Lambda_{\mathbf{y}}\mathbf{g}(t,\mathbf{p},\boldsymbol{\eta}_x,\boldsymbol{\eta}_y)\},$$

for some $a = \pm 1$, where $\Lambda_{\mathbf{y}}\mathbf{g}(t, \mathbf{p}, \boldsymbol{\eta}_x, \boldsymbol{\eta}_y)$ is the *combinatorial partial Jacobian* of $\mathbf{g}$ with respect to $\mathbf{y}$ at $(t, \mathbf{p}, \boldsymbol{\eta}_x, \boldsymbol{\eta}_y)$, defined as

$$\Lambda_{\mathbf{y}}\mathbf{g}(t, \mathbf{p}, \boldsymbol{\eta}_x, \boldsymbol{\eta}_y)$$
$$:= \left\{\mathbf{Y} \in \mathbb{R}^{n_y \times n_y} : \mathbf{Y}_i = \frac{\partial g_{(\delta_i),i}}{\partial \mathbf{y}}(t, \mathbf{p}, \boldsymbol{\eta}_x, \boldsymbol{\eta}_y), \forall i \in \{1, \ldots, n_y\},\ \boldsymbol{\delta} \in \{1, \ldots, n_{\text{ess}}\}^{n_y}\right\}.$$

(That is, the ith row in $\mathbf{Y}$ is equal to the ith row of the partial Jacobian matrix of the δ_ith selection function $\mathbf{g}_{(\delta_i)}$.) If every matrix in G_{R}^{CCO} has the same nonvanishing determinant sign (i.e., $\text{sign}(\det(\mathbf{Y})) = 1$ or -1), then $\mathbf{g}$ is called *completely coherently oriented* (CCO) with respect to $\mathbf{y}$ at $(t, \mathbf{p}, \boldsymbol{\eta}_x, \boldsymbol{\eta}_y)$, which implies the mapping $(t^*, \mathbf{p}^*, \boldsymbol{\eta}_x^*, \boldsymbol{\eta}_y^*) \mapsto (t^*, \mathbf{p}^*, \boldsymbol{\eta}_x^*, \mathbf{g}(t^*, \mathbf{p}^*, \boldsymbol{\eta}_x^*, \boldsymbol{\eta}_y^*))$ is a PC^1-homeomorphism on a neighborhood of $(t, \mathbf{p}, \boldsymbol{\eta}_x, \boldsymbol{\eta}_y)$ [49, Corollary 20], and therefore $G_{\mathrm{R}}^{CCO} \subset G_{\mathrm{R}}^{PC} \subset G_{\mathrm{R}}$.

Of note, if g is scalar-valued, $\Lambda_{\mathbf{y}}g(t, \mathbf{p}, \boldsymbol{\eta}_x, \boldsymbol{\eta}_y) = \partial_{\mathrm{B}}[\mathbf{g}(t, \mathbf{p}, \boldsymbol{\eta}_x, \cdot)](\boldsymbol{\eta}_y)$ (i.e., the *partial* B-subdifferential of $\mathbf{g}$ with respect to $\mathbf{y}$ at $(t, \mathbf{p}, \boldsymbol{\eta}_x, \boldsymbol{\eta}_y)$). If $\mathbf{g}$ is vector-valued but only has two selection functions at $(t, \mathbf{p}, \boldsymbol{\eta}_x, \boldsymbol{\eta}_y)$ and the partial Jacobian matrices $\frac{\partial \mathbf{g}_{(1)}}{\partial \mathbf{y}}(t, \mathbf{p}, \boldsymbol{\eta}_x, \boldsymbol{\eta}_y)$ and $\frac{\partial \mathbf{g}_{(2)}}{\partial \mathbf{y}}(t, \mathbf{p}, \boldsymbol{\eta}_x, \boldsymbol{\eta}_y)$ only differ in one row, then

$$\Lambda_{\mathbf{y}}\mathbf{g}(t, \mathbf{p}, \boldsymbol{\eta}_x, \boldsymbol{\eta}_y) = \left\{\frac{\partial \mathbf{g}_{(1)}}{\partial \mathbf{y}}(t, \mathbf{p}, \boldsymbol{\eta}_x, \boldsymbol{\eta}_y), \frac{\partial \mathbf{g}_{(2)}}{\partial \mathbf{y}}(t, \mathbf{p}, \boldsymbol{\eta}_x, \boldsymbol{\eta}_y)\right\}.$$

Lastly, consistent initialization of (3.1) is characterized through the *consistency set* and *initial consistency set*, which are defined as, respectively,

$$G_{\mathrm{C}} := \{(t, \mathbf{p}, \boldsymbol{\eta}_x, \boldsymbol{\eta}_y) \in D : \mathbf{g}(t, \mathbf{p}, \boldsymbol{\eta}_x, \boldsymbol{\eta}_y) = \mathbf{0}_{n_y}\},$$
$$G_{\mathrm{C}}^0 := \{(t, \mathbf{p}, \boldsymbol{\eta}_x, \boldsymbol{\eta}_y) \in G_{\mathrm{C}} : t = t_0, \boldsymbol{\eta}_x = \mathbf{f}_0(\mathbf{p})\}.$$

Then $\tilde{\mathbf{z}}$ is said to be a *(regular) solution of* (3.1) *on* $T \times \{\mathbf{p}_0\}$ *through* $\{(t_0, \mathbf{p}_0, \mathbf{x}_0, \mathbf{y}_0)\} \subset G_{\mathrm{C}}^0$ if $\tilde{\mathbf{z}}$ is a (regular) solution of (3.1) on $T \times \{\mathbf{p}_0\}$ and, in addition, $(\tilde{\mathbf{x}}(t_0, \mathbf{p}_0), \tilde{\mathbf{y}}(t_0, \mathbf{p}_0)) = (\mathbf{x}_0, \mathbf{y}_0)$.

Example 3.3 Consider the following IVP in semi-explicit DAEs:

$$\begin{aligned} \dot{x}(t, p) &= \text{sign}(t - 0.5), \\ 1 &= |x(t, p)| + |y(t, p)|, \\ x(t_0, p) &= \min(0, p), \end{aligned} \tag{3.4}$$

i.e., the right-hand side functions are given by

$$\begin{aligned} f &: \mathbb{R}^4 \to \mathbb{R} : (t, p, \eta_x, \eta_y) \mapsto \text{sign}(t - 0.5), \\ g &: \mathbb{R}^4 \to \mathbb{R} : (t, p, \eta_x, \eta_y) \mapsto |\eta_x| + |\eta_y| - 1, \\ f_0 &: \mathbb{R} \to \mathbb{R} : p \mapsto \min(0, p), \end{aligned}$$

which are PC^1 on their domains. The function g has points of nondifferentiability

$$Z_g = \{(t, p, \eta_x, \eta_y) \in \mathbb{R}^4 : \eta_x = 0 \text{ or } \eta_y = 0\}$$

and the following selection functions at $(t, p, 0, 0)$:

$$\begin{aligned}
g_{(1)}(t, p, \eta_x, \eta_y) &\equiv \eta_x + \eta_y - 1, \\
g_{(2)}(t, p, \eta_x, \eta_y) &\equiv -\eta_x + \eta_y - 1, \\
g_{(3)}(t, p, \eta_x, \eta_y) &\equiv -\eta_x - \eta_y - 1, \\
g_{(4)}(t, p, \eta_x, \eta_y) &\equiv \eta_x - \eta_y - 1.
\end{aligned}$$

Consequently,

$$\partial_{\mathrm{B}} g(t, p, 0, 0) = \{\mathbf{J}g_{(i)}(t, p, 0, 0)\} = \left\{ \begin{bmatrix} 1 \\ 1 \end{bmatrix}^{\mathrm{T}}, \begin{bmatrix} -1 \\ 1 \end{bmatrix}^{\mathrm{T}}, \begin{bmatrix} -1 \\ -1 \end{bmatrix}^{\mathrm{T}}, \begin{bmatrix} 1 \\ -1 \end{bmatrix}^{\mathrm{T}} \right\},$$

from which it follows that

$$\partial g(t, p, 0, 0) = \mathrm{conv}\{\mathbf{J}g_{(i)}(t, p, 0, 0)\} = \{[\lambda_1 \quad \lambda_2] : -1 \le \lambda_1 \le 1, -1 \le \lambda_2 \le 1\}.$$

Combining the above with the fact that the Clarke Jacobian is a singleton (i.e., the Jacobian matrix) for $(t, p, \eta_x, \eta_y) \in \mathbb{R}^4 \setminus Z_g$, it follows that for any $(t, p, \eta_x, \eta_y) \in \mathbb{R}^4$,

$$\pi_y \partial g(t, p, \eta_x, \eta_y) = \begin{cases} \{-1\}, & \text{if } \eta_y < 0, \\ [-1, 1], & \text{if } \eta_y = 0, \\ \{1\}, & \text{if } \eta_y > 0, \end{cases}$$

and

$$\Lambda_y g(t, p, \eta_x, \eta_y) = \begin{cases} \{-1\}, & \text{if } \eta_y < 0, \\ \{-1, 1\}, & \text{if } \eta_y = 0, \\ \{1\}, & \text{if } \eta_y > 0. \end{cases}$$

The consistency, initial consistency, and regularity sets are therefore given by

$$\begin{aligned}
G_{\mathrm{C}} &= \{(t, p, \eta_x, \eta_y) \in \mathbb{R}^4 : |\eta_x| + |\eta_y| = 1\}, \\
G_{\mathrm{C}}^0 &= \{(t, p, \eta_x, \eta_y) \in G_{\mathrm{C}} : t = t_0, \eta_x = \min(0, p)\}, \\
G_{\mathrm{R}}^{\pi} = G_{\mathrm{R}}^{CCO} &= \{(t, p, \eta_x, \eta_y) \in \mathbb{R}^4 : \eta_y \neq 0\}.
\end{aligned}$$

3.2 Existence and Uniqueness of Solutions of Nonsmooth DAEs

The well-posedness theory in [59] is collected in the next result.

Theorem 3.4 *Let Assumption 3.1 hold. Then the following statements hold:*

(i) If $(t_0, \mathbf{p}_0, \mathbf{x}_0, \mathbf{y}_0) \in G_C^0 \cap G_R$, *then there exist* $\alpha > 0$ *and a regular solution of* (3.1) *on* $[t_0 - \alpha, t_0 + \alpha] \times \{\mathbf{p}_0\}$ *through* $\{(t_0, \mathbf{p}_0, \mathbf{x}_0, \mathbf{y}_0)\}$.

(ii) If $\widetilde{\mathbf{z}}$ *is a regular solution of* (3.1) *on* $[t_0, t_f] \times \{\mathbf{p}_0\} \subset D_t \times D_p$ *through* $\{(t_0, \mathbf{p}_0, \mathbf{x}_0, \mathbf{y}_0)\}$ *and Assumption 3.2 holds, then* $\widetilde{\mathbf{z}}$ *is a unique regular solution*[1] *of* (3.1) *on* $[t_0, t_f] \times \{\mathbf{p}_0\}$ *through* $\{(t_0, \mathbf{p}_0, \mathbf{x}_0, \mathbf{y}_0)\}$.

(iii) If $\widetilde{\mathbf{z}}$ *is a regular solution of* (3.1) *on* $[t_0, t_f] \times \{\mathbf{p}_0\} \subset D_t \times D_p$ *through* $\{(t_0, \mathbf{p}_0, \mathbf{x}_0, \mathbf{y}_0)\}$, *then there exist* $\beta > 0$ *and a regular continuation*[2] *of* $\widetilde{\mathbf{z}}$ *on* $[t_0 - \beta, t_f + \beta]$. *Moreover, there exist* $t_L \in \mathbb{R} \cup \{-\infty\}$ *and* $t_U \in \mathbb{R} \cup \{+\infty\}$ *satisfying* $t_L < t_0 < t_f < t_U$ *and a maximal regular continuation of* $\widetilde{\mathbf{z}}$ *on* (t_L, t_U).

Example 3.5 Considering again (3.4) in Example 3.3 with $t_0 = 0.5$,

$$\widetilde{\mathbf{z}} \equiv (\widetilde{x}, \widetilde{y}) : [0.25, 0.75] \times \{-0.5\} \to \mathbb{R}^2 : (t, p) \mapsto \begin{cases} (-t, 1-t), & \text{if } t \in [0.25, 0.5], \\ (t-1, t), & \text{if } t \in (0.5, 0.75], \end{cases}$$

is a regular solution of (3.4) on $[0.25, 0.75]\times\{-0.5\}$ through $\{(0.5, -0.5, -0.5, 0.5)\}$; $\widetilde{y}(t, -0.5) > 0$ for all $t \in [0.25, 0.75]$. Since Assumption 3.2 holds on $D_t \times D_p \times D_x \times D_y = \mathbb{R} \times (-1, 1) \times (-1, 1) \times (-2, 2)$, $\widetilde{\mathbf{z}}$ is unique. Moreover, since Assumption 3.1 also holds on $D_t \times D_p \times D_x \times D_y$, there exists a maximal regular continuation, $\widetilde{\mathbf{z}}_{\max}$, of $\widetilde{\mathbf{z}}$ on $(t_L, t_U) = (-1, 2)$, given by

$$\widetilde{\mathbf{z}}_{\max} : (-1, 2) \times \{-0.5\} \to \mathbb{R}^2 : (t, p) \mapsto \begin{cases} (-t, 1+t), & \text{if } t \in (-1, 0], \\ (t-1, t), & \text{if } t \in (0, 2). \end{cases}$$

Note that $\widetilde{\mathbf{z}}_{\max}$ has no regular continuation for any superset of (t_L, t_U) by loss of regularity;

$$\lim_{t \to -1^+} (t, -0.5, \widetilde{\mathbf{z}}_{\max}(t, -0.5)) = (-1, -0.5, 1, 0) \notin G_R$$

---

[1] $\widetilde{\mathbf{z}}$ is *unique* in the sense that, given any other solution $\mathbf{z}^*$ of (3.1) on $T \times \{\mathbf{p}_0\}$ through $\{(t_0, \mathbf{p}_0, \mathbf{x}_0, \mathbf{y}_0)\}$ satisfying $T \cap [t_0, t_f] \neq \{t_0\}$, $\widetilde{\mathbf{z}}(t, \mathbf{p}_0) = \mathbf{z}^*(t, \mathbf{p}_0)$ for all $t \in T \cap [t_0, t_f]$.

[2] A *(regular) continuation* $\mathbf{z}^* : T \times \{\mathbf{p}_0\}$ of $\widetilde{\mathbf{z}}$ is a (regular) solution of (3.1) on $T \times \{\mathbf{p}_0\}$ through $\{(t_0, \mathbf{p}_0, \mathbf{x}_0, \mathbf{y}_0)\}$ satisfying $\mathbf{z}^*(t, \mathbf{p}_0) = \widetilde{\mathbf{z}}(t, \mathbf{p}_0)$ for all $t \in [t_0, t_f]$ and $T \subset D_t$ is a strict superset of $[t_0, t_f]$. $\mathbf{z}^*$ is a *maximal (regular) continuation* of $\widetilde{\mathbf{z}}$ if it has no (regular) continuation for any superset of T contained in D_t.

and

$$\lim_{t \to 2^-} (t, -0.5, \widetilde{\mathbf{z}}_{\max}(t, -0.5)) = (2, -0.5, 1, 0) \notin G_{\mathrm{R}}.$$

On the other hand, there exist two solutions of (3.4) on $[0, 1] \times \{-0.5\}$ through $\{(0, -0.5, -0.5, 0.5)\}$:

$$\mathbf{z}^{\dagger} \equiv (x^{\dagger}, y^{\dagger}) : [0, 1] \times \{-0.5\} \to \mathbb{R}^2 : (t, p) \mapsto \begin{cases} (-t - 0.5, 0.5 - t), & \text{if } t \in [0, 0.5], \\ (t - 1.5, t - 0.5), & \text{if } t \in (0.5, 1], \end{cases}$$

and

$$\mathbf{z}_{\dagger} \equiv (x_{\dagger}, y_{\dagger}) : [0, 1] \times \{-0.5\} \to \mathbb{R}^2 : (t, p) \mapsto \begin{cases} (-t - 0.5, 0.5 - t), & \text{if } t \in [0, 0.5], \\ (t - 1.5, 0.5 - t), & \text{if } t \in (0.5, 1]. \end{cases}$$

Non-uniqueness here is because of loss of regularity at $t = 0.5$; neither $\mathbf{z}^{\dagger}$ nor $\mathbf{z}_{\dagger}$ are regular since $y^{\dagger}(0.5, -0.5) = y_{\dagger}(0.5, -0.5) = 0$.

The following remarks are in order before returning to the nonsmooth flash model:

1. In summary, (3.1) is said to be generalized differentiation index one along a solution trajectory if the participating functions satisfy Assumption 3.1 and said solution is contained in G_{R} (i.e., the solution is regular). The latter condition can be checked in a global manner by evaluating the set $G_{\mathrm{R}}^{\pi} \subset G_{\mathrm{R}}$. If, in addition, $\mathbf{g}$ is PC^1 on its domain, then generalized differentiation index one corresponds to a solution being contained in G_{R}^{PC}, which can be verified by evaluating the set $G_{\mathrm{R}}^{CCO} \subset G_{\mathrm{R}}^{PC}$.
2. Well-posedness results of the IVP in DAEs (3.1) in [59] rely crucially on a nonsmooth extended implicit function theorem [59, Theorem 3.5]. This theorem allows for analyzing nonsmooth semi-explicit generalized differentiation index-one DAEs via nonsmooth ODE theory by providing an equivalence between said dynamical systems [57, Proposition 4.1]; if Assumptions 3.1 and 3.2 hold and $\widetilde{\mathbf{z}} \equiv (\widetilde{\mathbf{x}}, \widetilde{\mathbf{y}})$ is a regular solution of (3.1) on $[t_0, t_f] \times \{\mathbf{p}_0\}$ through $\{(t_0, \mathbf{p}_0, \mathbf{x}_0, \mathbf{y}_0)\}$, then there exist a neighborhood $N_{\mathbf{p}_0} \subset D_p$ of $\mathbf{p}_0$, a set $\Omega_0 \subset G_{\mathrm{C}}^{0}$ and a unique regular solution $\widetilde{\mathbf{z}}$ of (3.1) on $[t_0, t_f] \times N_{\mathbf{p}_0}$ through Ω_0. Moreover, there exist $\delta, \rho > 0$ and a Lipschitz continuous function

$$\mathbf{r} : B_{\delta}(\{(t, \mathbf{p}_0, \widetilde{\mathbf{x}}(t, \mathbf{p}_0)) : t \in [t_0, t_f]\}) \subset D_t \times D_p \times D_x \to \mathbb{R}^{n_y}$$

which satisfy

a. $\widetilde{\mathbf{y}}(t, \mathbf{p}) = \mathbf{r}(t, \mathbf{p}, \widetilde{\mathbf{x}}(t, \mathbf{p}))$ for all $(t, \mathbf{p}) \in [t_0, t_f] \times N_{\mathbf{p}_0}$;
b. $\{(t, \mathbf{p}, \widetilde{\mathbf{x}}(t, \mathbf{p})) : (t, \mathbf{p}) \in [t_0, t_f] \times N_{\mathbf{p}_0}\} \subset B_{\delta}(\{(t, \mathbf{p}_0, \widetilde{\mathbf{x}}(t, \mathbf{p}_0)) : t \in [t_0, t_f]\})$;
c. $\{(t, \mathbf{p}, \widetilde{\mathbf{z}}(t, \mathbf{p}) : (t, \mathbf{p}) \in [t_0, t_f] \times N_{\mathbf{p}_0}\} \subset B_{\rho}(\{(t, \mathbf{p}_0, \widetilde{\mathbf{z}}(t, \mathbf{p}_0)) : t \in [t_0, t_f]\})$.

Thus, the nonsmooth ODE

$$\dot{\mathbf{x}}(t, \mathbf{p}) = \mathbf{f}(t, \mathbf{p}, \mathbf{x}(t, \mathbf{p}), \mathbf{r}(t, \mathbf{p}, \mathbf{x}(t, \mathbf{p}))), \tag{3.5a}$$

$$\mathbf{x}(t_0, \mathbf{p}) = \mathbf{f}_0(\mathbf{p}), \tag{3.5b}$$

may be analyzed in a local neighborhood of $\{(t, \mathbf{p}, \widetilde{\mathbf{x}}(t, \mathbf{p})) : (t, \mathbf{p}) \in [t_0, t_f] \times N_{\mathbf{p}_0}\}$ to ascertain theoretical results for (3.1).

3. The regularity set in [59] is defined as G_{R}^{π} (denoted G_{R} in [59]), instead of the superset G_{R}, but the aforementioned Lipschitzian extended implicit function theorem [59, Theorem 3.5] holds if the Clarke Jacobian projection is replaced by the Lipschitz homeomorphism condition, allowing for such a generalization. Moreover, the PC^1 case is not considered in [59], but by the method of proof of [59, Theorem 3.5], a PC^1 extended implicit function theorem can be similarly proved by using the PC^1 local implicit function theorem [49, Corollary 20] in place of Clarke's locally Lipschitz implicit function theorem [13, Corollary to Theorem 7.1.1]. The various regularity sets defined above are therefore all suitable for guaranteeing (3.1) is generalized differentiation index one locally along a solution trajectory.
4. The uniqueness result in Theorem 3.4 states that the regular solution of (3.1) is unique in the sense that there cannot exist another solution, regular or not. Said result also holds for a regular solution defined on a non-singleton set (i.e., $P \subset D_p$). The existence and uniqueness results in [59] include other generalizations; the statement of the results here is motivated by facilitation of discussion.
5. Theorem 3.4 is applicable to nonlinear complementarity systems (NCSs) [23, 42, 54, 65]; the NCS

$$\dot{\mathbf{x}}(t, \mathbf{p}) = \mathbf{f}(t, \mathbf{p}, \mathbf{x}(t, \mathbf{p}), \mathbf{u}(t, \mathbf{p})), \tag{3.6a}$$

$$\mathbf{v}(t, \mathbf{p}) = \mathbf{h}(t, \mathbf{p}, \mathbf{x}(t, \mathbf{p}), \mathbf{u}(t, \mathbf{p})), \tag{3.6b}$$

$$v_i(t, \mathbf{p}) u_i(t, \mathbf{p}) = 0, \quad \forall i \in \{1, \dots, n_v\}, \tag{3.6c}$$

$$\mathbf{v}(t, \mathbf{p}) \geq \mathbf{0}_{n_v}, \mathbf{u}(t, \mathbf{p}) \geq \mathbf{0}_{n_v}, \tag{3.6d}$$

can be recast as (3.1) by letting

$$\mathbf{g} : (t, \mathbf{p}, \boldsymbol{\eta}_x, \boldsymbol{\eta}_v, \boldsymbol{\eta}_u) \mapsto \begin{bmatrix} \mathbf{h}(t, \mathbf{p}, \boldsymbol{\eta}_x, \boldsymbol{\eta}_u) - \boldsymbol{\eta}_v \\ \mathbf{min}(\boldsymbol{\eta}_v, \boldsymbol{\eta}_u) \end{bmatrix}$$

where $\mathbf{min}(\cdot, \cdot)$ is the component-wise vector-valued minimum function and the algebraic variable is $\mathbf{y} \equiv (\mathbf{u}, \mathbf{v})$. Mixed nonlinear complementarity systems can be similarly recast.
6. Continuous and Lipschitzian dependence of regular solutions of (3.1) on initial values and parameters is inherited from nonsmoothness of the right-hand side functions $\mathbf{f}$, $\mathbf{g}$ and $\mathbf{f}_0$ [59]; given a regular solution $\widetilde{\mathbf{z}}$ of (3.1) on $[t_0, t_f] \times \{\mathbf{p}_0\}$

through $\{(t_0, \mathbf{p}_0, \mathbf{x}_0, \mathbf{y}_0)\}$, Assumption 3.1 implies the existence of a neighborhood $N_{\mathbf{p}_0} \subset D_p$ of $\mathbf{p}_0$ and a regular solution $\widetilde{\mathbf{z}}$ of (3.1) on $[t_0, t_f] \times N_{\mathbf{p}_0}$ through $\Omega_0 \subset G_{\mathrm{C}}^0$ such that, for each $t \in [t_0, t_f]$, the mapping $\widetilde{\mathbf{z}}(t, \cdot)$ is continuous at $\mathbf{p}_0$. If in addition Assumption 3.2 holds, then the mapping $\widetilde{\mathbf{z}}(t, \cdot)$ is Lipschitz continuous on a neighborhood of $\mathbf{p}_0$, with a Lipschitz constant that is independent of t.

7. In the nonsmooth setting, the forward parametric sensitivity functions $\frac{\partial \widetilde{\mathbf{x}}}{\partial \mathbf{p}}(t, \mathbf{p}_0)$ and $\frac{\partial \widetilde{\mathbf{y}}}{\partial \mathbf{p}}(t, \mathbf{p}_0)$ may not be well-defined for all t. Instead, (local) sensitivity information for (3.1) can be characterized via elements of the Clarke Jacobians $\partial[\widetilde{\mathbf{x}}(t, \cdot)](\mathbf{p}_0)$ and $\partial[\widetilde{\mathbf{y}}(t, \cdot)](\mathbf{p}_0)$ (i.e., fixed t and varying $\mathbf{p}$). Such elements are computationally relevant by providing sensitivity information for state variables with respect to $\mathbf{p}$ at $\mathbf{p} = \mathbf{p}_0$ for use in dedicated nonsmooth optimization [33, 36] and equation-solving methods [16, 46], which exhibit attractive convergence properties. However, Clarke Jacobian elements are generally difficult to compute in an automatable way because the Clarke Jacobian satisfies calculus rules only as inclusions, among other reasons.

   A recent advancement in nonsmooth analysis provides a recourse here: the *lexicographic directional (LD-)derivative* [30], which is defined using lexicographic differentiation [41], is a new nonsmooth tool which provides an accurate, automatable and computationally tractable method for generalized derivative evaluation (e.g., via a nonsmooth vector forward mode of automatic differentiation [30]). Consequently, the generalized differentiation index-one DAE (3.1) now possesses a suitable sensitivity theory for use in, for example, dynamic optimization (e.g., via sequential methods such as single and multiple shooting) [58], in the form of a nonsmooth sensitivity DAE system [57, Theorem 4.1]. Mirroring the classical sensitivity theory, said nonsmooth sensitivity system admits a unique solution that can be used to furnish a computationally relevant generalized derivative element and simplifies to the classical sensitivity DAE system when the participating functions are C^1. (For a detailed account of theory and application of LD-derivatives to different classes of nonsmooth dynamical systems and optimization problems, the interested reader is directed to [6].)

4 Analysis and Simulation of Nonsmooth Flash Models

To solve the nonsmooth flash model introduced in Sect. 2, the thermodynamic properties of the phases must be determined, i.e., the specific molar volume of the vapor and liquid phases, v_V and v_L, respectively, and the molar enthalpy of the vapor and liquid, h_V and h_L, respectively. In this section, we describe three approaches to calculate such thermodynamic properties, and demonstrate their implications in simulating the full dynamic behavior of a flash process.

4.1 The Use of a Cubic Equation of State

A well-known approach to calculate the thermodynamic properties of both liquid and vapor phases is the use of a cubic equation of state, relating the specific volume v to the pressure, P, and the temperature, T, below the critical point. This algebraic equation can have up to three real roots, two corresponding to each of the phases (the smallest to the liquid phase and the largest to the vapor phase), and a third unphysical (and meaningless) one. The simplest example is the van der Waals (VDW) equation of state:

$$P = \frac{RT}{v-b} - \frac{a}{v^2}, \tag{4.1}$$

where R is the universal gas constant, and a and b are constants fitted to a specific pure species. To facilitate numerical solution, this equation can be rewritten in two different ways to solve for the vapor and liquid roots from an adequate initial guess; multiplying Eq. (4.1) by $f_V(v_V)$ where

$$f_V(v) \equiv \frac{v-b}{P}, \tag{4.2}$$

yields an equation suitable for calculating the vapor phase solution, while multiplying Eq. (4.1) by $f_L(v_L)$, where

$$f_L(v) \equiv \frac{(v-b)v^2}{a}, \tag{4.3}$$

yields an equation suitable for calculating the liquid phase solution. Dynamic versions of said equations are given by

$$v_V(t) = \frac{RT(t)}{P(t)} + b - \frac{a(v_V(t)-b)}{P(t)(v_V(t))^2}, \tag{4.4a}$$

$$v_L(t) = b + (v_L(t))^2 \frac{RT(t) + (b - v_L(t))P(t)}{a}. \tag{4.4b}$$

Note that both f_V and f_L are positive when evaluated at values near the solutions of the vapor and liquid roots of (4.1), respectively.

Defining the compressibility factor $Z \equiv Pv/RT$, $q \equiv a/bRT$ and $I \equiv b/v$, the following equation can be used to calculate the thermodynamically consistent enthalpy from the VDW equation of state using the definition of the residual enthalpy, h^R [56]:

$$\frac{h^R}{RT} \equiv \frac{h}{RT} - \frac{h^{\text{ig}}}{RT} = Z - 1 - qI,$$

so the vapor and liquid enthalpies follow:

$$h_V(t) = h^{\text{ig}}(T(t)) + P(t)v_V(t) - RT(t) - a/v_V(t), \tag{4.5a}$$

$$h_L(t) = h^{\text{ig}}(T(t)) + P(t)v_L(t) - RT(t) - a/v_L(t), \tag{4.5b}$$

where h^{ig} is the ideal gas molar enthalpy given by:

$$h^{\text{ig}}(T) = h_0^{\text{ig}} + \int_{T_0}^{T} C_p(s)ds, \tag{4.6}$$

where h_0^{ig} is the ideal gas molar enthalpy at the standard (reference) temperature ($T_0 = 298$ K), and C_p is the ideal gas heat capacity.

For a pure component system the saturation pressure is determined from the condition for equilibrium of two phases:

$$G^V(t) = G^L(t),$$

where G^V, G^L are the molar Gibbs free energy of the vapor and liquid phases, respectively. The residual Gibbs free energy, G^R, based on a cubic equation of state is given by:

$$\frac{G^R}{RT} \equiv \frac{G}{RT} - \frac{G^{\text{ig}}}{RT} = Z - 1 - \ln\left(Z - \frac{bP}{RT}\right) - qI,$$

where G^{ig} is the ideal gas Gibbs free energy of the pure component. Thus, the equilibrium condition equation reduces to the following form for the VDW equation of state:

$$P^{\text{sat}}v_V^{\text{sat}} - \frac{a}{v_V^{\text{sat}}} - RT\ln\left(\frac{P^{\text{sat}}(v_V^{\text{sat}} - b)}{RT}\right) = P^{\text{sat}}v_L^{\text{sat}} - \frac{a}{v_L^{\text{sat}}} - RT\ln\left(\frac{P^{\text{sat}}(v_L^{\text{sat}} - b)}{RT}\right), \tag{4.7}$$

where v_V^{sat} and v_L^{sat} are the saturated vapor and liquid molar volumes, respectively. It was not possible to get these equations to solve reliably with the typical DAE solver used in this paper. It is well known that vapor-liquid equilibrium equations, in particular with equations of state, are very difficult to converge with standard approaches such as Newton's method. This difficulty is often addressed by nesting tailored algorithms for the flash equations in the overall equation solve (applying the implicit function theorem). A class of these tailored algorithms has recently been extended to the nonsmooth flash formulation presented here [69, 70], but these have not yet been incorporated in the simulator used in this paper.

Alternatively, we use the empirical Antoine expression, Eq. (2.3), to solve for P^{sat} (with A, B, C fitted to empirical data). In doing so, P^{sat} has minor deviations from the one obtained by Eq. (4.7) in the temperature range of interest. In Sect. 4.3 the heat of vaporization is derived from the Antoine expression using the Clapeyron equation. This is not consistent with Eqs. (4.5a)–(4.5b), which will be replaced by other equations in Sect. 4.3.

The outlet flow rates of the vapor and liquid streams depend on the pressure and phase hold-up inside the vessel, expressed by the following nonsmooth equations:

$$F_V(t) = C_V^v \min\left(V^{\min}, V_V(t)\right) \max\left(\frac{P(t) - P_{\text{out}}}{\sqrt{|P(t) - P_{\text{out}}| + \epsilon}}, 0\right), \tag{4.8a}$$

$$F_L(t) = C_L^v \min\left(V^{\min}, V_L(t)\right) \max\left(\frac{P(t) + P_L^h(t) - P_{\text{out}}}{\sqrt{\left|P(t) + P_L^h(t) - P_{\text{out}}\right| + \epsilon}}, 0\right), \tag{4.8b}$$

where C_V^v and C_L^v are the valve constants, V_V and V_L are the total vapor phase and liquid phase hold-up volumes, respectively, and thus satisfy

$$V_V(t) = M_V(t)v_V(t), \quad V_L(t) = M_L(t)v_L(t).$$

P_L^h is the hydrostatic pressure exerted by the liquid phase (this is assumed to only affect the flow rate, otherwise the pressure is assumed to be uniform, as mentioned above), expressed by:

$$P_L^h(t) = 10^{-5} g M_L(t) mw/S,$$

where g is the gravitational constant, mw is the molecular weight, S is the cross-sectional area of the vessel and the factor 10^{-5} converts the units from Pa to bar. The first term in Eqs. (4.8a)–(4.8b) uses the nonsmooth min function to simulate the closing of the valve when the respective phase disappears, enforcing that liquid cannot flow through the vapor outlet and vice versa. The second term uses the nonsmooth max function to simulate a nonreturn check-valve, to prevent back flow when the pressure inside the vessel is lower than the downstream pressure, P_{out}. ϵ is a small regularization parameter to guarantee Lipschitz continuity.

4.2 *Analysis and Simulation of the Cubic Equation of State Model*

The simple flash model outlined thus far is thermodynamically consistent and is capable of simulating a transition between all three thermodynamic regimes (i.e., the vapor-only, two-phase and liquid-only regimes). Observe that Eqs. (4.8), (2.3) and (4.5) can be solved explicitly to eliminate the respective unknowns, P^{sat}, F_V, F_L, h_V and h_L. Also note that in this case after using (4.5a) and (4.5b), the terms Pv_V and Pv_L in (2.1d) cancel out. The simple flash model falls into the framework of the nonsmooth DAE system (3.1) with differential variables $\mathbf{x} \equiv (M, U)$, algebraic variables $\mathbf{y} \equiv (M_L, M_V, P, T, v_L, v_V)$, $\mathbf{f}$ corresponding to Eqs. (2.1a)–(2.1b) and $\mathbf{g}$ corresponding to Eqs. (2.1c), (2.1d), (2.1e), (2.2), (4.4). Note that $\mathbf{f}$ is C^1 on its domain, but $\mathbf{g}$ is PC^1 on its domain because the mid function is PC^1.

4.2.1 Structural Analysis

Structural analysis of the model can provide some insight into whether the problem may be high index in each of the phase regimes, as determined by Eq. (2.2). The occurrence matrices associated with the algebraic equations in each of the phase regimes (i.e., corresponding to the three selection functions of (2.2)) are presented below. In the liquid-only regime:

$$\begin{array}{c} \text{Eq.} \\ (2.1\text{c}) \\ (2.1\text{d}) \\ (2.1\text{e}) \\ (2.2) \\ (4.4\text{a}) \\ (4.4\text{b}) \end{array}
\begin{array}{cccccc} M_L & M_V & P & T & v_L & v_V \\ \left[\begin{array}{c} \textcircled{X} \\ \text{X} \\ \text{X} \\ \\ \\ \\ \end{array}\right. & \begin{array}{c} 0 \\ 0 \\ 0 \\ \textcircled{0} \\ \\ \\ \end{array} & \begin{array}{c} \\ \\ \\ \\ \text{X} \\ \textcircled{X} \end{array} & \begin{array}{c} \\ \textcircled{X} \\ \\ \\ \text{X} \\ \text{X} \end{array} & \begin{array}{c} \\ \text{X} \\ \textcircled{X} \\ \\ \\ \text{X} \end{array} & \left.\begin{array}{c} \\ \text{X} \\ \text{X} \\ \\ \textcircled{X} \\ \\ \end{array}\right] \end{array}$$

In the two-phase regime:

$$\begin{array}{c} \text{Eq.} \\ (2.1\text{c}) \\ (2.1\text{d}) \\ (2.1\text{e}) \\ (2.2) \\ (4.4\text{a}) \\ (4.4\text{b}) \end{array}
\begin{array}{cccccc} M_L & M_V & P & T & v_L & v_V \\ \left[\begin{array}{c} \textcircled{X} \\ \text{X} \\ \text{X} \\ \\ \\ \\ \end{array}\right. & \begin{array}{c} \text{X} \\ \text{X} \\ \textcircled{X} \\ \\ \\ \\ \end{array} & \begin{array}{c} \\ \\ \\ \textcircled{X} \\ \text{X} \\ \text{X} \end{array} & \begin{array}{c} \\ \textcircled{X} \\ \\ \text{X} \\ \text{X} \\ \text{X} \end{array} & \begin{array}{c} \\ \text{X} \\ \text{X} \\ \\ \\ \textcircled{X} \end{array} & \left.\begin{array}{c} \\ \text{X} \\ \text{X} \\ \\ \textcircled{X} \\ \\ \end{array}\right] \end{array}$$

In the vapor-only regime:

$$
\begin{array}{r}
\text{Eq.} \\
(2.1c) \\
(2.1d) \\
(2.1e) \\
(2.2) \\
(4.4a) \\
(4.4b)
\end{array}
\begin{array}{c}
\begin{array}{cccccc} M_L & M_V & P & T & v_L & v_V \end{array} \\
\left[\begin{array}{cccccc}
0 & \textcircled{X} & & & & \\
0 & X & & \textcircled{X} & X & X \\
0 & X & & & X & \textcircled{X} \\
\textcircled{0} & & & & & \\
 & & \textcircled{X} & X & & X \\
 & & X & X & \textcircled{X} &
\end{array}\right]
\end{array}
$$

The six unknowns are denoted above the matrices with the corresponding equations listed on the left. Variables that are identically zero (by Eq. (2.2)) are denoted by 0 in the matrices to emphasize that certain equations cannot be used to solve for certain variables (e.g., Eq. (2.1d) cannot be used to solve for v_V in the liquid-only regime). The problem structure changes as the nonsmooth Eq. (2.2) switches between three selection functions, where only the fourth component differs:

$$g_{(1),4} \equiv M_V, \quad g_{(2),4} \equiv P - P^{\text{sat}}(T), \quad g_{(3),4} \equiv -M_L.$$

The equation used to solve for a particular unknown is denoted by a circle. When only one phase exists Eq. (2.2) sets the corresponding phase hold-up to zero, leading to a qualitative change in the structure of the problem. The main difference is that the pressure has to be determined from Eq. (4.4a) or (4.4b) instead of from Eq. (2.2).

4.2.2 Regularity Analysis

Nevertheless, it is apparent that this model is structurally regular in each of the phase regimes, thus it may be generalized differentiation index one in all three regimes. In particular, the partial Jacobian matrices associated with each selection function (and thus each of the three regimes), $\mathbf{Y}_{(i)} \equiv \frac{\partial \mathbf{g}_{(i)}}{\partial \mathbf{y}}$ are given as follows: for the liquid-only regime,

$$
\mathbf{Y}_{(1)} = \begin{bmatrix}
1 & 1 & 0 & 0 & 0 & 0 \\
\mathscr{H}_L & \mathscr{H}_V & 0 & \mathscr{A} & a\frac{M_L}{v_L^2} & a\frac{M_V}{v_V^2} \\
v_L & v_V & 0 & 0 & M_L & M_V \\
0 & 1 & 0 & 0 & 0 & 0 \\
0 & 0 & \mathscr{T}_V & \mathscr{R}_V & 0 & \mathscr{C}_V \\
0 & 0 & \mathscr{T}_L & \mathscr{R}_L & \mathscr{C}_L & 0
\end{bmatrix},
$$

where

$$\begin{aligned}
\mathscr{H}_L &\equiv h_0^{\mathrm{ig}} + C_p(T - T_0) - \frac{a}{v_L} - RT, \\
\mathscr{H}_V &\equiv h_0^{\mathrm{ig}} + C_p(T - T_0) - \frac{a}{v_V} - RT, \\
\mathscr{A} &\equiv (C_p - R)(M_L + M_V), \\
\mathscr{C}_V &\equiv -1 + \frac{a(v_V - 2b)}{Pv_V^3}, \\
\mathscr{C}_L &\equiv -1 + \frac{2v_L(RT + Pb - 1.5Pv_L)}{a}, \\
\mathscr{R}_V &\equiv \frac{R}{P}, \\
\mathscr{R}_L &\equiv \frac{Rv_L^2}{a}, \\
\mathscr{T}_V &\equiv \frac{a(v_V - b)}{P^2 v_V^2} - \frac{RT}{P^2}, \\
\mathscr{T}_L &\equiv -\frac{v_L^2(v_L - b)}{a}.
\end{aligned}$$

For the two-phase regime,

$$\mathbf{Y}_{(2)} = \begin{bmatrix} 1 & 1 & 0 & 0 & 0 & 0 \\ \mathscr{H}_L & \mathscr{H}_V & 0 & \mathscr{A} & a\frac{M_L}{v_L^2} & a\frac{M_V}{v_V^2} \\ v_L & v_V & 0 & 0 & M_L & M_V \\ 0 & 0 & 1 & \mathscr{P} & 0 & 0 \\ 0 & 0 & \mathscr{T}_V & \mathscr{R}_V & 0 & \mathscr{C}_V \\ 0 & 0 & \mathscr{T}_L & \mathscr{R}_L & \mathscr{C}_L & 0 \end{bmatrix},$$

where

$$\mathscr{P} \equiv -\frac{10^{A-B/(C+T)} B \ln(10)}{(C+T)^2}.$$

For the vapor-only regime,

$$\mathbf{Y}_{(3)} = \begin{bmatrix} 1 & 1 & 0 & 0 & 0 & 0 \\ \mathscr{H}_L & \mathscr{H}_V & 0 & \mathscr{A} & a\frac{M_L}{v_L^2} & a\frac{M_V}{v_V^2} \\ v_L & v_V & 0 & 0 & M_L & M_V \\ -1 & 0 & 0 & 0 & 0 & 0 \\ 0 & 0 & \mathscr{T}_V & \mathscr{R}_V & 0 & \mathscr{C}_V \\ 0 & 0 & \mathscr{T}_L & \mathscr{R}_L & \mathscr{C}_L & 0 \end{bmatrix}.$$

As expected, only the fourth row of each matrix differs because of Eq. (2.2). (i.e., **g** has three selection functions whose partial Jacobian matrices only differ in the fourth row). Since at most two selection functions can be active at a domain point of interest, it follows that

$$\Lambda_{\mathbf{y}}\mathbf{g} = \begin{cases} \{\mathbf{Y}_{(1)}^{\text{liquid-only}}\} & \text{if } (\mathbf{x},\mathbf{y}) \in \text{liquid-only regime}, \\ \{\mathbf{Y}_{(1)}^{\text{liquid-only}}, \mathbf{Y}_{(2)}^{\text{liquid-only}}\} & \text{if } (\mathbf{x},\mathbf{y}) \in \text{liquid-only and two-phase transition}, \\ \{\mathbf{Y}_{(2)}^{\text{two-phase}}\} & \text{if } (\mathbf{x},\mathbf{y}) \in \text{two-phase regime}, \\ \{\mathbf{Y}_{(2)}^{\text{vapor-only}}, \mathbf{Y}_{(3)}^{\text{vapor-only}}\} & \text{if } (\mathbf{x},\mathbf{y}) \in \text{vapor-only and two-phase transition}, \\ \{\mathbf{Y}_{(3)}^{\text{vapor-only}}\} & \text{if } (\mathbf{x},\mathbf{y}) \in \text{vapor-only regime}, \end{cases}$$

where, for example, the notation $\mathbf{Y}_{(1)}^{\text{liquid-only}}$ denotes the matrix-valued function $\mathbf{Y}_{(1)}$ evaluated at realizable values of state variables in the liquid-only regime.

Recall that a solution trajectory $(\widetilde{\mathbf{x}}, \widetilde{\mathbf{y}})$ associated with reference parameter $\mathbf{p}_0$ (see Table 1) is regular at time t if

$$(\widetilde{\mathbf{x}}(t,\mathbf{p}_0), \widetilde{\mathbf{y}}(t,\mathbf{p}_0)) \in G_{\text{R}}^{CCO} = \{(\mathbf{x},\mathbf{y}) : \text{sign}(\det(\mathbf{Y})) = a\ \forall \mathbf{Y} \in \Lambda_{\mathbf{y}}\mathbf{g}\},$$

for some $a = \pm 1$, which is sufficient for generalized differentiation index one. Regularity of solutions can therefore be shown by analyzing the signs of the determinants of $\mathbf{Y}_{(1)}, \mathbf{Y}_{(2)}, \mathbf{Y}_{(3)}$ at realizable values of state variables: since $M_V = 0$ in the liquid-only regime, it follows that

$$\det(\mathbf{Y}_{(1)}^{\text{liquid-only}}) = -M_L \mathscr{A} \mathscr{C}_V \mathscr{T}_L.$$

Similarly, since $M_L = 0$ in the vapor-only regime,

$$\det(\mathbf{Y}_{(3)}^{\text{vapor-only}}) = -M_V \mathscr{A} \mathscr{C}_L \mathscr{T}_V.$$

Table 1 Model parameters for condensation of n-Butane

Constant	Value	Units	Constant	Value	Units
mw	0.0581	kg/mol	A	6.83	
P_{out}	1	bar	B	945.8	K
a	13.93	$\text{barL}^2/\text{mol}^2$	C	-33	K
b	0.1168	L/mol	C_p	0.093	kJ/mol/K
V	9	L	C_V^v	100	mol/s/L/$\sqrt{\text{bar}}$
$V^{\min}$	0.01	L	C_L^v	50	mol/s/L/$\sqrt{\text{bar}}$
S	0.01	m^2	F_{in}	1	mol/s
ϵ	10^{-6}		h_{in}	-125	kJ/mol
			h_0^{ig}	-125.6	kJ/mol

Consider the two non-classical cases (i.e., at the two phase boundaries): for state variable values in the liquid-only regime, $M_L > 0$, $\mathscr{A} > 0$ and $\mathscr{T}_L < 0$. Recall that

$$\mathscr{C}_V = \frac{\partial g_{(1),5}}{\partial v_V}$$

where $g_{(1),5}$ is the right-hand side function associated with the algebraic equation (4.4a) evaluated at the state variables of the system (and equals $g_{(2),5}$ and $g_{(3),5}$ since said function is C^1 on its domain). Equation (4.4a) was obtained by multiplying the VDW equation of state (4.1) by $f_V(v)$, as defined in (4.2). That is,

$$0 = \left(\frac{RT}{v-b} - \frac{a}{v^2} - P \right) f_V(v).$$

The partial derivative of the right-hand side of this equation with respect to v, and evaluated at a solution of the dynamic system, equals $\mathscr{C}_V$ by definition. Therefore,

$$\mathscr{C}_V = \frac{\partial P}{\partial v}(v_V) f_V(v_V) + f'_V(v_V)\left(P(v_V) - P\right) = \frac{\partial P}{\partial v}(v_V) f_V(v_V),$$

where the argument v_V appearing above is understood to be the vapor root of the equation of state and

$$P(v) \equiv \frac{RT}{v-b} - \frac{a}{v^2}.$$

Moreover, since $(P(v_V) - P) = 0$ at points where Eq. (4.1) is satisfied, $f_V(v_V) > 0$ (as mentioned at the beginning of Sect. 4.1), and the fact that $\frac{\partial P}{\partial v}(v_V) < 0$ holds from physical consideration, it follows that $\mathscr{C}_V < 0$. From this it follows that

$$\det(\mathbf{Y}_{(1)}^{\text{liquid-only}}) = \det\left(\frac{\partial \mathbf{g}_{(1)}}{\partial \mathbf{y}}(\widetilde{\mathbf{x}}(t, \mathbf{p}_0), \widetilde{\mathbf{y}}(t, \mathbf{p}_0)) \right) = -M_L \mathscr{A} \mathscr{C}_V \mathscr{T}_V < 0. \quad (4.9)$$

Therefore, regularity holds at the transition between liquid-only and two-phase regimes if

$$\text{sign}(\det(\mathbf{Y}_{(2)}^{\text{liquid-only}})) = -1.$$

Since $M_V > 0$ and $\mathscr{A} > 0$, a similar analysis as above can be used to show that $\mathscr{C}_L < 0$ with state variables assuming values at the vapor-only and two-phase boundary, so that

$$\det(\mathbf{Y}_{(3)}^{\text{vapor-only}}) = \det\left(\frac{\partial \mathbf{g}_{(3)}}{\partial \mathbf{y}}(\widetilde{\mathbf{x}}(t, \mathbf{p}_0), \widetilde{\mathbf{y}}(t, \mathbf{p}_0)) \right) = -M_V \mathscr{A} \mathscr{C}_L \mathscr{T}_L < 0. \quad (4.10)$$

Hence, regularity holds at the transition between vapor-only and two-phase regimes if

$$\mathrm{sign}(\det(\mathbf{Y}_{(2)}^{\text{vapor-only}})) = -1.$$

Away from the phase boundaries (i.e., where the model is C^1), regularity implies classical differentiation index one and holds for values of state variables in the interior of the liquid-only and vapor-only regimes by Eqs. (4.9)–(4.10). Regularity holds in the interior of the two-phase regime if

$$\det(\mathbf{Y}_{(2)}^{\text{two-phase}}) = \det\left(\frac{\partial \mathbf{g}_{(2)}}{\partial \mathbf{y}}(\widetilde{\mathbf{x}}(t, \mathbf{p}_0), \widetilde{\mathbf{y}}(t, \mathbf{p}_0))\right) \neq 0.$$

Evaluation of $\det(\mathbf{Y}_{(2)}^{\text{liquid-only}})$, $\det(\mathbf{Y}_{(2)}^{\text{two-phase}})$ and $\det(\mathbf{Y}_{(2)}^{\text{vapor-only}})$ would therefore yield definitive conclusions with respect to the model being of (generalized) differentiation index one.

4.2.3 Simulation

A dynamic process of condensing n-Butane in a constant volume flash vessel is simulated by solving Eqs. (2.1)–(2.3) and (4.4)–(4.8). The simulation was performed in Jacobian (Res Group Inc.) using DSL48SE [61, 62] as the solver, which is based on DASSL [45]. The duty is assumed to follow:

$$Q(t) \equiv \max(-10, -0.25t) \quad \text{kW}.$$

The model parameters are given in Table 1. The Antoine expression constants A, B, and C correspond to P^{sat} in units of mmHg. The feed flow rate is constant at constant conditions. Furthermore, C_p in Eq. (4.6) is assumed to be constant. The model requires two initial conditions to be specified; assume that the vessel is initially filled with vapor:

$$M(0) = 0.5 \text{ mol, and } T(0) = 313.15 \text{ K}.$$

Figure 2 shows the solution for the pressure, hold-up volumes of the phases and the temperature in the flash vessel. The initial state of the system corresponds to the vapor-only regime. As cooling starts the temperature, and thus also P^{sat}, decrease, while a slight increase in the pressure is observed due to the accumulation of moles in the vessel. A transition to the two-phase regime occurs when $P = P^{\text{sat}}$ at $t \approx 6$ s, Fig. 2a, leading to the formation of a liquid phase, Fig. 2b. Another transition to the liquid-only regime occurs at $t \approx 191$ s, when P increases above P^{sat}. The increase in P is very sharp because now liquid is forced into the vessel, which is filled entirely with liquid, and the weak dependency of the liquid phase density on

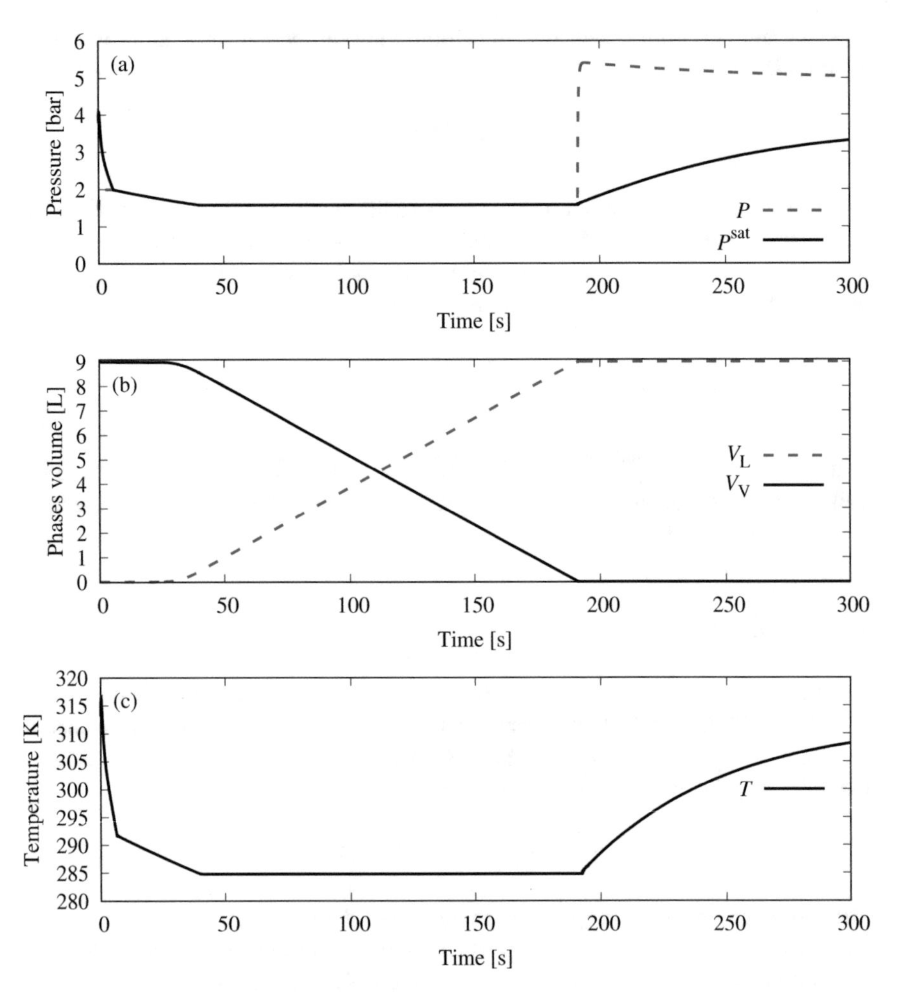

Fig. 2 Simulation results of a flash vessel fed by n-Butane modeled by nonsmooth DAEs. The initial state is in the vapor-only regime. A transition to the two-phase regime occurs when $P = P^{\text{sat}}$ at $t \approx 6$ s (**a**). Another transition to the liquid-only regime occurs at $t \approx 191$ s. The hold-up volumes of the corresponding phases are depicted in (**b**), and the temperature in (**c**)

the pressure results in a rapid pressure increase (practically, this will only be feasible if a high enough upstream pressure is applied).

Figure 3 illustrates the simulation results on a PV diagram (phase plane plot), and highlights the reasoning behind Eq. (2.2). The vapor state, (P, v_V, T), is represented by the rightmost red curve (large molar volume values), and the liquid state, (P, v_L, T), is represented by the leftmost red curve (small molar volume values), magnified to the right for clarity. The isotherms corresponding to the highest and lowest temperatures obtained, 316.5 and 284.8 K, are marked by dashed black lines.

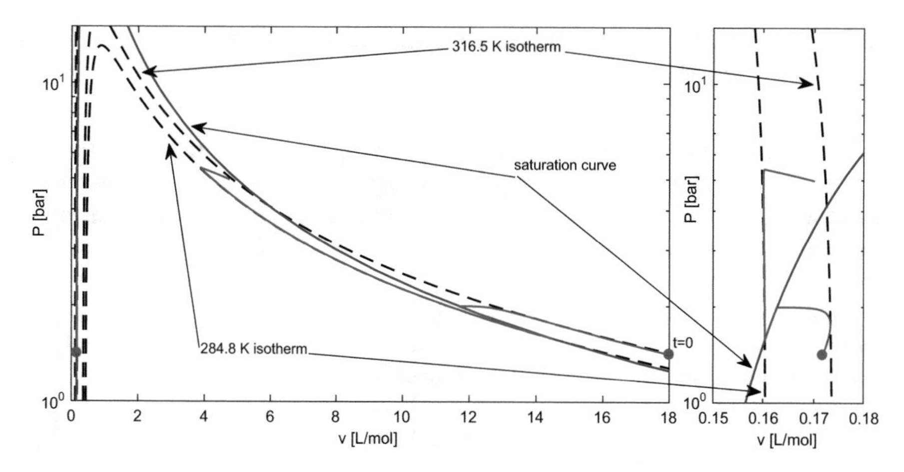

Fig. 3 Simulation results on a PV diagram (red curves): the initial state is marked by red dots; the isotherms for the lowest and highest temperatures are marked by dashed black lines; and the saturation curve, $P^{sat}(T)$, is marked by a blue line. The liquid molar volume curve is magnified on the right for clarity. The molar volumes are calculated from the VDW equation of state for all three regimes

These are the function $P(v)$, according to Eq. (4.1), for constant temperatures. The saturation curve, $P^{sat}(T)$, calculated from Eqs. (2.3), (4.4a) and (4.4b) is depicted by a blue curve.

Starting from the initial state in Fig. 3 (marked by red dots), the system is initialized in the vapor-only regime: the vapor molar volume is to the right of the saturation curve, *below* the saturation pressure at the system temperature (the intersection of the saturation curve with the isotherm), implying that $P(t) - P^{sat}(T(t)) < 0$. Since M_L and M_V are always non-negative Eq. (2.2) sets $M_L(t) = 0$ in this case. In this state the liquid molar volume calculated by (4.4b) is below the saturation curve, but is not used in the total hold-up and enthalpy calculation because $M_L(t) = 0$ is enforced by Eq. (2.2). Eventually, the temperature of the system decreases, the pressure increases and the state variables T, P, v_L, v_V realize values such that the solution trajectory (red curves) intersect the saturation curve. This corresponds to the system transitioning to the two-phase regime, where $P(t) - P^{sat}(T(t)) = 0$ is enforced by Eq. (2.2), and neither $M_V(t) = 0$ nor $M_L(t) = 0$ are enforced. The system remains in this regime until transitioning to the liquid-only regime, corresponding to the left of the saturation curve where the pressure is *above* the saturation pressure for this temperature, so that $P(t) - P^{sat}(T(t)) > 0$ and Eq. (2.2) enforces $M_V(t) = 0$. It is clear from this diagram that it is possible to solve for both the liquid and vapor phase molar volumes even when the system is in a single phase regime. Extrapolations of the state equations might be necessary for extreme conditions [67], where the equation of state has only one solution. However, this was not necessary in this example.

4.3 Unphysical Behavior from Common Modeling Simplifications

Engineers often use simplifying assumptions when appropriate to make the problem clearer, more readily solvable or require less empirical parameters. Here, we wish to demonstrate problems that may arise when trying to simulate phase changes modeled by nonsmooth DAEs using certain common simplifications.

A simplification of the model could make the acceptable assumption of an incompressible liquid, i.e., its specific volume, v_L, is constant, and an *ideal gas* vapor phase, i.e., a and b in (4.1) are zero. In this case $h_L = u_L$ (and $C_p = C_v$). Thus, for this model equation (4.4a) is replaced with

$$v_V(t) = RT(t)/P(t), \tag{4.11}$$

and Eq. (4.5) are replaced with:

$$h_V(t) = h^{ig}(T(t)), \tag{4.12a}$$

$$h_L(t) = h_V(t) - \Delta h^{\text{vap}}(T(t)), \tag{4.12b}$$

where Δh^{vap} is the heat of vaporization calculated from the Clapeyron equation using Eq. (2.3)

$$\Delta h^{\text{vap}}(T(t)) = -R\Delta Z^{vap}(t)\frac{d\ln P^{\text{sat}}}{d(1/T)} = R\Delta Z^{vap}(t)\ln(10)\frac{BT(t)^2}{(T(t)+C)^2}, \tag{4.13}$$

where $\Delta Z^{vap} = 1 - P^{\text{sat}}(T)v_L/RT$.

Equation (2.1d) becomes:

$$U(t) = M_L(t)h_L(t) + M_V(t)\left(h_V(t) - P(t)v_V(t)\right), \tag{4.14}$$

and v_L becomes a parameter, thus the reduced system of algebraic equations has five equations with five unknowns instead of six previously.

However, if we consider the liquid-only regime and analyze the model structure as before we find that it has become structurally singular. Equation (2.2) enforces $M_V(t) = 0$, and the following occurrence matrix is obtained:

$$\begin{array}{c} \text{Eq.} \\ (2.1c) \\ (4.14) \\ (2.1e) \\ (4.11) \end{array} \begin{array}{cccc} M_L & P & T & v_V \\ \otimes & & & \\ X & & \otimes & \\ X & & & \\ & X & X & \otimes \end{array}$$

Clearly, the model is ill-defined, as the pressure and the vapor specific volume cannot be simultaneously determined. This issue would be alleviated if it were possible to determine v_V from Eq. (2.1e), because Eqs. (4.11) and (4.14) could be used to solve for P and T. However, this is not possible because M_V is identically zero in the liquid-only regime. In other words, the fact that the liquid is incompressible and there is no vapor phase makes it impossible to calculate the pressure from this system of equations. This structural singularity indicates that the model is not index one anywhere in the liquid-only regime where the model is C^1 (away from the phase boundary). An index reduction procedure is possible but involves differentiating potentially very complex (and even nondifferentiable [67]) thermodynamic property models that are often implemented as complex libraries of nested subroutines. Consequently, index reduction is not very practical in the context of how these models might be implemented and used in practice. At the phase boundary, the structural singularity implies that the determinant of one B-subdifferential element is zero. This violates both given sufficient conditions to be index one, but is not definitive proof that the model is not index one at such points.

To address the issue above, a different simplification of the model is to assume that the liquid phase specific volume is a function of temperature but not a function of pressure. In this case we replace Eqs. (4.4b) and (4.12b) with [56]:

$$v_L(t) = v_L^0 \exp\left(\beta(T(t) - T_0)\right), \tag{4.15}$$

$$h_L(t) = h_V(t) - \Delta h^{\text{vap}}(T(t)) + (1 - \beta T(t))v_L(t)(P(t) - P^{\text{sat}}(T(t))), \tag{4.16}$$

where v_L^0 is the liquid molar volume at some reference temperature T_0 and β is the *volume expansivity* of the liquid, defined by

$$\beta \equiv \frac{1}{V}\left(\frac{\partial V}{\partial T}\right)_P,$$

where the subscript P denotes that P is held constant. It is apparent that this model is not structurally singular in the liquid-only regime:

$$\begin{array}{c} \text{Eq.} \\ (2.1c) \\ (2.1d) \\ (2.1e) \\ (4.15) \\ (4.11) \end{array} \begin{array}{c} \begin{array}{ccccc} M_L & P & T & v_L & v_V \end{array} \\ \left[\begin{array}{ccccc} \textcircled{X} & & & & \\ X & \textcircled{X} & X & X & \\ X & & & \textcircled{X} & \\ & & \textcircled{X} & X & \\ & X & X & & \textcircled{X} \end{array}\right] \end{array}$$

However, this is a thermodynamically inconsistent model because any real fluid is characterized by a finite ratio of β/κ [56]. κ is the *isothermal compressibility*, defined by

$$\kappa \equiv \frac{1}{V}\left(\frac{\partial V}{\partial P}\right)_T,$$

which is zero in this model for the liquid phase, from which it follows that β/κ is infinite. Nevertheless, this approximation is adopted by some commercial simulation packages such as ASPEN (Aspen Technology Inc.). Indeed, this model behaves unphysically, as we explain hereafter.

Focusing on the liquid-only regime, we express the flow rates as:

$$F_L(t) = C_L^v\sqrt{P(t) + P_L^h(t) - P_{\text{out}}}, \tag{4.17}$$

$$F_{\text{in}}(t) = k\sqrt{P_{\text{in}} - P(t)}. \tag{4.18}$$

This allows the system to be expressed by the following DAEs, where $\mathbf{x} \equiv (M, U)$ are the differential variables and $\mathbf{y} \equiv (T, P)$ are the algebraic variables:

$$\frac{dM}{dt}(t) = F_{\text{in}}(t) - F_L(t) \equiv f_1(\mathbf{y}(t)) \tag{4.19a}$$

$$\frac{dU}{dt}(t) = F_{\text{in}}(t)h_{\text{in}} - F_L(t)h_L(t) \equiv f_2(\mathbf{y}(t)), \tag{4.19b}$$

$$M(t) = V/v_L(t) \equiv g_1(\mathbf{y}(t)), \tag{4.19c}$$

$$U(t) = (V/v_L(t))\,(h_L(t) - P(t)v_L(t)) \equiv g_2(\mathbf{y}(t)), \tag{4.19d}$$

where h_{in} is calculated from Eq. (4.16) based on the known inlet temperature and pressure, $T_{\text{in}} = 260$ K and $P_{\text{in}} = 10$ bar. Other values used are as follows: $T_0 = 260$ K, $v_L^0 = 0.09455$ L/mol, $C_L^v = k = 4.5$ mol/s/$\sqrt{\text{bar}}$ and $\beta = 0.0017\,\text{K}^{-1}$. A steady-state solution for this system, representing a vessel full of liquid only, is given by

$$\mathbf{x}_{\text{SS}} = \begin{bmatrix} M_{\text{SS}} \\ U_{\text{SS}} \end{bmatrix} = \begin{bmatrix} 94.66 \\ -12,354.57 \end{bmatrix}, \quad \mathbf{y}_{\text{SS}} = \begin{bmatrix} T_{\text{SS}} \\ P_{\text{SS}} \end{bmatrix} = \begin{bmatrix} 263.24 \\ 5.47 \end{bmatrix}.$$

However, this state is unstable as verified by linearizing the system [15, 27, 35, 47, 50]; Eq. (4.19) can be written as

$$\dot{\mathbf{x}}(t) = \mathbf{f}(\mathbf{y}(t)), \tag{4.20a}$$

$$\mathbf{x}(t) = \mathbf{g}(\mathbf{y}(t)), \tag{4.20b}$$

which can be linearized about the steady-state solution $(\mathbf{x}_{\text{SS}}, \mathbf{y}_{\text{SS}})$ to yield the linear DAE system

$$\frac{\mathrm{d}\Delta\mathbf{x}}{\mathrm{d}t}(t) = \mathbf{Jf}(\mathbf{y}_{\text{SS}})\Delta\mathbf{y}(t),$$

$$\Delta\mathbf{x}(t) = \mathbf{Jg}(\mathbf{y}_{\text{SS}})\Delta\mathbf{y}(t),$$

which possesses the underlying ODE system

$$\begin{bmatrix} \frac{\mathrm{d}\Delta \mathbf{x}}{\mathrm{d}t}(t) \\ \frac{\mathrm{d}\Delta \mathbf{y}}{\mathrm{d}t}(t) \end{bmatrix} = \mathbf{D} \begin{bmatrix} \Delta \mathbf{x}(t) \\ \Delta \mathbf{y}(t) \end{bmatrix}, \tag{4.22}$$

where

$$\mathbf{D} = \begin{bmatrix} \mathbf{0}_{2\times 2} & \mathbf{Jf}(\mathbf{y}_{\mathrm{SS}}) \\ \mathbf{0}_{2\times 2} & (\mathbf{Jg}(\mathbf{y}_{\mathrm{SS}}))^{-1}\,\mathbf{Jf}(\mathbf{y}_{\mathrm{SS}}) \end{bmatrix}.$$

The eigenvalues of the right-hand side matrix $\mathbf{D}$ are 0, 0, 23.03 and -0.10, implying instability of the steady-state solution, from which it follows that this model displays unphysical behavior. The two zero eigenvalues are indicative of the decoupled structure of the system (i.e., $\mathbf{f}$ and $\mathbf{g}$ do not depend on $\mathbf{x}$ explicitly).

The underlying ODEs associated with (4.20) are given by (4.20a) and

$$\dot{\mathbf{y}}(t) = (\mathbf{Jg}(\mathbf{y}(t)))^{-1}\mathbf{f}(\mathbf{y}(t)), \tag{4.23}$$

which can be written as

$$\begin{aligned} \dot{T}(t) &= \gamma_1(T(t), P(t)), \\ \dot{P}(t) &= \gamma_2(T(t), P(t)), \end{aligned} \tag{4.24}$$

since $\mathbf{y} \equiv (T, P)$ (i.e., ODEs that are decoupled from (4.20a)). The instability is illustrated in Fig. 4, where the vector field associated with Eq. (4.24) is plotted. It is evident that the steady-state solution of this ODE system is unstable as a small disturbance in pressure is amplified (i.e., the pressure continues increasing or decreasing depending on the direction of the disturbance). Note that in Fig. 2 the system appears to be approaching a steady state in the liquid-only regime.

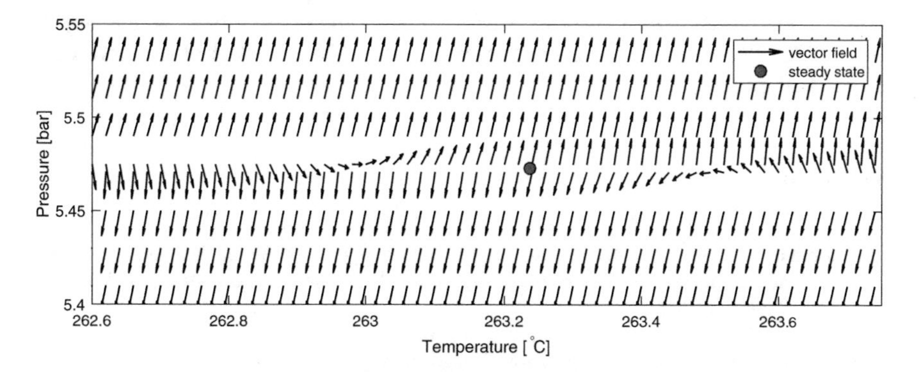

Fig. 4 A graph of the (normalized) vector field associated with Eq. (4.24) (black arrows). The steady-state solution $\mathbf{y}(t) = \mathbf{y}_{\mathrm{SS}}$ is also plotted (red dot)

5 Conclusions

The interaction of continuous and discrete phenomena encountered in a variety of chemical engineering problems is correctly captured by a nonsmooth modeling framework. Process operations modeled as nonsmooth DAEs include phase equilibrium problems, crystallization kinetics, and flow control devices (such as weirs), among others [60]. Other examples include nonsmooth models used for dynamic simulation of separation processes [52], with non-ideal mixtures considered in [69]; modeling and design of multistream heat exchangers [66, 68]; and dynamic flux balance analysis modeling of microbial consortia [20, 21, 24, 25]. Additional complex nonsmooth models are expected to arise in the future.

Thanks to advanced simulation technology (e.g., event detection and discontinuity locking), practically implementable dynamic simulation and optimization methods are possible for nonsmooth DAEs; dynamic simulation only requires extension of function libraries to include nonsmooth elemental functions (e.g., the absolute-value function, min, max, etc.). For example, the Jacobian software (Res Group Inc.), used for the process simulations above, supports such elemental functions. Extensions of sequential methods (e.g., single and multiple shooting method) for dynamic optimization of nonsmooth DAE systems (see [58]) is enabled by parametric regularity of nonsmooth generalized differentiation index-one DAEs (i.e., the sensitivity analysis theory [57]). Numerical treatment for large optimization problems with nonsmooth ODEs embedded is underway, based on the theory in [31] and [28, Chapter 7].

Areas for future work include application of the nonsmooth theoretical tools [57, 59] discussed in this article to "high-index" nonsmooth DAEs with special structures, investigating the adjoint sensitivities case (i.e., extending [51]), and extension of deterministic global optimization methods to nonsmooth DAEs (i.e., using McCormick's framework [32, 40, 55, 63]). Other possible directions for future work include the following: developing simultaneous methods [8] in this setting, possibly by using moving finite elements [11, 12] to adjust mesh points as dictated by the presence of nonsmoothness; establishing optimality conditions for nonsmooth optimal control problems with nonsmooth DAEs embedded; and applying the newly developed theoretical tools to analyze sensitivities of discontinuous dynamical systems.

References

1. Anitescu, M., Tseng, P., Wright, S.J.: Elastic-mode algorithms for mathematical programs with equilibrium constraints: global convergence and stationarity properties. Math. Program. **110**(2), 337–371 (2007)
2. Ascher, U.M., Petzold, L.R.: Computer Methods for Ordinary Differential Equations and Differential-Algebraic Equations. SIAM, Philadelphia (1998)
3. Barton, P.I., Pantelides, C.C.: Modeling of combined discrete/continuous processes. AIChE J. **40**(6), 966–979 (1994)

4. Barton, P.I., Banga, J.R., Galán, S.: Optimization of hybrid discrete/continuous dynamic systems. Comput. Chem. Eng. **24**(9), 2171–2182 (2000)
5. Barton, P.I., Lee, C.K.: Modeling, simulation, sensitivity analysis, and optimization of hybrid systems. ACM Trans. Model. Comput. Simul. **12**(4), 256–289 (2002)
6. Barton, P.I., Khan, K.A., Stechlinski, P., Watson, H.A.J.: Computationally relevant generalized derivatives: theory, evaluation and applications. Optim. Methods Softw. In Press
7. Baumrucker, B.T., Biegler, L.T.: MPEC strategies for optimization of a class of hybrid dynamic systems. J. Process Control **19**(8), 1248–1256 (2009)
8. Biegler, L.T.: An overview of simultaneous strategies for dynamic optimization. Chem. Eng. Process. Process Intensif. **46**(11), 1043–1053 (2007)
9. Brenan, K.E., Campbell, S.L., Petzold, L.R.: Numerical Solution of Initial-Value Problems in Differential-Algebraic Equations. SIAM, Philadelphia (1996)
10. Bullard, L., Biegler, L.: Iterated linear programming strategies for non-smooth simulation: a penalty based method for vapor-liquid equilibrium applications. Comput. Chem. Eng. **17**(1), 95–109 (1993)
11. Chen, W., Biegler, L.T.: Nested direct transcription optimization for singular optimal control problems. AIChE J. **62**(10), 3611–3627 (2016)
12. Chen, W., Shao, Z., Biegler, L.T.: A bilevel NLP sensitivity-based decomposition for dynamic optimization with moving finite elements. AIChE J. **60**(3), 966–979 (2014)
13. Clarke, F.H.: Optimization and Nonsmooth Analysis. SIAM, Philadelphia (1990)
14. Coddington, E.A., Levinson, N.: Theory of Ordinary Differential Equations. McGraw-Hill, New York (1955)
15. Collins, D.M.: Tools to analyse cell signaling models. PhD thesis, Massachusetts Institute of Technology (2003)
16. Facchinei, F., Fischer, A., Herrich, M.: An LP-Newton method: nonsmooth equations, KKT systems, and nonisolated solutions. Math. Program. **146**, 1–36 (2014)
17. Filippov, A.F.: Differential Equations with Discontinuous Righthand Sides. Kluwer Academic Publishers, Dordrecht (1988)
18. Galán, S., Feehery, W.F., Barton, P.I.: Parametric sensitivity functions for hybrid discrete/continuous systems. Appl. Numer. Math. **31**, 17–47 (1999)
19. Goebel, R., Sanfelice, R., Teel, A.R.: Hybrid dynamical systems. IEEE Control Syst. **29**, 28–93 (2009)
20. Gomez, J.A., Höffner, K., Barton, P.I.: From sugars to biodiesel using microalgae and yeast. Green Chem. **18**, 461–475 (2015)
21. Gomez, J.A., Höffner, K., Khan, K.A., Barton, P.I.: Generalized derivatives of lexicographic linear programs. J. Optim. Theory Appl. **178**(2), 477–501 (2018)
22. Gopal, V., Biegler, L.: Nonsmooth dynamic simulation with linear programming based methods. Comput. Chem. Eng. **21**(7), 675–689 (1997)
23. Heemels, W.P.M.H., Schumacher, J.M., Weiland, S.: Linear complementarity systems. SIAM J. Appl. Math. **60**(4), 1234–1269 (2000)
24. Höffner, K., Barton, P.I.: Design of microbial consortia for industrial biotechnology. Comput. Aided Chem. Eng. **34**, 65–74 (2014)
25. Höffner, K., Khan, K.A., Barton, P.I.: Generalized derivatives of dynamic systems with a linear program embedded. Automatica **63**, 198–208 (2016)
26. Johansson, K.H., Egerstedt, M., Lygeros, J., Sastry, S.: On the regularization of Zeno hybrid automata. Syst. Control Lett. **38**, 141–150 (1999)
27. Khalil, H.K.: *Nonlinear Systems*, 3rd edn. (Prentice-Hall, New Jersey, 2002)
28. Khan, K.A.: Sensitivity analysis for nonsmooth dynamic systems. PhD thesis, Massachusetts Institute of Technology, 2014
29. Khan, K.A., Barton, P.I.: Generalized derivatives for solutions of parametric ordinary differential equations with non-differentiable right-hand sides. J. Optim. Theory Appl. **163**, 355–386 (2014)
30. Khan, K.A., Barton, P.I.: A vector forward mode of automatic differentiation for generalized derivative evaluation. Optim. Methods Softw. **30**(6), 1185–1212 (2015)

31. Khan, K.A., Barton, P.I.: Generalized derivatives for hybrid systems. IEEE Trans. Autom. Control **62**(7), 3193–3208 (2017)
32. Khan, K.A., Watson, H.A.J., Barton, P.I.: Differentiable McCormick relaxations. J. Glob. Optim. **67**(4), 687–729 (2017)
33. Kiwiel, K.C.: Methods of Descent for Nondifferentiable Optimization. Lecture Notes in Mathematics. Springer, Berlin (1985)
34. Kunkel, P., Mehrmann, V.: Differential-Algebraic Equations: Analysis and Numerical Solution. European Mathematical Society, Zurich (2006)
35. Kunkel, P., Mehrmann, V.: Stability properties of differential-algebraic equations and spin-stabilized discretizations. Electron. Trans. Numer. Anal. **26**, 385–420 (2007)
36. Lemaréchal, C., Strodiot, J.J., Bihain, A.: On a bundle algorithm for nonsmooth optimization. In: Mangasarian, O.L., Meyer, R.R., Robinson, S.M. (eds.) Nonlinear Programming 4. Academic Press, New York (1981)
37. Lukšan, L., Vlček, J.: A bundle-Newton method for nonsmooth unconstrained minimization. Math. Program. **83**, 373–391 (1998)
38. Lygeros, J., Johansson, K.H., Sastry, S., Egerstedt, M.: On the existence of executions of hybrid automata. In: 38th IEEE Conference on Decision and Control, pp. 2249–2254, 1999
39. Lygeros, J., Johansson, K., Simic, S., Sastry, S.: Dynamical properties of hybrid automata. IEEE Trans. Autom. Control **48**(1), 2–17 (2003)
40. McCormick, G.P.: Computability of global solutions to factorable nonconvex programs: part I – convex underestimating problems. Math. program. **10**(1), 147–175 (1976)
41. Nesterov, Y.: Lexicographic differentiation of nonsmooth functions. Math. Program. **104**, 669–700 (2005)
42. Pang, J.S., Shen, J.: Strongly regular differential variational systems. IEEE Trans. Autom. Control **52**(2), 242–255 (2007)
43. Patrascu, M., Barton, P.I.: Optimal campaigns in end-to-end continuous pharmaceuticals manufacturing. Part 1: nonsmooth dynamic modeling. Chem. Eng. Process. Process Intensif. **125**, 298–310 (2018)
44. Patrascu, M., Barton, P.I.: Optimal campaigns in end-to-end continuous pharmaceuticals manufacturing. Part 2: dynamic optimization. Chem. Eng. Process. Process Intensif. **125**, 124–132 (2018)
45. Petzold, L.R.: A description of DASSL: a differential/algebraic system solver. In: *Scientific computing (Montreal, Quebec, 1982)*, pp. 65–68. Sandia National Labs., Livermore (1983)
46. Qi, L., Sun, J.: A nonsmooth version of Newton's method. Math. Program. **58**, 353–367 (1993)
47. Rabier, P.J., Rheinboldt, W.C.: Theoretical and numerical analysis of differential-algebraic equations. Handb. Numer. Anal. **8**, 183–540 (2002)
48. Raghunathan, A.U., Diaz, M.S., Biegler, L.T.: An MPEC formulation for dynamic optimization of distillation operations. Comput. Chem. Eng. **28**(10), 2037–2052 (2004)
49. Ralph, D., Scholtes, S.: Sensitivity analysis of composite piecewise smooth equations. Math. Program. **76**, 593–612 (1997)
50. Reich, S.: On the local qualitative behavior of differential-algebraic equations. Circuits Syst. Signal Process. **14**(4), 427–443 (1995)
51. Ruban, A.I.: Sensitivity coefficients for discontinuous dynamic systems. J. Comput. Syst. Sci. Int. **36**(4), 536–542 (1997)
52. Sahlodin, A.M., Watson, H.A.J., Barton, P.I.: Nonsmooth model for dynamic simulation of phase changes. AIChE J. **62**(9), 3334–3351 (2016)
53. Scholtes, S.: Introduction to Piecewise Differentiable Equations. Springer, New York (2012)
54. Schumacher, J.M.: Complementarity systems in optimization. Math. Program. Ser. B **101**, 263–295 (2004)
55. Scott, J.K., Barton, P.I.: Convex and concave relaxations for the parametric solutions of semi-explicit index-one differential-algebraic equations. J. Optim. Theory Appl. **156**(3), 617–649 (2013)
56. Smith, J.M., Van Ness, H.C., Abbott, M.M.: Introduction to Chemical Engineering Thermodynamics. McGraw-Hill Chemical Engineering Series, 7th edn. McGraw-Hill, Boston (2005)

57. Stechlinski, P.G., Barton, P.I.: Generalized derivatives of differential–algebraic equations. J. Optim. Theory Appl. **171**(1), 1–26 (2016)
58. Stechlinski, P.G., Barton, P.I.: Generalized derivatives of optimal control problems with nonsmooth differential-algebraic equations embedded. In: 55th IEEE Conference on Decision and Control, pp. 592–597, 2016
59. Stechlinski, P.G., Barton, P.I.: Dependence of solutions of nonsmooth differential–algebraic equations on parameters. J. Differ. Equ. **262**(3), 2254–2285 (2017)
60. Stechlinski, P., Patrascu, M., Barton, P.I.: Nonsmooth differential-algebraic equations in chemical engineering. Comput. Chem. Eng. **114**, 52–68 (2018)
61. Tolsma, J., Barton, P.I.: DAEPACK: an open modeling environment for legacy models. Ind. Eng. Chem. Res. **39**(6), 1826–1839 (2000)
62. Tolsma, J.E., Barton, P.I.: Hidden discontinuities and parametric sensitivity calculations. SIAM J. Sci. Comput. **23**(6), 1861–1874 (2002)
63. Tsoukalas, A., Mitsos, A.: Multivariate McCormick relaxations. J. Glob. Optim. **59**(2–3), 633–662 (2014)
64. Ullman, J., Hopcroft, J.: Introduction to Automata Theory, Languages, and Computation. Addison-Wesley, Massachusetts (1979)
65. van der Schaft, A.J., Schumacher, J.M.: Complementarity modeling of hybrid systems. IEEE Trans. Autom. Control **43**(4), 483–490 (1998)
66. Watson, H.A.J., Barton, P.I.: Modeling phase changes in multistream heat exchangers. Int. J. Heat Mass Transf. **105**, 207–219 (2017)
67. Watson, H.A.J., Barton, P.I.: Reliable flash calculations: Part 3. A nonsmooth approach to density extrapolation and pseudoproperty evaluation. Ind. Eng. Chem. Res. **56**(50), 14832–14847 (2017)
68. Watson, H.A.J., Khan, K.A., Barton, P.I.: Multistream heat exchanger modeling and design. AIChE J. **61**(10), 3390–3403 (2015)
69. Watson, H.A.J., Vikse, M., Gundersen, T., Barton, P.I.: Reliable flash calculations: Part 1. Nonsmooth inside-out algorithms. Ind. Eng. Chem. Res. **56**(4), 960–973 (2017)
70. Watson, H.A.J., Vikse, M., Gundersen, T., Barton, P.I.: Reliable flash calculations: Part 2. Process flowsheeting with nonsmooth models and generalized derivatives. Ind. Eng. Chem. Res. **56**(50), 14848–14864 (2017)

Continuous, Semi-discrete, and Fully Discretised Navier-Stokes Equations

R. Altmann and J. Heiland

Abstract The Navier-Stokes equations are commonly used to model and to simulate flow phenomena. We introduce the basic equations and discuss the standard methods for the spatial and temporal discretisation. We analyse the semi-discrete equations – a semi-explicit nonlinear DAE – in terms of the strangeness index and quantify the numerical difficulties in the fully discrete schemes, that are induced by the strangeness of the system. By analysing the Kronecker index of the difference-algebraic equations, that represent commonly and successfully used time stepping schemes for the Navier-Stokes equations, we show that those time-integration schemes factually remove the strangeness. The theoretical considerations are backed and illustrated by numerical examples.

Keywords DAEs · Difference-algebraic equations · Navier-Stokes equations · Strangeness index

Mathematics Subject Classification (2010) 65L80, 65M12, 35Q30

1 Introduction

The *Navier-Stokes equations* (NSE) are a system of nonlinear partial-differential equations that have been commonly used to model fluid flows for more than a century. The NSE are believed to describe all kinds of incompressible flows sufficiently well as long as the setup supports the hypothesis that the fluid is a

R. Altmann
Department of Mathematics, University of Augsburg, Augsburg, Germany
e-mail: robert.altmann@math.uni-augsburg.de

J. Heiland (✉)
Max Planck Institute for Dynamics of Complex Technical Systems, Magdeburg, Germany
e-mail: heiland@mpi-magdeburg.mpg.de

S. Campbell et al. (eds.), *Applications of Differential-Algebraic Equations: Examples and Benchmarks*, Differential-Algebraic Equations Forum,
https://doi.org/10.1007/11221_2018_2

continuum. Indeed, comparisons of numerical simulations with experiments show arbitrarily good agreement of the model with the observations over a long range from slowly moving flows in small geometries like a pipe up to highly turbulent flows over large spatial extensions like the flow around an airplane or even weather phenomena. Nevertheless, the mere equations and the extent of their applicability have not been fully deciphered by now and a substantial progress in this respect will be eligible for a Clay price.[1]

Under the assumption of continuity of the observed quantities, the NSE can be derived from fundamental laws of physics; see [41] and [12]. These considerations are well backed for a macroscopic viewpoint, from which a fluid like water appears as a continuum. On a microscopic level, where discrete molecular structures define the states, the NSE cannot capture the physics right, as it is well-known, e.g., for capillary flows.

On the molecular level, fluids are better described by the *Boltzmann equations*, which model molecular interactions. This fact seems undisputed the more that the NSE can also be interpreted and derived through a limiting process of the Boltzmann equations in the sense of averaging the microscopic quantities for a macroscopic description [54].

As a mathematical object the NSE have ever been subject to fundamental investigations and led to its own research field and its own subject definitions in the *MSC* classification scheme.[2] The research on the NSE has focused on the analysis of the equations and their numerical approximation. Early results on the existence of solutions are due to Leray [43] (weak solutions) and Fujita and Kato [19] (smooth solutions). The first textbooks on the functional and on the numerical analysis were written by Ladyzhenskaya [40] and Temam [58], respectively.

For the numerical analysis of the spatial discretisation, one may distinguish two lines of development. The mathematical line focuses on Galerkin methods in the realm of variational formulations whereas the engineering orientated line has been advancing finite volume methods (FVM) as they appear well-suited for simulations. On the side of Galerkin methods, and in particular finite element methods (FEM), there have been many efforts in designing stable elements like the famous *Taylor-Hood* elements [59] as well as for general convergence results; see the textbooks [21, 40, 49] for the numerical analysis and [60] for an application oriented overview. On the side of FVM that are the method of choice in most general purpose flow solvers, there have been general developments in view of discretising conservation laws [44] and particular progress in view of stable approximation of fluid flow [18].

A numerical analysis of approximations to the time-dependent NSE with FEM semi-discretisations has been carried out by Heywood and Rannacher [33]; see also the textbook [24], that covers implementation issues. For time marching schemes for FVM formulations we refer to [18]. Strategies for the iterative solution of the arising linear systems can be found in [15] (FEM) or [18] (FVM).

[1] See http://www.claymath.org/millennium-problems/navier-stokes-equation.

[2] See http://www.ams.org/mathscinet/msc/msc.html?t=35Q30&btn=Current and related.

In this work we revisit the NSE for incompressible flows from a differential-algebraic equations (DAE) perspective. This includes the modelling where the incompressibility is treated as an algebraic constraint in an abstract space and the spatial semi-discretisation, that has to be handled with care to respect the incompressibility constraint and to lead to a well-posed classical DAE. It also includes the temporal discretisation, in which the DAE properties of the NSE become evident and (hopefully not) problematic. Such a pure DAE perspective has been taken on by Weickert [62] who analysed finite difference approximations and certain time-stepping schemes for the NSE and by Emmrich and Mehrmann [16] who provided conditions and solution representations for linearised NSE in abstract spaces in line with linear time-invariant DAEs in finite-dimensional state-spaces. In [31] the nonlinear NSE and its Galerkin approximations have been analysed in view of consistency of reformulations of semi-discrete DAE approximations with the infinite-dimensional model.

The paper is organized as follows. In Sect. 2 we derive the NSE from first principles and formulate the weak form as an *operator DAE*. The direct connection to DAEs is then made in Sect. 3, in which we report on several spatial discretisation schemes and that commonly used FEM schemes lead to systems of *strangeness index* one. In Sect. 4 we analyse the time approximation schemes in terms of the index of the resulting *difference-algebraic equations* (ΔAE). We show that the straight-forward temporal discretisation leads to a scheme of higher index than that of well-established time-stepping schemes for incompressible flows. We further confirm, that schemes with a ΔAE of lower index can also be obtained from a standard time-discretisation applied to a reformulation of the DAE with lower index. In the numerical examples in Sect. 5, we confirm the superiority of the lower index ΔAE approximations of the NSE over the straight-forward time-discretisation. As a benchmark for time-integration schemes for the considered class of nonlinear semi-explicit DAEs of strangeness index 1, we provide reference trajectories and the system coefficients and nonlinear inhomogeneities for direct realization in *Python*. We conclude this paper with summarizing remarks in Sect. 6.

2 Continuous Model

2.1 Derivation of the Navier-Stokes Equations for Incompressible Flows

The NSE provide a model of a flow as a continuum. The basic assumption for their derivation and, thus, the validity of the model is that the flow under consideration forms a continuous entity of flow particles in a spatial domain $\Omega \subset \mathbb{R}^3$ and a time interval $\mathcal{I}$ such that the functions

$$\mathfrak{v}\colon \mathcal{I} \times \Omega \to \mathbb{R}^3, \quad \mathfrak{p}\colon \mathcal{I} \times \Omega \to \mathbb{R}, \quad \text{and} \quad \rho\colon \mathcal{I} \times \Omega \to \mathbb{R}, \tag{2.1}$$

describing the velocity, the pressure, and the density as measured at the position $x \in \Omega$ at time $t \in \mathcal{I}$ are continuous functions. Here, continuous means that we can apply differential calculus in order to derive basic partial differential equations. Later, when we derive the weak formulation of the NSE, the needed continuity will be specified further.

As mentioned in [12, p. 2], these assumptions lead to a model that is believed to provide accurate descriptions of common macroscopic flow phenomena. In, e.g., setups of small geometric scales like in capillary flows or under vacuum-like conditions [35], the discrete microscopic molecular structure of the fluid that constitutes the flow has to be taken into account.

The basic assumption of continuity of the matter allows for the consideration of a possibly infinitesimal small control volume $W \subset \Omega$, an open bounded domain in $\mathbb{R}^3$ that contains a given agglomerate of fluid particles in the considered flow. Continuity also implies that a fixed W is deformed and convected by the flow but always consists of the same fluid particles.

Under this continuity assumption, one can call on the *Reynolds Transport Theorem*, that relates the temporal change of an integral quantity over W to the convection velocity $\mathfrak{v}$.

Theorem 2.1 (See [52], Eq.(16) and [12], p. 10) *Let $W\colon I \to \mathbb{R}^3$ describe a smoothly moving control volume and let $f\colon t \times W(t) \mapsto \mathbb{R}$ be a sufficiently smooth function, then*

$$\frac{d}{dt}\int_W f \,\mathrm{d}V = \int_W \frac{\partial f}{\partial t} + \operatorname{div}(f\mathfrak{v})\,\mathrm{d}V. \tag{2.2}$$

2.1.1 Incompressibility and Mass Conservation

The *Reynolds Transport Theorem* can be used to show that a flow is *incompressible*, which means that the volume of any agglomerate W is constant over time, if and only if the velocity field v is divergence free. In fact, if (2.2) applies, then one has that

$$\frac{\mathrm{d}}{\mathrm{d}t}\int_W \mathrm{d}V = 0 \quad \text{if, and only if,} \quad \int_W \operatorname{div}\mathfrak{v}\,\mathrm{d}V = 0. \tag{2.3}$$

With a well-defined density function ρ, the mass of a control volume is defined as the integral over the (mass) density ρ and, with the assumption that in the flow there are neither mass sinks nor mass sources, an application of (2.2) gives

$$\frac{\mathrm{d}}{\mathrm{d}t}\int_W \rho\,\mathrm{d}V = 0 \quad \text{if, and only if,} \quad \int_W \frac{\partial\rho}{\partial t} + \operatorname{div}(\rho\mathfrak{v})\,\mathrm{d}V = 0. \tag{2.4}$$

Remark 2.1 The conservation of mass and the incompressibility of a flow are closely related and only equivalent in the case that the density ρ is constant in space and time. Flow models that assume incompressibility and varying density functions ρ are applied, e.g., in oceanography [57].

Remark 2.2 If the volume of a fluid parcel changes over time, the flow is called *compressible*. In this case, the mass density ρ is modelled as an unknown function and related to the pressure p through constitutive or so-called *state equations*, like the *ideal gas law*; see [17].

2.1.2 Balance of Momentum

Under the continuum assumption, the momentum of a fluid agglomerate W can be expressed as the integral of the mass density times velocity and the temporal change equated with volume and surface forces on W:

$$\frac{\mathrm{d}}{\mathrm{d}t}\int_W \rho \mathfrak{v} \,\mathrm{d}V = \int_W \rho g \,\mathrm{d}V + \int_{\partial W} \sigma n \,\mathrm{d}S, \tag{2.5}$$

where $g\colon \mathcal{I}\times\Omega \to \mathbb{R}^3$ is the density of a body force, and where $\sigma\colon \mathcal{I}\times\Omega\to\mathbb{R}^{3,3}$ is a tensor such that σn, where n is the normal field on the boundary ∂W of W, represents the density of the forces acting on the surface. Note that the expression in (2.5) is vector valued and that integration and differentiation is performed componentwise.

Applying the *Divergence Theorem* componentwise, one can write the term with the surface forces as a volume integral

$$\int_{\partial W} \sigma n \,\mathrm{d}S = \int_W \operatorname{div}\sigma \,\mathrm{d}V, \tag{2.6}$$

with the divergence of a tensor defined accordingly.

So far, all assumptions have based on first principles. For the mathematical modelling of the tensor σ, however, ad hoc assumptions and heuristics are employed. First of all, it is assumed that σ can be written as

$$\sigma = -\mathfrak{p} I + \tau \tag{2.7}$$

where $\mathfrak{p}$ – the pressure – is a smooth scalar function and $\tau\colon \mathcal{I}\times\Omega\to\mathbb{R}^{3,3}$ is the tensor of shear stresses. It is assumed that $\tau(t,x)$ is symmetric and invariant under rigid body rotation. Furthermore, τ is a linear function of the velocity gradient $\nabla\mathfrak{v} := \left[\frac{\partial \mathfrak{v}_j}{\partial x_i}\right]_{i,j=1,2,3}$. Under the additional assumption that the fluid under consideration is a *Newtonian fluid*, the tensor τ is defined by the velocity field and two further parameters (see [12, Ch. 1.3]) and commonly written as

$$\tau = \mu\left[\nabla\mathfrak{v} + \nabla\mathfrak{v}^{\mathsf{T}} - \frac{2}{3}(\operatorname{div}\mathfrak{v})\mathrm{I}\right] + \zeta(\operatorname{div}\mathfrak{v})\mathrm{I}. \tag{2.8}$$

The parameter μ is the *(first coefficient of the) viscosity* and has been experimentally determined and tabulated for many gases and fluids. The parameter ζ is called the *bulk viscosity* and, in line with the *Stokes Hypothesis*, often set to zero.

As for the left hand side in (2.5), one proceeds as follows. Let $\mathfrak{v}_i$ denote the i-th component of $\mathfrak{v}$, $i = 1, 2, 3$. Then, an application of (2.2) gives that

$$\frac{\mathrm{d}}{\mathrm{d}t}\int_W \rho\mathfrak{v}_i \,\mathrm{d}V = \int_W \frac{\partial\rho\mathfrak{v}_i}{\partial t} + \mathrm{div}(\rho\mathfrak{v}_i\mathfrak{v}) \,\mathrm{d}V = \int_W \frac{\partial\rho\mathfrak{v}_i}{\partial t} + \rho\mathfrak{v}_i\mathrm{div}\mathfrak{v} + \nabla(\rho\mathfrak{v}_i)\cdot\mathfrak{v} \,\mathrm{d}V \tag{2.9}$$

where basic vector calculus has been applied. If one considers ∇ as the formal column vector of the three space derivatives, relation (2.9) for all components of v can be written in compact form as

$$\frac{\mathrm{d}}{\mathrm{d}t}\int_W \rho\mathfrak{v} \,\mathrm{d}V = \int_W \frac{\partial\rho\mathfrak{v}}{\partial t} + \rho\mathfrak{v}\mathrm{div}\mathfrak{v} + (\mathfrak{v}\cdot\nabla)(\rho\mathfrak{v}) \,\mathrm{d}V. \tag{2.10}$$

2.1.3 The Navier-Stokes Equations for Incompressible Flows

In the preceding derivations, integral quantities over a flow agglomerate W were considered. If the underlying *continuity assumption* includes that all relations hold on arbitrary (small) control volumes W, instead of equating the integrals, one can equate the integrands pointwise in space, which leads to partial-differential equations.

Thus, putting together all assumptions and derivations, the balance of momentum (2.5) for an isotropic Newtonian fluid under the *Stokes Hypothesis* defines the (NSE) as

$$\frac{\partial\rho\mathfrak{v}}{\partial t} + \rho\mathfrak{v}\mathrm{div}\mathfrak{v} + (\mathfrak{v}\cdot\nabla)(\rho\mathfrak{v}) = \rho g - \nabla\mathfrak{p} + \mathrm{div}\big(\mu\big[\nabla\mathfrak{v} + \nabla\mathfrak{v}^{\mathsf{T}} - \frac{2}{3}(\mathrm{div}\mathfrak{v})\mathrm{I}\big]\big). \tag{2.11}$$

In the case that the flow is incompressible with a constant density $\rho \equiv \rho^*$ and a constant viscosity $\mu \equiv \mu^*$, the combination of (2.4) and (2.11) results in the system

$$\rho^*\big(\frac{\partial\mathfrak{v}}{\partial t} + (\mathfrak{v}\cdot\nabla)\mathfrak{v}\big) + \nabla\mathfrak{p} - \mu^*\Delta\mathfrak{v} = \rho^* g, \tag{2.12a}$$

$$\mathrm{div}\mathfrak{v} = 0. \tag{2.12b}$$

Remark 2.3 Note that in the derivation, the pressure $\mathfrak{p}$ is not a variable but a function which is assumed to be known and to describe the normal forces on a fluid element, cf. (2.7) and [12, Ch. 1.1.ii]. Thus, the NSE (2.11) has only the velocity $\mathfrak{v}$ as an unknown and the divergence free formulation (2.12) can be seen as an abstract ODE for $\mathfrak{v}$ with an invariant [29]. Nonetheless, since it turns out that the divergence constraint defines the function $\mathfrak{p}$, system (2.12) is commonly considered as an abstract differential-algebraic equation with $\mathfrak{v}$ and $\mathfrak{p}$ as unknowns.

For the numerical treatment and for similarity considerations, one relates all dependent and independent variables to a characteristic length L and characteristic velocity V via

$$\mathfrak{v}' = \frac{\mathfrak{v}}{V}, \quad \mathfrak{p}' = \frac{\mathfrak{p}}{\rho^* V^2}, \quad x' = \frac{x}{L}, \quad \text{and} \quad t' = \frac{tV}{L} \tag{2.13}$$

With this and with $f' := \frac{V^2}{L} g$, Eq. (2.12) can be rewritten in dimensionless form

$$\frac{\partial}{\partial t'}\mathfrak{v}' + (\mathfrak{v}' \cdot \nabla)\mathfrak{v}' - \frac{1}{\mathrm{Re}}\Delta \mathfrak{v}' + \nabla \mathfrak{p}' = f', \tag{2.14a}$$

$$\operatorname{div} \mathfrak{v}' = 0. \tag{2.14b}$$

Thus, the system is completely parameterized by only one parameter $\mathrm{Re} := \frac{VL\rho^*}{\mu^*}$, the *Reynolds number*. In what follows, we will always consider the dimensionless NSE (2.14) but drop the dashes of the dimensionless variables.

2.1.4 Boundary Conditions

For the considered domain Ω, let Γ denote the boundary in the abstract and in the physical sense. If the domain is bounded by nonpermeable walls, then the *no-slip* condition

$$v = g \quad \text{on } \Gamma \tag{2.15}$$

applies, where g is the velocity of the wall. The no-slip conditions align well with experiments and macroscopic considerations; see [42, Ch. 5.3] for references but also for examples where no-slip conditions seem insufficient to describe the flow at walls.

If the domain is not fully bounded by walls, typically because the computational domain needs to be bounded whereas the physical domain of the flow is unbounded, *artificial boundary* conditions are needed. If the direction of the flow is known, one may decompose these artificial boundaries into a part Γ_{i} for the incoming flow and Γ_{o} for the part where the flow leaves the domain. On Γ_{i} we can prescribe a velocity profile by

$$\mathfrak{v} = \mathfrak{v}_i \quad \text{on } \Gamma_{\mathrm{i}}, \tag{2.16}$$

whereas for Γ_{o}, there is no immediate physical insight into what should be the mathematical conditions. Accordingly, outflow boundary conditions are motivated as being *useful for the implementation of downstream boundary conditions* (see, e.g., [22, 49]) and equipped with the advice to put the outlet sufficiently far away from the region of interest; see [18, Ch.8.10.2]. The most common outflow boundary conditions are the *no stress* conditions:

$$\sigma n = 0 \quad \text{on } \Gamma_{\text{o}},$$

where σ is the stress tensor as defined in its general form in (2.7), the *do-nothing* conditions:

$$\frac{1}{\mathrm{Re}} \frac{\partial \mathfrak{v}}{\partial n} - \mathfrak{p} n = 0 \quad \text{on } \Gamma_{\text{o}}, \tag{2.17}$$

that are formulated for the nondimensional NSE (2.14), or the *no gradient* conditions:

$$\frac{\partial \mathfrak{v}}{\partial n} = 0 \quad \text{and} \quad \frac{\partial \mathfrak{p}}{\partial n} = 0 \quad \text{on } \Gamma_{\text{o}}.$$

Remark 2.4 The *do-nothing* conditions have been extended; see [7], to the case of backflow, i.e., when some, possibly spurious, inflow occurs at the boundary Γ_{o}.

2.2 Formulation as Operator DAE

In this subsection we provide yet another formulation of the dimensionless NSE (2.14), namely in the weak form as an operator DAE. This is a DAE in an abstract setting, where the solution is an element of a Sobolev space instead of a vector. We consider here homogeneous Dirichlet boundary conditions. More general boundary conditions with in- and outflow may modeled in a similar way, cf. [48]. We define the spaces

$$\mathcal{V} := [H_0^1(\Omega)]^n, \quad \mathcal{H} := [L^2(\Omega)]^n, \quad \text{and } \mathcal{Q} := L^2(\Omega)/\mathbb{R}.$$

By $\mathcal{V}'$ we denote the dual space of $\mathcal{V}$. Note that the spaces $\mathcal{V}, \mathcal{H}, \mathcal{V}'$ form a Gelfand or evolution triple [64, Ch. 23.4]. Furthermore, we define $\mathcal{W}(0, T)$ as the space of functions $u \in L^2(0, T; \mathcal{V})$, which have a weak time derivative $\dot{u} \in L^2(0, T; \mathcal{V}')$. Well-known embedding results then imply $u \in C([0, T]; \mathcal{H})$, cf. [53, Lem. 7.3].

We now consider the weak formulation of (2.14) in operator form. This means that for given right-hand sides $\mathcal{F} \in L^2(0, T; \mathcal{V}')$, $\mathcal{G} \in L^2(0, T; \mathcal{Q}')$ and an initial condition $a \in \mathcal{H}$, we seek for a pair $(\mathfrak{v}, \mathfrak{p}) \in \mathcal{W}(0, T) \times L^2(0, T; \mathcal{Q})$ satisfying

$$\dot{\mathfrak{v}}(t) + \mathcal{K}(\mathfrak{v}(t)) - \mathcal{B}'\mathfrak{p}(t) = \mathcal{F}(t) \quad \text{in } \mathcal{V}', \tag{2.18a}$$

$$\mathcal{B}\mathfrak{v}(t) = \mathcal{G}(t) \quad \text{in } \mathcal{Q}', \tag{2.18b}$$

$$\mathfrak{v}(0) = a \qquad \text{in } \mathcal{H} \tag{2.18c}$$

a.e. on $(0, T)$. The derivative of $\mathfrak{v}$ should be understood in the weak sense. The operators $\mathcal{K}\colon \mathcal{V} \to \mathcal{V}'$ and $\mathcal{B}\colon \mathcal{V} \to \mathcal{Q}'$ are defined via

$$\langle \mathcal{K}(\mathfrak{v}), w\rangle = \int_\Omega (\mathfrak{v} \cdot \nabla)\mathfrak{v} \cdot w \, dx + \frac{1}{\text{Re}} \int_\Omega \nabla \mathfrak{v} \cdot \nabla w \, dx \tag{2.19}$$

and

$$\langle \mathcal{B}\mathfrak{v}, q\rangle = \int_\Omega (\text{div}\mathfrak{v})\mathfrak{q} \, dx = \langle \mathfrak{v}, \mathcal{B}'\mathfrak{q}\rangle, \tag{2.20}$$

respectively, for a given $\mathfrak{v} \in \mathcal{V}$ and for all test functions $w \in \mathcal{V}$ and $q \in \mathcal{Q}$. We emphasize that system (2.18) not only covers the NSE but also systems with time-dependent Dirichlet boundary conditions, since we have introduced an inhomogeneity $\mathcal{G}$, see [6].

For results on the existence solutions to the weak formulation of the NSE, we refer to [57]. For a compact summary that also considers nonzero $\mathcal{G}$ in (2.18b) see [1, 2]. The differential-algebraic structure of (2.18) and the possible decoupling of differential and algebraic parts and variables has been discussed in [31].

3 Semi-discrete Equations

In this section, we consider the DAE, which results from a spatial discretisation of system (2.14) or (2.18):

$$M\dot{v} + K(v) - B^T p = f, \tag{3.1a}$$

$$Bv = g. \tag{3.1b}$$

Here, v and p denote the finite-dimensional approximations of $\mathfrak{v}$ and $\mathfrak{p}$, respectively. In view of solvability, we assume additionally a consistent initial condition of the form $v(0) = v_0$, i.e., we demand $Bv_0 = g(0)$.

The appearance of B and B^T in (3.1) reflects the duality of the divergence and the gradient in the infinite-dimensional model that is preserved by mixed finite element discretisations of (2.18) (see [15]) and finite difference discretisations of (2.14) (see [62]). If finite volumes are used for the spatial discretisation, typically, the coefficients in (3.1) are not explicitly assembled; see [18].

There are many reasons to include a right-hand side g in Eq. (3.1b) rather than a zero as it seems naturally for incompressible flows. Firstly, a nonzero g in the continuous continuity equations (2.14b) may appear also in generalizations of the NSE model to fluid-structure interactions (see, e.g., [50]) or in optimal control setups; see, e.g., [34]. Secondly, a nonzero g maybe a numerical artifact from the semi-discretisation of flows with nonzero Dirichlet boundary conditions as in the example we will provide below. Finally, in view of analysing the DAE, this inclusion of a nonzero g leads to a better understanding as one can track where the derivatives of the right-hand sides appear within the solution. This is of importance also in the case $g = 0$, since inexact solves lead to errors in (3.1b) that act like a non-smooth inhomogeneity.

3.1 Spatial Discretisation by Finite Elements

We recall the basic principles of a discretisation of the incompressible Navier-Stokes equations by finite elements; see the text books [15] for examples and [21] for the numerical analysis. Let V_h and Q_h be finite-dimensional subspaces of $\mathcal{V}$ and $\mathcal{Q}$, respectively, that are associated with a (shape) regular triangulation $\mathcal{T}$ of the polygonal Lipschitz domain Ω, cf. [13]. Given a basis $\{\varphi_1, \ldots, \varphi_n\}$ of V_h, we can identify the finite-dimensional approximation of the velocity $\mathfrak{v}(t)$ by the coefficient vector $v(t) \in \mathbb{R}^n$. The discrete representative of the pressure $\mathfrak{p}$(t) is denoted by $p(t) \in \mathbb{R}^m$ and corresponds to a basis $\{\psi_1, \ldots, \psi_m\}$ of Q_h.

With the basis functions of V_h we define the symmetric and positive definite mass matrix $M \in \mathbb{R}^{n,n}$ by

$$M := [m_{jk}] \in \mathbb{R}^{n\times n}, \qquad m_{jk} := \int_\Omega \varphi_j \cdot \varphi_k \,\mathrm{d}x.$$

The discretisation of the nonlinearity of the Navier-Stokes equation, i.e., the discretisation of the operator $\mathcal{K}$ in Sect. 2.2, is denoted by $K\colon \mathbb{R}^n \to \mathbb{R}^n$. Note that the given model also includes linearisations of the Navier-Stokes equation such as the unsteady Stokes or Oseen equation. In this case, K can be written as a $n \times n$ matrix. In view of the index analysis of system (3.1), however, this is not of importance; see Appendix 1. We define for $v \in \mathbb{R}^n$ and its representative $\tilde{v} = \sum_{j=1}^n v_j \varphi_j \in V_h$,

$$K_j(v) := \int_\Omega (\tilde{v} \cdot \nabla)\tilde{v} \cdot \varphi_j \,\mathrm{d}x + \frac{1}{\mathrm{Re}} \int_\Omega \nabla\tilde{v} \cdot \nabla\varphi_j \,\mathrm{d}x. \tag{3.2}$$

Finally, we define the matrix B, which corresponds to the divergence operator $\mathcal{B}$, i.e.,

$$B := [b_{ij}] \in \mathbb{R}^{m\times n}, \qquad b_{ij} := \langle \mathcal{B}\varphi_j, \psi_i \rangle = \int_\Omega \psi_i \mathrm{div}\varphi_\mathrm{j} \,\mathrm{dx}.$$

By duality of the continuous operators, the discretisation of $\mathcal{B}'$ corresponds to the transpose B^T. This yields the saddle point structure of system (3.1).

Especially for the stable approximation of the pressure, it is necessary that the chosen finite element spaces are compatible [9, Ch. VI.3]. Let V_h and Q_h denote again finite-dimensional (sub)spaces of $\mathcal{V}$ and $\mathcal{Q}$, respectively, with

$$\dim V_h = n, \qquad \dim Q_h = m < n.$$

The spaces V_h and Q_h are compatible if they satisfy a so-called *inf-sup* or *Ladyzhenskaya-Babuška-Brezzi* condition [9, Ch. VI.3]. This means that there exists a constant $\beta > 0$, independent of the chosen mesh size, such that

$$\inf_{p_h \in Q_h} \sup_{v_h \in V_h} \frac{|\langle \operatorname{div} v_h, p_h \rangle|}{\|v_h\|_{\mathcal{V}} \|p_h\|_{\mathcal{Q}}} \geq \beta.$$

The inf-sup condition, with β independent of h, is a necessary condition for the convergence of the FEM; see, e.g., [40] or [21]. For a fixed spatial discretisation, this condition implies that the matrix B resulting from the discretisation scheme V_h, Q_h is of full rank.

Remark 3.1 In the case of an internal flow, i.e., if there are no inflow or outflow boundaries, the pressure solution to (2.18) takes on values in $L^2(\Omega)/\mathbb{R}$ meaning that $\mathfrak{p}(t)$ is defined up to a constant. In a finite element discretisation that does not *fix* this constant explicitly, this leads to a rank deficit of B; see [15, Ch. 5.3].

3.2 Finite Volumes and Finite Differences

In this section we briefly touch the spatial discretisation of the incompressible NSE by the methods of *finite volumes* (FVM) and *finite differences* (FDM). If applied to (2.14) in a straight forward manner, both approaches lead to a DAE of type (3.1).

Finite volume approximations base on the integral formulation of the conservation laws as it reads for the momentum equation

$$\int_W \frac{\partial \rho \mathfrak{v}}{\partial t}\, \mathrm{d}V + \int_{\partial W} \rho \mathfrak{v}\mathfrak{v} \cdot n \, \mathrm{d}S = -\int_{\partial W} \mathfrak{p} n \, \mathrm{d}S + \int_{\partial W} \tau \cdot n \, \mathrm{d}S + \int_W \rho g \, \mathrm{d}V, \tag{3.3}$$

cf. (2.5) and (2.7). Balances, like (3.3) in particular, hold for the whole domain of computation. For the discretisation, the domain of the flow is subdivided into small volumes often referred to as cells. The velocities and the pressure are assumed to be, say, constant over the cells, and their approximated values are determined by evaluating and equating the volume and surface integrals associated with every cell; see [18, Ch. 8.6].

Because of its flexibility in the discretisation, because of its variants that provide unconditional stability properties, and since typical models for the effective treatment of turbulence are formulated as conservation laws too, the FVM is the method of choice in most general purpose solvers. As for the discussion in the DAE context we note that, typically, the DAE (3.1) is never assembled but rather decoupled during the time discretisation; see [18, Ch. 7] and Sect. 4.3.2 below.

Approximations of the NSE via FDM come with the known flaws of finite difference approximations like high regularity requirements and confinement to regular grids. Nonetheless, FDM discretisations have been successfully used in flow discretisations and are probably still in use in certain specified, say, single-purpose codes. Also, FDM are often well suited for teaching the fundamentals of flow simulation of laminar and turbulent setups; see [25].

3.3 *Index of the DAE*

As argued in Remark 3.1, the coefficient matrix B may be rank-deficient and, thus, the *differentiation index* may be not well-defined [61]. However, the *strangeness index* [36, 37, 39] that applies to over- and underdetermined DAE systems and, thus, also can be determined in the case of a rank-deficient coefficient matrix B. A rough index analysis has been realized in [62] with the result that under general and reasonable assumptions, system (3.1) has strangeness index 1. See also Appendix 1, where the strangeness index has been determined in a rigorous way. This matches the, say, observations in [27, Ch. VII.1] that the system is of differentiation index 2 if the matrix B is of full rank.

In the sequel we always assume that the discretisation scheme is chosen in such a way that the matrix B is of full rank. Note that this can always be realized, choosing linearly independent finite element basis functions with respect to the space $\mathcal{Q}$, i.e., keeping in mind that constant functions are in the same equivalence class as the zero-function. If the ansatz functions form a partition of unity, then this means nothing else than eliminating one of these ansatz functions. Numerical schemes used in practice usually satisfy this condition as the discrete ansatz space for the pressure $\mathcal{Q}_h$ is chosen appropriately.

4 Fully Discrete Approximation Schemes

The final step of a numerical approximation of the infinite-dimensional NSE (2.14) is the numerical time integration of the semi-discrete approximation (3.1). For this, one discretises the time interval $(0, T)$ via the grid

$$t_0 := 0 < t_1 < t_2 < \cdots < t_N = T$$

and computes a sequence $(v^k, p^k)_{k\in\mathbb{N}}$ of values that are supposed to approximate the solution of (3.1) at the discrete time instances, i.e., $v^k \approx v(t^k)$ and $p^k \approx p(t^k)$.

On the one hand side, this approximating sequence can be defined via a numerical time integration scheme applied to the DAE (3.1); see [39, Ch. 5] for a general introduction and [3, 8] for so-called *half-explicit* methods as we will consider them below. In this case, the resulting fully discrete scheme inherits the properties of the associated time-continuous DAE. In particular, if the DAE was strangeness-free, then suitable methods will deliver accurate and stable time-discrete approximations, whereas certain difficulties arise if the DAE is of higher index; cp. [39, Introduction to Ch. 6].

On the other hand side, the discrete sequence (v^k, p^k) may be defined as an approximation to (3.1) on a time grid without resorting to a time-discretisation scheme. In this case, the resulting fully-discrete scheme cannot be assessed in terms of, say, the index of the associated time-continuous equations.

Therefore, to provide a qualitative analysis of fully discrete schemes without resorting to the continuous-time system, we employ the Kronecker index for discrete-time systems.

4.1 Index and Causality of Discrete Systems

For the analysis, we will restrict our considerations to a linear time-invariant setup. This is no restriction, since in the particular semi-linear semi-explicit form, the DAE structure is not affected by the nonlinearity. Further we assume that the time grid is equidistantly spaced and of size $\tau = t^{k+1} - t^k$.

We will cast the fully-discrete schemes into the standard form

$$\mathcal{E}x^{k+1} = \mathcal{A}x^k + h^k, \tag{4.1}$$

with coefficient matrices $\mathcal{E}$ and $\mathcal{A}$ and x^k containing all variables like $x^k = [v^k; p^k]$. We call such a fully discrete approximation scheme (4.1) a *difference-algebraic equation* (ΔAE).

Remark 4.1 This recast of the scheme, which will basically amount to a shift of the time indices for some variables, is needed to avoid ambiguities. In fact, for continuous time DAEs with delays, one can deliberately apply the operations of time-shift and differentiation to produce equivalent formulations of different indices [26].

In what follows, we argue that the matrix pair $(\mathcal{E}, \mathcal{A})$ can be assumed to be regular (cf. [39, Def. 2.5]) such that it can be brought into a particular Kronecker form

$$(\mathcal{E}, \mathcal{A}) \sim \left(\begin{bmatrix} I & 0 \\ 0 & N \end{bmatrix}, \begin{bmatrix} J & 0 \\ 0 & I \end{bmatrix} \right), \tag{4.2}$$

where J is a matrix in Jordan form and N is a nilpotent matrix, cf. [39, Thm. 2.7].

Definition 4.1 The index of nilpotency – that integer ν for which $N^\nu = 0$ while $N^{\nu-1} \neq 0$ – is called the *index of the matrix pair* $(\mathcal{E}, \mathcal{A})$; see [39, Def. 2.9], or the *Kronecker index* of the ΔAE (4.1), cf. [51, p. 39].

Equivalently, we can classify the discrete schemes using the notion of *causality* that attributes systems with states depending only on the past or current inputs.

Definition 4.2 ([14, Def. 8-1.1]) The sequence x^k, $k = 1, 2, \ldots$, defined through (4.1) is called *causal*, if x^k is determined completely by an initial condition x^0 and former (and the current) inputs $h^0, h^1, \ldots, h^k$.

Conversely, in a *noncausal* system, the current state x^k depends on future inputs like f^{k+1}. We will analyse time-stepping schemes for causality, which in the considered case is equivalent to being of Kronecker index 1 [14, Thm. 8-1.1] and discuss how and why a noncausal system poses difficulties in the numerical approximation.

To introduce the procedure, to fix the notation, and to have a benchmark for further comparisons, we start with analysing an *half-explicit Euler* discretisation of a linearised version of (3.1), namely

$$\tfrac{1}{\tau} M v^{k+1} = (\tfrac{1}{\tau} M + A) v^k + B^T p^k + f^k, \tag{4.3a}$$

$$B v^{k+1} = g^{k+1}. \tag{4.3b}$$

This defines the difference equations for v^k and p^k, which approximate the velocity and pressure at the discrete time instances t^k, $k = 1, 2, \ldots$. Similarly, f^k and g^k stand for the approximations of f and g at the time instances.

The difference scheme is of the standard form (4.1) with $x^k = [v^k; p^k]$, $h^k = [f^k; g^k]$, a shift of the index in (4.3b), and

$$(\mathcal{E}, \mathcal{A}) = \left(\begin{bmatrix} \frac{1}{\tau} M & 0 \\ 0 & 0 \end{bmatrix}, \begin{bmatrix} \frac{1}{\tau} M + A & B^T \\ B & 0 \end{bmatrix} \right). \tag{4.4}$$

In Appendix 2 it is shown that under standard conditions and for τ sufficiently small, the pair $(\mathcal{E}, \mathcal{A})$ in (4.4) is regular and equivalent to

$$\left(\begin{bmatrix} I & 0 & 0 \\ 0 & 0 & 0 \\ 0 & I & 0 \end{bmatrix}, \begin{bmatrix} * & 0 & 0 \\ 0 & I & 0 \\ 0 & 0 & I \end{bmatrix} \right).$$

This means that the difference scheme based on an *half-explicit Euler* discretisation is of *Kronecker index* 2. In fact, a straight forward calculation reveals that in (4.3) the state p^k depends on g^{k+1}.

Remark 4.2 The index k for the pressure p in (4.3) is consistent with f^k, as can be directly derived from the case that $A = 0$ and $g = 0$. A fully implicit scheme, i.e., considering f^{k+1} and p^{k+1} in (4.3a), cannot be brought into the standard form (4.1), since a shift in the index, $k+1 \leftarrow k$, would lead to the appearance of v^{k-1}.

4.2 *Inherent Instabilities of ΔAEs of Higher Index*

In this section, we illustrate a mechanism that leads to a numerical instability and, thus, possibly to divergence of the approximation of a dynamical system through a time discretisation, i.e. a ΔAE, with a Kronecker index greater than 1. Consider a DAE in Kronecker form

$$N\dot{x} = x + g$$

with $N \neq 0$ and $N^2 = 0$ and it's time discrete approximation through an Euler scheme:

$$\frac{1}{\tau}Nx^{k+1} = (\frac{1}{\tau}N + I)x^k + g^k. \tag{4.5}$$

The ΔAE (4.5) is of Kronecker index 2, according to Definition 4.1 and as it can be read off after a premultiplication by $(I - N)$, and it has the solution

$$x^k = -g^k - \frac{1}{\tau}(Ng^{k+1} - Ng^k),$$

as it follows from an adaption of the arguments in [39, Lem. 2.8] to the discrete case. From this solution representation one can conclude, that the solution to a ΔAE of higher index that discretises a DAE may depend on numerical differentiations and that any error in the computation may be amplified by the factor τ^{-1}. Note that for index-1 ΔAE s, where $N = 0$, this derivative is not present and that for even higher indices higher (numerical) derivatives will appear in the solution. Also note, that for systems that are not in Kronecker form such as (4.3), these derivations will be realized implicitly; see [1].

4.3 *Common Time-Stepping Schemes as Index-1 ΔAEs*

In this subsection, we discuss the different strategies, which are used in practice, to solve the spatially discretised NSE (3.1). In practical applications one typically uses schemes that decouple pressure and velocity computations. Although, as we have argued from a DAE perspective [1], this may lead to instabilities, the advantages that

- one has to solve two smaller systems rather than one large and
- one basically solves Poisson equations and convection-diffusion rather than saddle-point problems

seem to prevail despite of the active research on iterative solvers for saddle-point system. In particular, for the symmetric case of Stokes flow, theory provides efficient and well-understood preconditioning approaches. These approaches serve as a base for preconditioning techniques for coupled systems with highly non-normal coefficient matrices as they appear in the case of Navier-Stokes flow with higher Reynolds numbers; see, e.g., the numerical results in [15, Ch. 8] or [30] for the scaling of the performance of iterative solvers with respect to the Reynolds number.

Another common feature of the schemes used in practice is their explicit approach to the momentum equation which avoids the repeated assembling of Jacobians. The error is then either controlled through a small time-step or through some fixed point iterations.

We consider the schemes *Projection* as it was described in [23] and *SIMPLE* and *artificial compressibility*; see [18]. We will use a description and formulation general enough, to also accommodate numerous variants of the methods.

4.3.1 Projection

The principle of these methods is to solve for an intermediate velocity approximation that does not need to be divergence-free, and project it onto the divergence-free constraint in a second step. As far as the velocity approximation is concerned, *Projection* methods can be formulated both in infinite and finite dimensions. Since the first work by Chorin [11], a number of variants have been developed mainly proposing different approaches to the approximation of the pressure tackling or circumventing the need of solving a *Poisson equation* for the pressure, which requires certain regularity assumptions (cf. [24, p. 642]). Another problem is the requirement of boundary conditions for the pressure update that do not have a physical motivation and may cause inaccuracies close to the boundary. Nonetheless, these schemes have been extensively studied and certain heuristics ensure satisfactory convergence behaviour; see, e.g, [23, 24].

As an example for a *Projection* scheme, we present the variant proposed in [23]:

1. Solve for intermediate velocity with the old pressure

$$\tfrac{1}{\tau} M \tilde{v}^{k+1} = (\tfrac{1}{\tau} M + A) v^k + B^T p^k + f^k. \tag{4.6}$$

2. Determine the new velocity v^{k+1} as the projection of the intermediate velocity onto $\ker B$ by solving

$$M v^{k+1} - B^T \phi^{k+1} = M \tilde{v}^{k+1}, \tag{4.7a}$$

$$B v^{k+1} = g^{k+1}. \tag{4.7b}$$

3. Update the pressure via

$$p^{k+1} = p^k + \frac{2}{\tau}\phi^{k+1}. \tag{4.8}$$

Remark 4.3 Instead of solving the saddle-point problem (4.7) as a whole and in order to avoid the division of a numerically computed quantity by τ in (4.8), one can decouple the system. Then, one solves for $\tilde{\phi}^{k+1} := \frac{2}{\tau}\phi^{k+1}$ through

$$-BM^{-1}B^T\phi^{k+1} = \frac{2}{\tau}(B\tilde{v}^{k+1} - g^{k+1})$$

and obtains the updates via

$$v^{k+1} = \tilde{v}^{k+1} + \frac{\tau}{2}M^{-1}B^T\tilde{\phi}^{k+1}$$

and

$$p^{k+1} = p^k + \tilde{\phi}^{k+1}.$$

Remark 4.4 The formula for the update of the pressure is derived from the relation

$$\phi(t+\tau) = -\frac{\tau^2}{2}\dot{p}(t) + \mathcal{O}(\tau^3),$$

cf. [23, p. 595], that holds under certain regularity assumptions. Note also that, if formulated for the space-continuous problem, this *Projection 2* algorithm requires a sophisticated treatment of the boundary conditions. The presented variant is a simplification for spatially discretised equations; see [24, Ch. 3.16.6c].

As a single system, this projection scheme defines a ΔAE in the form of (4.1) with $x^k := [\tilde{v}^k; \phi^k; v^k; p^k]$ and

$$(\mathcal{E}, \mathcal{A}) = \left(\begin{bmatrix} \frac{1}{\tau}M & 0 & 0 & 0 \\ 0 & 0 & 0 & 0 \\ -M & -\frac{\tau}{2}B^T & M & 0 \\ 0 & -I & 0 & I \end{bmatrix}, \begin{bmatrix} 0 & \frac{1}{\tau}M + A & 0 & B^T \\ -\frac{2}{\tau}B & -BM^{-1}B^T & 0 & 0 \\ 0 & 0 & 0 & 0 \\ 0 & 0 & 0 & I \end{bmatrix} \right), \tag{4.9}$$

which is a matrix pair of Kronecker index 1; see Appendix 2.

4.3.2 SIMPLE Scheme: Implicit Pressure Correction

The SIMPLE scheme and its variants are based on the decomposition

$$v^{k+1} = \tilde{v}^{k+1} + v_\Delta^{k+1} \quad \text{and} \quad p^{k+1} = p^k + p_\Delta^{k+1}, \tag{4.10}$$

where $\tilde{v}^{k+1}$ is the tentative velocity computed by means of the old pressure. We present the basic variant, in which the velocity correction v_Δ^{k+1} is discarded when solving for p_Δ^{k+1}:

1. Solve for the intermediate velocity $\tilde{v}^{k+1}$ with the old pressure as in (4.6).
2. Compute p_Δ^{k+1} through

$$B(\tfrac{1}{\tau}M + A)^{-1}B^T p_\Delta^{k+1} = -B\tilde{v}^{k+1} + g^{k+1} \tag{4.11}$$

   as the correction to p^k such that the
3. updates of the velocity and pressure defined through

$$v^{k+1} = \tilde{v}^{k+1} + (\tfrac{1}{\tau}M + A)^{-1}B^T p_\Delta^{k+1}, \tag{4.12a}$$

$$p^{k+1} = p^k + p_\Delta^{k+1}. \tag{4.12b}$$

   jointly fulfill the time discrete momentum and the continuity equation.

Remark 4.5 Step (2) of the presented SIMPLE algorithm computes p_Δ^{k+1} under the temporary assumption that $v_\Delta^{k+1} = 0$. This is hardly justified and probably a reason for slow convergence, when applied in a fixed-point iteration within fully implicit schemes. Certain variants of the SIMPLE scheme try to approximate v_Δ^{k+1} at this step, e.g., through interpolation; see [18, Ch. 7.3.4].

As a single system, the SIMPLE scheme defines a ΔAE in the form of (4.1) with $x^k := [\tilde{v}^k; p_\Delta^k; v^k; p^k]$ and

$$(\mathcal{E}, \mathcal{A}) = \left(\begin{bmatrix} \frac{1}{\tau}M & 0 & 0 & 0 \\ 0 & 0 & 0 & 0 \\ I & (\frac{1}{\tau}M + A)^{-1}B^T & -I & 0 \\ 0 & -I & 0 & I \end{bmatrix}, \begin{bmatrix} 0 & \frac{1}{\tau}M + A & 0 & B^T \\ -B & -B(\frac{1}{\tau}M + A)^{-1}B^T & 0 & 0 \\ 0 & 0 & 0 & 0 \\ 0 & 0 & 0 & I \end{bmatrix} \right), \tag{4.13}$$

which is a matrix pair of Kronecker index 1; see Appendix 2.

4.3.3 Artificial Compressibility

In this class of methods, the divergence-free constraint (2.14b) is relaxed by adding a scaled time-derivative of the pressure, i.e.,

$$\frac{1}{\beta}\frac{\partial \mathfrak{p}}{\partial t} + \operatorname{div} \mathfrak{v} = 0. \tag{4.14}$$

The parameter is assumed to satisfy $\beta \gg 1$. The corresponding spatial semi-discretisation (3.1) then reads

$$M\dot{v} + K(v) - B^T p = f, \tag{4.15a}$$

$$M_p \dot{p} + Bv = g \tag{4.15b}$$

with M_p denoting the mass matrix of the pressure approximation.

In theory, system (4.15) is an ODE and could be solved by standard time-stepping schemes. In practice, however, the common solution approaches to System (4.15) decouple pressure and velocity computations through introducing auxiliary quantities and, thus, a DAE structure. A possible method, that defines a pressure update similar to the SIMPLE scheme (4.12), can be interpreted and implemented as follows, cf. [18, Ch. 7.4.3]. As in the SIMPLE approach a first velocity approximation $\tilde{v}^{k+1}$ is computed via (4.6) on the base of the old pressure value p^k.

Next, one subtracts the contribution of the old pressure,

$$\bar{v}^{k+1} := \tilde{v}^{k+1} - (\frac{1}{\tau}M + A)^{-1}B^T p^k \tag{4.16}$$

and defines the new velocity v^{k+1} through the linear expansion of $\bar{v}^{k+1}$ in terms of the pressure gradient, i.e.,

$$v^{k+1} = \bar{v}^{k+1} + \frac{\partial(\bar{v}^{k+1})}{\partial(B^T p)}[B^T p^{k+1} - B^T p^k] = \bar{v}^{k+1} + (\frac{1}{\tau}M + A)^{-1}B^T p_\Delta^{k+1}. \tag{4.17}$$

Note that one uses (4.16) to determine the Jacobian $\frac{\partial(\bar{v}^{k+1})}{\partial(B^T p)}$ and that we have defined $p_\Delta^{k+1} = p^{k+1} - p^k$. Expression (4.17) is then inserted in a, e.g., *implicit Euler* discretisation of (4.15b), which gives an equation for p_Δ^{k+1}. Accordingly, the fully discrete scheme using artificial compressibility reads

$$(\tfrac{1}{\tau}M + A)\tilde{v}^{k+1} = \tfrac{1}{\tau}Mv^k + B^T p^k + f^k, \tag{4.18a}$$

$$\bar{v}^{k+1} = \tilde{v}^{k+1} - (\tfrac{1}{\tau}M + A)^{-1}B^T p^k, \tag{4.18b}$$

$$\frac{1}{\beta\tau}p_\Delta^{k+1} + B\bar{v}^{k+1} + B(\tfrac{1}{\tau}M + A)^{-1}B^T p_\Delta^{k+1} = g^{k+1}, \tag{4.18c}$$

$$v^{k+1} = \bar{v}^{k+1} - (\tfrac{1}{\tau}M + A)^{-1}B^T p_\Delta^{k+1}, \tag{4.18d}$$

$$p^{k+1} = p^k + p_\Delta^{k+1}. \tag{4.18e}$$

The corresponding ΔAE equals the ΔAE of the SIMPLE scheme up to a slight modification of the equation for the pressure update such that we can use the arguments laid out in Appendix 2 to conclude that it is of Kronecker index 1.

Remark 4.6 Note that in practice there may be nonlinear solves to determine, e.g., $\tilde{v}^{k+1}$ such that, e.g., the correction $\bar{v}^{k+1}$ is possibly different from $\tilde{v}^{k+1}$ computed with $p^k = 0$.

4.4 Time-Stepping Schemes Resulting from Index Reduction

Besides the presented schemes in Sect. 4.3, one may also apply an index reduction to system (3.1) and then discretise in time. This then also leads to matrix pairs $(\mathcal{E}, \mathcal{A})$ of Kronecker index 1.

4.4.1 Penalty Methods

Similar to Sect. 4.3.3, we may reduce the index of the DAE (3.1) by relaxing the divergence-free constraint or, in other words, add a penalty term [55]. With the penalty parameter $\beta \gg 1$ we replace the incompressibility condition by

$$\mathfrak{p} = -\beta \operatorname{div} \mathfrak{v}.$$

In the semi-discrete case this corresponds to the constraint equation $M_p p + Bv = g$. It can be shown that this then leads to a DAE of differentiation index 1. However, this approach changes the solution of the system. In order to keep this difference of reasonable size, β should be choosen relatively large. On the other hand, the condition number of the involved matrices increase with β and thus, lead to numerical difficulties. The difficulty of a reasonable choice of β is one drawback of this approach. A second disadvantage is that small velocities of order β^{-1} or less cannot be resolved [32]. Further modifications of the penalty method, which are also applicable for slightly compressible fluids (Newtonian fluids), are discussed, e.g., in [32, Sect. 5].

4.4.2 Derivative of the Constraint

Another very simple possibility to reduce the index of the given DAE is to replace the constraint by its derivative, the so-called *hidden constraint*. Instead of (3.1) we then consider the system

$$M\dot{\tilde{v}} + K(\tilde{v}) - B^T\tilde{p} = f, \tag{4.19a}$$

$$B\dot{\tilde{v}} = \dot{g}. \tag{4.19b}$$

The initial condition remains unchanged, i.e., $\tilde{v}(0) = v_0$. It is well-known that this system has differentiation index 1. Nevertheless, numerical simulations, which rely on this formulation, show a linear drift from the solution manifold given by $Bv = g$. This can be seen as follows.

Although the two systems are equivalent, numerical errors are integrated over time and thus amplified. Solving the constraints only up to a small error, i.e.,

$$Bv(t) = g(t) + \varepsilon, \qquad B\dot{\tilde{v}}(t) = \dot{g}(t) + \eta$$

for small and constant ε and η, we calculate for $\tilde{v}$ that

$$B\tilde{v}(t) = Bv_0 + \int_0^t B\dot{\tilde{v}}(s)\,\mathrm{d}s = Bv_0 + \int_0^t \big(\dot{g}(s) + \eta\big)\,\mathrm{d}s = g(t) + t\,\eta.$$

Note that the last step holds because of the assumed consistency condition $Bv_0 = g(0)$. This shows that a constant error in the constraint of $\tilde{v}$ leads to an error which grows linearly in time. Because of this, the method of replacing the constraint by its derivative is – although it is of lower index – not advisable.

4.4.3 Minimal Extension

Finally, we present the index reduction technique of *minimal extension* [39, Ch. 6.4]. A general framework for an index reduction based on derivative arrays is given in [39, Ch. 6]; see also [10]. Because of the special saddle point structure of system (3.1), in which the constraint is explicitly given, this procedure can be simplified by using so-called *dummy variables*, cf. [38, 46].

Since we assume that B is of full rank, there exists an invertible matrix $Q \in \mathbb{R}^{n,n}$ such that BQ has the block structure $BQ = [B_1, \; B_2]$ with an invertible matrix $B_2 \in \mathbb{R}^{m,m}$. Note that the choice of Q is not unique. We then use this transformation to partition the variable v, namely

$$\begin{bmatrix} v_1 \\ v_2 \end{bmatrix} := Q^{-1} v,$$

with according dimensions $v_1 \in \mathbb{R}^{n-m}$ and $v_2 \in \mathbb{R}^m$. With this, the constraint and its derivative may be written as

$$B_2 v_2 = g - B_1 v_1, \qquad B_2 \dot{v}_2 = \dot{g} - B_1 \dot{v}_1.$$

Together with the differential equation (3.1a) this yields an overdetermined system. Instead of an (expensive) search for projectors [39, Ch. 6.2] or selectors [56], which would bring the system back to its original size, we introduce a dummy variable $w_2 := \dot{v}_2$ to get rid of the redundancy. Note, however, that this leads to a slightly bigger system. More precisely, we extend the system dimensions from $n + m$ to

$n+2m$, which is still moderate, since we usually have $m \ll n$. The extended system is again square and has the form

$$M Q \begin{bmatrix} \dot{v}_1 \\ w_2 \end{bmatrix} + K(v_1, v_2) - B^T p = f,$$

$$B_2 v_2 = g - B_1 v_1,$$

$$B_2 w_2 = \dot{g} - B_1 \dot{v}_1.$$

This DAE is of differentiation index 1 and has the same solution set as the original system, cf. [1].

The drawback of this method is the need of a transformation matrix Q. With a suitable reordering of the variables, however, we can choose Q to be the identity matrix. In this case, the needed variable transformation is just a permutation and thus, all variables keep their physical meaning. For some specific finite element schemes an algorithm to find such a permutation (which leads to an invertible B_2 block) is given in [1]. This paper also considers the half-explicit Euler scheme applied to the minimally extended system, which leads to a pair $(\mathcal{E}, \mathcal{A})$ of Kronecker index 1.

5 Numerical Experiments

To illustrate the performance and particular issues of the time-stepping schemes for the NSE, we consider the numerical simulation of the flow passing a cylinder in two space dimensions. This problem, also known as *cylinder wake*, is a popular flow benchmark problem and a test field for flow control [5, 47, 63].

We consider the incompressible NSE (2.14) on the domain as illustrated in Fig. 1 with boundary Γ and boundary conditions as follows. At the inflow Γ_i, we prescribe a parabolic velocity profile through the function

$$g(s) = 4\big(1 - \frac{s}{0.41}\big)\frac{s}{0.41}.$$

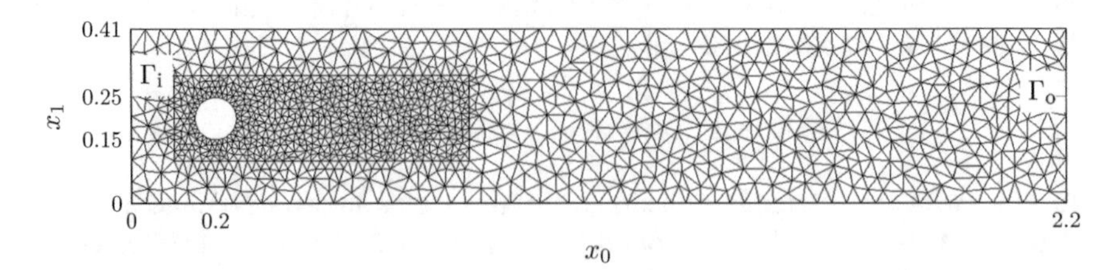

Fig. 1 Illustration of the geometrical setup including the domains of distributed control and observation and of the velocity magnitude for the cylinder wake. Figure taken from [4]

At the outflow Γ_o, we impose *do-nothing* conditions, cf. (2.17). At the upper and the lower wall of the channel and at the cylinder periphery, we employ *no-slip*, i.e., zero Dirichlet conditions. Thus, we set

$$\gamma(v,p)\colon \begin{cases} v=[g(x_1),0]^T & \text{on } \Gamma_\mathrm{i},\\ p\vec{n}-\frac{1}{\mathrm{Re}}\frac{\partial v}{\partial \vec{n}}=[0,0]^T & \text{on } \Gamma_\mathrm{o},\\ v=[0,0]^T & \text{elsewhere on the boundary.}\end{cases}$$

We set $\mathrm{Re} = 60$ and consider a $\mathcal{P}_2 - \mathcal{P}_1$ (Taylor-Hood) finite element discretisation of (2.14) on the grid depicted in Fig. 1 with 9, 356 degrees of freedom in the velocity and 1, 289 degrees of freedom in the pressure approximation. The result of the semi-discretisation is a DAE of the form (3.1), which we write as

$$M\dot{v}+Av+N(v)-B^Tp=f, \tag{5.1a}$$

$$Bv=g \tag{5.1b}$$

on the time interval $(0,1]$. Here, A is the Laplacian or the linear part of K as defined in (3.2) and N denotes the convection part. The right-hand sides f and g account for the (static) boundary conditions. As initial value we take the corresponding steady-state *Stokes* solution v_S, which is part of the solution to

$$\begin{bmatrix} A & -B^T\\ B & 0\end{bmatrix}\begin{bmatrix} v_\mathrm{S}\\ p_\mathrm{S}\end{bmatrix}=\begin{bmatrix} f\\ g\end{bmatrix}. \tag{5.2}$$

For the schemes that need an initial value for the pressure, we provide it as p_NS solving (5.1) at $t=0$ with $v(0)=v_\mathrm{S}$ and $\dot{v}(0)=0$. Note that this gives a consistent initial pressure, since g is constant, and thus, $B\dot{v}=0$.

We consider the time-discretisation of (5.1) by means of the half-explicit Euler scheme and compare it to the time-discretisation via the SIMPLE scheme as described in Sect. 4.3.2. Both schemes treat the nonlinearity explicitly. Comparing the computed approximations to a reference, obtained by the time-discretisation via the implicit trapezoidal rule on a fine grid, we show that both schemes are convergent of order 1 as long as the resulting linear systems are solved with sufficient precision. For inexact solves, we show that the pressure approximation in the half-explicit Euler scheme diverges unlike for the SIMPLE approximation. For snapshots of the approximate velocity solution, see Fig. 2.

The finite element implementation uses *FEniCS, Version 2017.2* [45]. For the iterative solutions of the linear system, we employ *Krypy* [20]. The code used for the numerical investigations is freely available for reproducing the reported results and as a benchmark for further developments in the time integration of semi-explicit DAEs of strangeness index 1; see Fig. 3 for a stable link to the online repository.

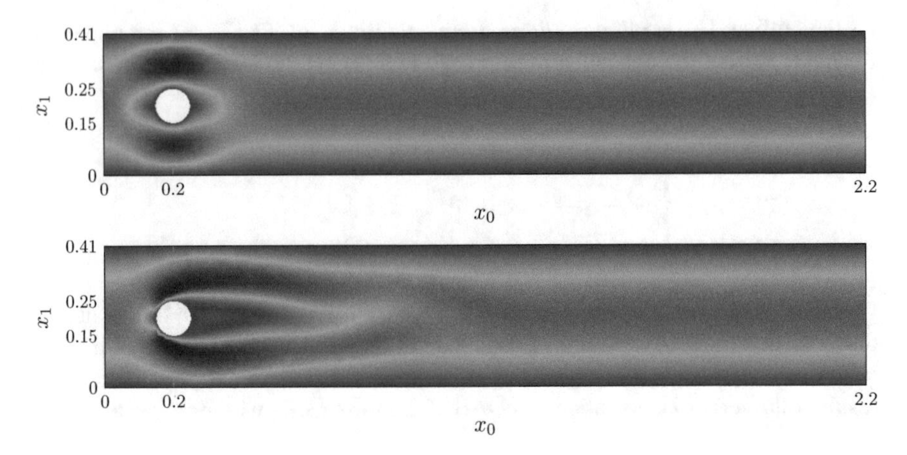

Fig. 2 Snapshots of the velocity magnitude computed with SIMPLE with $\tau = 1/1{,}024$ and exact solves taken at $t = 0$ (top) and $t = 1$ (bottom)

Code and Data Availability

The source code and the data of the implementations used to compute the presented results are available from:

`doi:10.5281/zenodo.998909`

Contact the author Jan Heiland for licensing information.

Fig. 3 Link to code and data

5.1 *Time Integration with Half-Explicit Euler*

With the half-explicit Euler time discretisation, at every time step the linear system

$$\begin{bmatrix} \frac{1}{\tau}M + A & B^T \\ B & 0 \end{bmatrix} \begin{bmatrix} v^{k+1} \\ -p^k \end{bmatrix} = \text{rhs} := \begin{bmatrix} \frac{1}{\tau}M - N(v^k) + f \\ g \end{bmatrix} \tag{5.3}$$

is solved for the velocity and the pressure approximations. For the approximate solution of the linear systems, we use *MinRes* iterations, which are stopped as soon as the relative residual drops below a given tolerance `tol`, i.e.,

$$\| \begin{bmatrix} \frac{1}{\tau}M + A & B^T \\ B & 0 \end{bmatrix} \begin{bmatrix} v^{k+1} \\ -p^k \end{bmatrix} - \text{rhs} \, \|_{(M^{-1}, M_Q^{-1})} \leq \texttt{tol} \; \|\text{rhs}\|_{(M^{-1}, M_Q^{-1})},$$

where we use the norm induced by the inverses of the mass matrix M and of the mass matrix of the pressure approximation space M_Q.

It can be seen from the error plots in Fig. 4, for exact solves, that the half-explicit Euler converges linearly in the velocity and the pressure approximation, which is

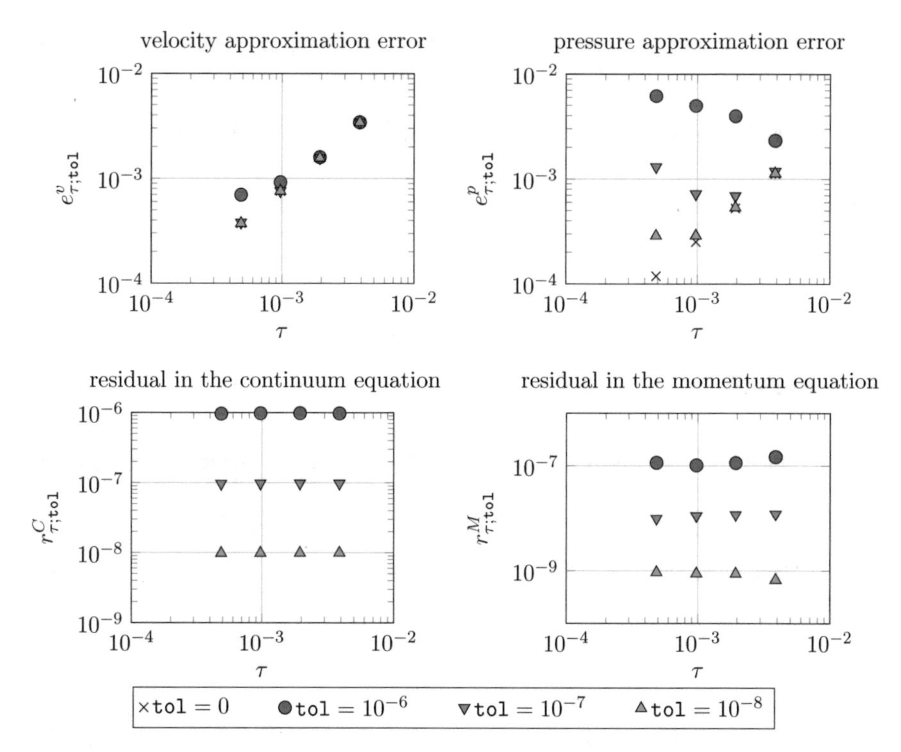

Fig. 4 Error in the velocity and the pressure approximation provided by the half-explicit Euler algorithm for iterative solves with varying tolerances. The crosses are the errors obtained with direct solves

in line with the theory; see [28, Tab. 2.3] and [8]. For inexact solves, however, for smaller time-steps, the pressure approximation diverges linearly. As we have shown in [1] this is an inherent instability of the index-2 formulation.

The reported residuals and errors are defined as follows:

$$e^v_{\tau;\mathtt{tol}} = \operatorname{trp}(\|v_{\tau;\mathtt{tol}} - v_{\mathtt{ref}}\|_M)$$

where the subscript $\mathtt{ref}$ denotes the reference solution, the subscripts $\tau;\mathtt{tol}$ denote the approximation that is computed on the grid of size τ with the linear equations solved with tolerance $\mathtt{tol}$, and where trp(s) denotes the approximation to the integral $\int_0^{t_N} s(t)\,\mathrm{d}t$ by means of the piecewise trapezoidal rule with step size τ. Analogously, we define the error in the pressure approximation

$$e^p_{\tau;\mathtt{tol}} = \operatorname{trp}(\|p_{\tau;\mathtt{tol}} - p_{\mathtt{ref}}\|_{M_P})$$

and the integrated residuals in the momentum equation $r^M_{\tau;\mathtt{tol}}$ as the integral of the function

$$t^k \mapsto \|(\tfrac{1}{\tau}M + A)v^{k+1}_{\tau;\mathtt{tol}} - \tfrac{1}{\tau}Mv^k_{\tau;\mathtt{tol}} - B^T p^k_{\tau;\mathtt{tol}} + N(v^k_{\tau;\mathtt{tol}}) - f^k\|_{M^{-1}}$$

and in the continuum equation as

$$r^C_{\tau;\mathtt{tol}} := \operatorname{trp}(\|Bv_{\tau;\mathtt{tol}} - g\|_{M_Q^{-1}}).$$

5.2 *Time Integration with SIMPLE*

For this particular time discretisation, we have to solve three linear systems, namely

$$(\tfrac{1}{\tau}M + A)\tilde{v}^{k+1} = \mathrm{rhs}_1 := \tfrac{1}{\tau}Mv^k + B^T p^k + f^k, \tag{5.4a}$$

$$B(\tfrac{1}{\tau}M + A)^{-1}B^T p^{k+1}_\Delta = \mathrm{rhs}_2 := -B\tilde{v}^{k+1} + g^{k+1}, \tag{5.4b}$$

and

$$(\tfrac{1}{\tau}M + A)v^{k+1}_\Delta = \mathrm{rhs}_3 := B^T p^{k+1}_\Delta \tag{5.4c}$$

to compute the updates as $v^{k+1} = \tilde{v}^{k+1} + v^{k+1}_\Delta$ and $p^{k+1} = p^k + p^{k+1}_\Delta$. For the approximate solution, we use *CG* iterations until the relative residuals, measured in the M^{-1} norm (or M_Q^{-1} for (5.4b)), drop below a given tolerance `tol`.

As one can see from the error plots in Fig. 5, the inexact solves affect the approximation only for a rough tolerance $\mathtt{tol} = 10^{-4}$, which is only one order of magnitude smaller than the actual approximation error. Thus the breakdown in the convergence observed in Fig. 5 is probably due to the accumulation of errors that every single step scheme suffers from. For smaller tolerances, the approximations almost achieve the accuracy of the direct solves. Interestingly, the continuity equation (3.1b), which is only an implicit part of the SIMPLE scheme, is fulfilled at much better accuracy than the choices of the tolerances suggest; see Fig. 5.

6 Conclusion

In this paper, we have discussed the incompressible NSE from a DAE point of view, in which the incompressibility is interpreted as an algebraic constraint. Thus, a spatial discretisation leads to a DAE. If the time is discretised as well, a ΔAE is obtained – a sequence of equations, that define the numerical approximations.

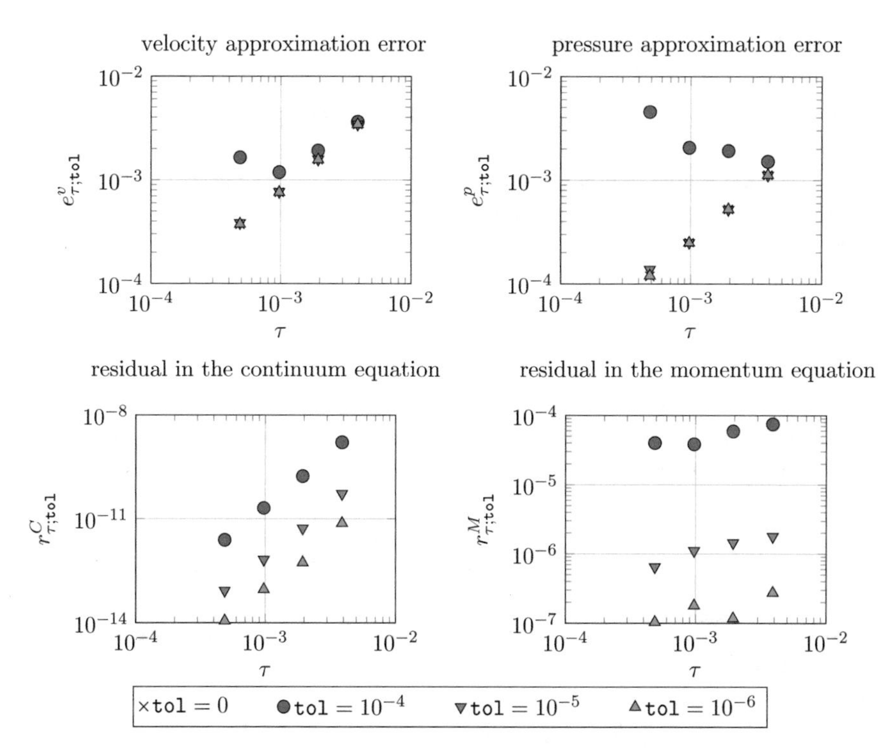

Fig. 5 Error in the velocity and the pressure approximation provided by the SIMPLE algorithm for exact solves and iterative solves with varying tolerances `tol`

We have discussed suitable approaches to well-posed semi-discrete approximations and investigated the ΔAE stemming from different time-discretisations in terms of the Kronecker index. By means of the Kronecker index it became possible to qualitatively compare the time discretised schemes with established numerical schemes that may not have time continuous counterpart. It turned out that commonly and successfully used time integration schemes like the SIMPLE algorithm define a ΔAE of index 1, whereas a time-discretisation of the semi-discrete NSE by an half-explicit Euler scheme leads to a ΔAE of index 2. If, however, an index reduction on the semi-discrete level was applied first, then the time marching schemes led to a ΔAE of lower index.

The advantage of the discrete index-1 formulations is that they avoid implicit derivations that amplify computational errors. This mechanism will likely lead to larger errors, in particular if the equations are solved with limited accuracy. We have illustrated the origin of this behavior in a small analytical example and verified it in a numerical simulation. The code of the numerical test case, an implementation of the 2D cylinder wake, is provided to serve as a benchmark for future developments of time integration schemes for DAEs of Navier-Stokes type.

Appendix 1: Strangeness Index of Eq. (3.1)

We analyse in detail the strangeness index of the DAE (3.1); see [39, Def. 4.4] for the precise definition. Note that we do not ask for any assumptions on the nonlinearity K and that we allow the matrix B to be rank-deficient. This then also implies that the differentiation index of (3.1) equals 2 if it is well-defined, i.e., if B is of full rank.

Linear Case

Considering any linearisation of the Navier-Stokes equations, i.e., $K(u) = Ku$ in (3.1), we deal with the matrix pair

$$(E, A) = \left(\begin{bmatrix} M & 0 \\ 0 & 0 \end{bmatrix}, \begin{bmatrix} K & B^T \\ B & 0 \end{bmatrix} \right).$$

Following [39, Th. 3.11], we can construct a to (E, A) (globally) equivalent pair $(\tilde{E}, \tilde{A})$ of the form

$$\tilde{E} = \left[\begin{array}{cc|c} I_b & & \\ & I_{n-b} & \\ \hline & & 0 \\ & & 0 \end{array} \right], \qquad \tilde{A} = \left[\begin{array}{cc|c} 0 & A_{12} & A_{13} \\ 0 & 0 & A_{23} \\ \hline I_b & & 0 \\ & 0 & 0 \end{array} \right]$$

with $A_{13} \in \mathbb{R}^{b,m}$ being of full rank. Thus, the original system (3.1) is equivalent to a system of the form

$$\dot{x}_1 = A_{12}x_2 + A_{13}x_3 + f_1, \qquad \dot{x}_2 = A_{23}x_3 + f_2, \qquad 0 = x_1 + f_3, \qquad 0 = f_4$$

with dimensions $x_1(t) \in \mathbb{R}^b$, $x_2(t) \in \mathbb{R}^{n-b}$, $x_3(t) \in \mathbb{R}^m$. Since we have a differential and an algebraic equation for x_1 (this causes the 'strangeness'), we use the derivative of $0 = x_1 + f_3$ in order to eliminate $\dot{x}_1$ in the first equation. Hence, we consider the pair $(E_{\text{mod}}, A_{\text{mod}})$ with

$$E_{\text{mod}} = \left[\begin{array}{cc|c} 0 & & \\ & I_{n-b} & \\ \hline & & 0 \\ & & 0 \end{array} \right], \qquad A_{\text{mod}} = \left[\begin{array}{cc|c} 0 & A_{12} & A_{13} \\ 0 & 0 & A_{23} \\ \hline I_b & & 0 \\ & 0 & 0 \end{array} \right].$$

Since A_{13} is of full rank, one can show that system $(E_{\text{mod}}, A_{\text{mod}})$ is strangeness-free, cf. the calculation in [39, Th. 3.7]. Since we have obtained a strangeness-free system with only one differentiation, system (3.1) has strangeness index one.

Nonlinear Case

The general form of a nonlinear DAE is given by

$$F(t, x, \dot{x}) = 0.$$

In regard of system (3.1) we set $x := [q^T, \ p^T]^T$ and define

$$F(t, x, \dot{x}) := \begin{bmatrix} M\dot{q} - K(q) - B^T p - f \\ -Bq + g \end{bmatrix} = E\dot{x} - A(x) - h$$

with

$$E := \begin{bmatrix} M & 0 \\ 0 & 0 \end{bmatrix}, \qquad A_x := \frac{\partial A(x)}{\partial x} = \begin{bmatrix} K_q & B^T \\ B & 0 \end{bmatrix}.$$

In the sequel we show that (3.1) has strangeness index 1 also in the nonlinear case. For this, we assume that B has full rank such that there are no vanishing equations and the pressure variable is uniquely defined. In the case $\operatorname{rank} B = b < m$, we consider the following transformation.

Let $C_0 \in \mathbb{R}^{m,m-b}$ be the matrix of full rank satisfying $B^T C_0 = 0$. Furthermore, $C' \in \mathbb{R}^{m,b}$ defines any matrix such that $C = [C_0 \ C'] \in \mathbb{R}^{m,m}$ is invertible. With this, we obtain the relation

$$B^T C = \begin{bmatrix} B^T C_0 & B^T C' \end{bmatrix} = \begin{bmatrix} 0 & \tilde{B}^T \end{bmatrix}$$

with $\tilde{B} \in \mathbb{R}^{b,n}$ having full rank. With the matrix C in hand, we first introduce the new pressure variable $\tilde{p} := C^{-1} p$. Thus, we consider the pair $z := [q^T, \tilde{p}^T]^T$. As a second step, we multiply equation (3.1) by the block-diagonal matrix $\operatorname{diag}(\mathrm{I_n}, \mathrm{C^T})$ from the left. In total, this yields the equivalent DAE

$$M\dot{q} = K(q) + \begin{bmatrix} 0 & \tilde{B}^T \end{bmatrix} \tilde{p} + f, \qquad \begin{bmatrix} 0 \\ \tilde{B} \end{bmatrix} q = C^T g.$$

Note that the constraint contains $(m - b)$ consistency equations of the form $0 = g_1$. Assuming that system (3.1) is solvable, we suppose that these are in fact vanishing equations. Thus, they have no influence on the index of the system. Furthermore, the first $(m - b)$ components of the transformed pressure $\tilde{p}$ do not influence the system. These components are underdetermined and may be omitted, again without changing the index. Leaving out the underdetermined parts as well as the vanishing equations, we obtain a system of the form (3.1) with a full rank matrix B.

In the sequel, we assume that $\operatorname{rank} B = m$ and show that [39, Hyp. 4.2] is satisfied for $\mu = 1$. Note that this hypothesis is not satisfied for $\mu = 0$, i.e., the system is not strangeness-free. We define the matrices

$$M_1 := \begin{bmatrix} E & 0 \\ -A_x & E \end{bmatrix}, \qquad N_1 := \begin{bmatrix} A_x & 0 \\ 0 & 0 \end{bmatrix}.$$

We now pass through the list of points of the hypothesis in [39, Hyp. 4.2]:

1. First, we note that the rank of M_1 equals $2n$ and we set $a := 2(n+m) - 2n = 2m$. Thus, the system contains $2m$ algebraic variables (the pressure and the part of q, which is not divergence-free). Furthermore, we define $Z_2 \in \mathbb{R}^{2(n+m),2m}$ by $Z_2^T M_1 = 0$, i.e.,

$$Z_2 = \left[\begin{array}{c|c} 0 & M^{-1}B^T \\ I_m & 0 \\ \hline 0 & 0 \\ 0 & I_m \end{array}\right].$$

2. As a second step we define $\hat{A}_2 := Z_2^T N_1 [I_{n+m},\ 0]^T$, which yields

$$\hat{A}_2 = \left[\begin{array}{cc|cc} 0 & I_m & 0 & 0 \\ \hline BM^{-1} & 0 & I_m & 0 \end{array}\right] \left[\begin{array}{cc} K_q & B^T \\ B & 0 \\ \hline 0 & 0 \\ 0 & 0 \end{array}\right] = \left[\begin{array}{c|c} B & 0 \\ \hline BM^{-1}K_u & BM^{-1}B^T \end{array}\right].$$

   This matrix has rank $2m$, since the full-rank property of B implies that $BM^{-1}B^T$ is invertible. We define $d := n - m$ as the number of differential variables and $T_2 \in \mathbb{R}^{n+m,n-m}$ by $\hat{A}_2 T_2 = 0$. Let $C \in \mathbb{R}^{n,n-m}$ be a matrix of full rank with $BC = 0$ and $C_2 := -(BM^{-1}B^T)^{-1} BM^{-1} K_u C \in \mathbb{R}^{m,n-m}$. Then, we set

$$T_2 := \begin{bmatrix} C \\ C_2 \end{bmatrix}.$$

3. Finally, we compute the rank of ET_2. Since C has full rank, this equals $\operatorname{rank} MC = n - m = d$. The matrix $Z_1^T := [C^T\ 0] \in \mathbb{R}^{n-m,n+m}$ satisfies

$$\operatorname{rank} Z_1^T E T_2 = \operatorname{rank} C^T M C = n - m = d.$$

Thus, the hypothesis in [39, Hyp. 4.2] is satisfied for $\mu = 1$, which implies that the nonlinear DAE (3.1) has strangeness index one.

Appendix 2: Difference-Algebraic Equation Index of the Considered Systems

In this appendix, we derive the Kronecker index for the discrete schemes considered in Sect. 4.3.

Half-Explicit Euler

We start with the half-explicit Euler discretisation, that gives a scheme $\mathcal{E}x^{k+1} = \mathcal{A}^k x^k + h^k$ with the matrix pair

$$(\mathcal{E}, \mathcal{A}) = \left(\begin{bmatrix} \frac{1}{\tau}M & 0 \\ 0 & 0 \end{bmatrix}, \begin{bmatrix} \frac{1}{\tau}M + A & B^T \\ B & 0 \end{bmatrix} \right)$$

as in (4.4). For sufficiently small τ, due to the definiteness of M and the full-rank property of B, the matrix $\mathcal{A}$ is invertible and thus, the pair $(\mathcal{E}, \mathcal{A})$ is regular. Let S denote the matrix $BM^{-1}B^T$. If one applies

$$\begin{bmatrix} M^{-\frac{1}{2}} & B^T S \\ 0 & I \end{bmatrix} \begin{bmatrix} I & 0 \\ BM^{-1} & I \end{bmatrix} \to (\mathcal{E}, \mathcal{A}) \leftarrow \begin{bmatrix} M^{-\frac{1}{2}} & 0 \\ -S^{-1}BM^{-1}(\frac{1}{\tau}M + A) & I \end{bmatrix}$$

from the left and the right, one finds that $(\mathcal{E}, \mathcal{A})$ is similar to

$$\left(\begin{bmatrix} \frac{1}{\tau}(I - M^{-\frac{1}{2}}B^T SBM^{-\frac{1}{2}}) & 0 \\ \frac{1}{\tau}B & 0 \end{bmatrix}, \right.$$
$$\left. \begin{bmatrix} (I - M^{-\frac{1}{2}}B^T SBM^{-\frac{1}{2}})(\frac{1}{\tau}I + M^{-\frac{1}{2}}AM^{-\frac{1}{2}}) - M^{-\frac{1}{2}}B^T SBM^{-\frac{1}{2}} & 0 \\ 0 & S \end{bmatrix} \right).$$

Since B is of full rank, there exists an orthogonal matrix Q and an invertible matrix R such that $BM^{-\frac{1}{2}}Q = \begin{bmatrix} 0 & R \end{bmatrix}$ and, in particular,

$$Q^T(I - M^{-\frac{1}{2}}B^T SBM^{-\frac{1}{2}})Q = \begin{bmatrix} I & 0 \\ 0 & 0 \end{bmatrix}.$$

Thus, the corresponding similarity transformation transforms $(\mathcal{E}, \mathcal{A})$ into

$$
(\begin{bmatrix} \frac{1}{\tau}I & 0 & 0 \\ 0 & 0 & 0 \\ 0 & \frac{1}{\tau}R & 0 \end{bmatrix}, \begin{bmatrix} \tilde{a}_{11} & 0 & 0 \\ \tilde{a}_{21} & I & 0 \\ 0 & 0 & S \end{bmatrix}),
$$

where $\tilde{a}_{11}$ and $\tilde{a}_{21}$ stand for unspecified but possibly nonzero block matrix entries. With another few regular row and column transformations, one can eliminate the entry $\tilde{a}_{21}$ and read off the Kronecker index of $(\mathcal{E}, \mathcal{A})$ as the index of nilpotency of $\begin{bmatrix} 0 & 0 \\ \frac{1}{\tau}R & 0 \end{bmatrix}$ which is 2.

Projection Scheme

The matrix coefficient pair of the Projection scheme (4.9) reads

$$
(\mathcal{E}, \mathcal{A}) = (\begin{bmatrix} \frac{1}{\tau}M & 0 & 0 & 0 \\ 0 & 0 & 0 & 0 \\ -M & -\frac{\tau}{2}B^T & M & 0 \\ 0 & -I & 0 & I \end{bmatrix}, \begin{bmatrix} 0 & \frac{1}{\tau}M + A & 0 & B^T \\ -\frac{2}{\tau}B & -BM^{-1}B^T & 0 & 0 \\ 0 & 0 & 0 & 0 \\ 0 & 0 & 0 & I \end{bmatrix}). \tag{6.1}
$$

If we define $S := BM^{-1}B^T$, if we move the second row and column to the left and bottom, respectively, and if we rescale certain rows and columns, we find that the pair is equivalent to

$$
(\mathcal{E}, \mathcal{A}) \sim (\begin{bmatrix} I & 0 & 0 & 0 \\ \tilde{e}_{21} & I & 0 & \tilde{e}_{24} \\ 0 & 0 & I & \tilde{e}_{34} \\ 0 & 0 & 0 & 0 \end{bmatrix}, \begin{bmatrix} 0 & 0 & \tilde{a}_{13} & \tilde{a}_{14} \\ 0 & 0 & 0 & 0 \\ 0 & 0 & \tilde{a}_{33} & 0 \\ \tilde{a}_{41} & 0 & 0 & -S \end{bmatrix}),
$$

where the $\tilde{e}$'s and $\tilde{a}$'s stand for unspecified but possibly nonzero entries. Since, in particular, S is invertible, one can eliminate the entries $\tilde{e}_{24}$, $\tilde{e}_{34}$, $\tilde{a}_{41}$, and $\tilde{a}_{14}$ by regular row and column manipulations without affecting the invertibility of the left upper 3×3 block in the transformed $\mathcal{E}$ and read off the Kronecker index of (4.9) as the index of nilpotency of 0 which is 1.

SIMPLE

The matrix coefficient pair of the SIMPLE scheme (4.13) reads

$$(\mathcal{E}, \mathcal{A}) = \left(\begin{bmatrix} \frac{1}{\tau}M & 0 & 0 & 0 \\ 0 & 0 & 0 & 0 \\ I & (\frac{1}{\tau}M + A)^{-1}B^T & -I & 0 \\ 0 & -I & 0 & I \end{bmatrix}, \begin{bmatrix} 0 & \frac{1}{\tau}M + A & 0 & B^T \\ -B & -B(\frac{1}{\tau}M + A)^{-1}B^T & 0 & 0 \\ 0 & 0 & 0 & 0 \\ 0 & 0 & 0 & I \end{bmatrix} \right).$$

If we define $S_A := B(\frac{1}{\tau}M^{-1} + A)^{-1}B^T$, move the second row and column to the left and bottom, respectively, and rescale certain rows and columns, then we find that the pair is equivalent to

$$(\mathcal{E}, \mathcal{A}) \sim \left(\begin{bmatrix} I & 0 & 0 & 0 \\ \tilde{e}_{21} & I & 0 & \tilde{e}_{24} \\ 0 & 0 & I & \tilde{e}_{34} \\ 0 & 0 & 0 & 0 \end{bmatrix}, \begin{bmatrix} 0 & 0 & \tilde{a}_{13} & \tilde{a}_{14} \\ 0 & 0 & 0 & 0 \\ 0 & 0 & \tilde{a}_{33} & 0 \\ \tilde{a}_{41} & 0 & 0 & -S_A \end{bmatrix} \right),$$

where, again, the $\tilde{e}$'s and $\tilde{a}$'s stand for unspecified but possibly nonzero entries. Since S_A is invertible for sufficiently small τ, we find that this matrix pair has the very same structure as the one of the projection scheme (see section "Projection Scheme" in Appendix) and, thus, is of index 1.

References

1. Altmann, R., Heiland, J.: Finite element decomposition and minimal extension for flow equations. ESAIM: Math. Model. Numer. Anal. **49**(5):1489–1509 (2015)
2. Altmann, R., Heiland, J.: Regularization and Rothe discretization of semi-explicit operator DAEs. Int. J. Numer. Anal. Model. **15**(3), 452–477 (2018)
3. Arnold, M., Strehmel, K., Weiner, R.: Half-explicit Runge-Kutta methods for semi-explicit differential-algebraic equations of index 1. Numer. Math. **64**(1), 409–431 (1993)
4. Behr, M., Benner, P., Heiland, J.: Example setups of Navier-Stokes equations with control and observation: spatial discretization and representation via linear-quadratic matrix coefficients. Technical Report (2017). arXiv:1707.08711
5. Benner, P., Heiland, J.: LQG-balanced truncation low-order controller for stabilization of laminar flows. In: King, R. (ed.) Active Flow and Combustion Control 2014, pp. 365–379. Springer, Berlin (2015)
6. Benner, P., Heiland, J.: Time-dependent Dirichlet conditions in finite element discretizations. ScienceOpen Research, 1–18 (2015)
7. Braack, M., Mucha, P.B.: Directional do-nothing condition for the Navier-Stokes equations. J. Comput. Math. **32**(5), 507–521 (2014)
8. Brasey, V., Hairer, E.: Half-explicit Runge–Kutta methods for differential-algebraic systems of index 2. SIAM J. Numer. Anal. **30**(2), 538–552 (1993)
9. Brezzi, F., Fortin, M.: Mixed and Hybrid Finite Element Methods. Springer, New York (1991)

10. Campbell, S.: A general form for solvable linear time varying singular systems of differential equations. SIAM J. Math. Anal. **18**(4), 1101–1115 (1987)
11. Chorin, A.J.: Numerical solution of the Navier-Stokes equations. Math. Comput. **22**, 745–762 (1968)
12. Chorin, A.J., Marsden, J.E.: A Mathematical Introduction to Fluid Mechanics, 3rd edn. Springer, New York (1993)
13. Ciarlet, P.G.: The Finite Element Method for Elliptic Problems. North-Holland, Amsterdam (1978)
14. Dai, L.: Singular Control Systems. Lecture Notes in Control and Information Sciences, vol. 118. Springer, Berlin (1989)
15. Elman, H.C., Silvester, D.J., Wathen, A.J.: Finite Elements and Fast Iterative Solvers: With Applications in Incompressible Fluid Dynamics. Oxford University Press, Oxford (2005)
16. Emmrich, E., Mehrmann, V.: Operator differential-algebraic equations arising in fluid dynamics. Comp. Methods Appl. Math. **13**(4), 443–470 (2013)
17. Feireisl, E., Karper, T.G., Pokorný, M.: Mathematical Theory of Compressible Viscous Fluids. Analysis and Numerics. Birkhäuser/Springer, Basel (2016)
18. Ferziger, J.H., Perić, M.: Computational Methods for Fluid Dynamics, 3rd edn. Springer, Berlin (2002)
19. Fujita, H., Kato, T.: On the Navier-Stokes initial value problem. I. Arch. Ration. Mech. Anal. **16**, 269–315 (1964)
20. Gaul, A.: Krypy – a Python toolbox of iterative solvers for linear systems, commit: 36e40e1d (2017). https://github.com/andrenarchy/krypy
21. Girault, V., Raviart, P.-A.: Finite Element Methods for Navier–Stokes Equations. Theory and Algorithms. Springer, Berlin (1986)
22. Glowinski, R.: Finite element methods for incompressible viscous flow. In: Numerical Methods for Fluids (Part 3). Handbook of Numerical Analysis, vol. 9, pp. 3–1176. Elsevier, Burlington (2003)
23. Gresho, P.M.: On the theory of semi-implicit projection methods for viscous incompressible flow and its implementation via a finite element method that also introduces a nearly consistent mass matrix. I: Theory. Int. J. Numer. Methods Fluids **11**(5), 587–620 (1990)
24. Gresho, P.M., Sani, R.L.: Incompressible Flow and the Finite Element Method. Vol. 2: Isothermal Laminar Flow. Wiley, Chichester (2000)
25. Griebel, M., Dornseifer, T., Neunhoeffer, T.: Numerical Simulation in Fluid Dynamics. A Practical Introduction. SIAM, Philadelphia (1997)
26. Ha, P.: Analysis and numerical solutions of delay differential algebraic equations. Ph.D. thesis, Technische Universität Berlin (2015)
27. Hairer, E., Wanner, G.: Solving Ordinary Differential Equations II: Stiff and Differential-Algebraic Problems, 2nd edn. Springer, Berlin (1996)
28. Hairer, E., Lubich, C., Roche, M.: The numerical solution of differential-algebraic systems by Runge-Kutta methods. Springer, Berlin (1989)
29. Hairer, E., Lubich, C., Wanner, G.: Geometric Numerical Integration. Structure-Preserving Algorithms for Ordinary Differential Equations, 2nd edn. Springer Series in Computational Mathematics. Springer, Berlin (2006)
30. He, X., Vuik, C.: Comparison of some preconditioners for the incompressible Navier-Stokes equations. Numer Math. Theory Methods Appl **9**(2), 239–261 (2016)
31. Heiland, J.: Decoupling and optimization of differential-algebraic equations with application in flow control. Ph.D. thesis, TU Berlin (2014). http://opus4.kobv.de/opus4-tuberlin/frontdoor/index/index/docId/5243
32. Heinrich, J.C., Vionnet, C.A.: The penalty method for the Navier-Stokes equations. Arch. Comput. Method E **2**, 51–65 (1995)
33. Heywood, J.G., Rannacher, R.: Finite-element approximation of the nonstationary Navier–Stokes problem. IV: Error analysis for second-order time discretization. SIAM J. Numer. Anal. **27**(2), 353–384 (1990)

34. Hinze, M.: Optimal and instantaneous control of the instationary Navier-Stokes equations. Habilitationsschrift, Institut für Mathematik, Technische Universität Berlin (2000)
35. Karniadakis, G., Beskok, A., Narayan, A.: Microflows and Nanoflows. Fundamentals and Simulation. Springer, New York (2005)
36. Kunkel, P., Mehrmann, V.: Canonical forms for linear differential-algebraic equations with variable coefficients. J. Comput. Appl. Math. **56**(3), 225–251 (1994)
37. Kunkel, P., Mehrmann, V.: Analysis of over- and underdetermined nonlinear differential-algebraic systems with application to nonlinear control problems. Math. Control Signals Syst. **14**(3), 233–256 (2001)
38. Kunkel, P., Mehrmann, V.: Index reduction for differential-algebraic equations by minimal extension. Z. Angew. Math. Mech. **84**(9), 579–597 (2004)
39. Kunkel, P., Mehrmann, V.: Differential-Algebraic Equations. Analysis and Numerical Solution. European Mathematical Society Publishing House, Zürich (2006)
40. Ladyzhenskaya, O.A.: The Mathematical Theory of Viscous Incompressible Flow. Gordon and Breach Science Publishers, New York (1969)
41. Landau, L.D., Lifshits, E.M.: Fluid Mechanics. Course of Theoretical Physics, vol. 6, 2nd edn. Elsevier, Amsterdam (1987). Transl. from the Russian by J. B. Sykes and W. H. Reid.
42. Layton, W.: Introduction to the Numerical Analysis of Incompressible Viscous Flows. SIAM, Philadelphia (2008)
43. Leray, J.: étude de diverses équations intégrales non linéaires et de quelques problèmes que pose l'hydrodynamique. J. Math. Pures Appl. **12**, 1–82 (1933)
44. LeVeque, R.J.: Numerical Methods for Conservation Laws, 2nd edn. Birkhäuser, Basel (1992)
45. Logg, A., Ølgaard, K.B., Rognes, M.E., Wells, G.N.: FFC: the FEniCS form compiler. In: Automated Solution of Differential Equations by the Finite Element Method, pp. 227–238. Springer, Berlin (2012)
46. Mattsson, S.E., Söderlind, G.: Index reduction in differential-algebraic equations using dummy derivatives. SIAM J. Sci. Comput. **14**(3), 677–692 (1993)
47. Noack, B.R., Afanasiev, K., Morzyński, M., Tadmor, G., Thiele, F.: A hierarchy of low-dimensional models for the transient and post-transient cylinder wake. J. Fluid Mech. **497**, 335–363 (2003)
48. Nguyen, P.A., Raymond, J.-P.: Boundary stabilization of the Navier–Stokes equations in the case of mixed boundary conditions. SIAM J. Control. Optim. **53**(5), 3006–3039 (2015)
49. Pironneau, O.: Finite Element Methods for Fluids. Wiley/Masson, Chichester/Paris (1989) Translated from the French
50. Raymond, J.-P.: Feedback stabilization of a fluid-structure model. SIAM J. Cont. Optim. **48**(8), 5398–5443 (2010)
51. Reis, T.: Systems Theoretic Aspects of PDAEs and Applications to Electrical Circuits. Shaker, Aachen (2006)
52. Reynolds, O.: Papers on Mechanical und Physical Subjects. Volume III. The Sub-mechanics of the Universe. Cambridge University Press, Cambridge (1903)
53. Roubíček, T.: Nonlinear Partial Differential Equations with Applications. Birkhäuser, Basel (2005)
54. Saint-Raymond, L.: Hydrodynamic Limits of the Boltzmann Equation. Springer, Berlin (2009)
55. Shen, J.: On error estimates of the penalty method for unsteady Navier-Stokes equations. SIAM J. Numer. Anal. **32**(2), 386–403 (1995)
56. Steinbrecher, A.: Numerical Solution of Quasi-Linear Differential-Algebraic Equations and Industrial Simulation of Multibody Systems. Ph.D. thesis, Technische Universität Berlin (2006)
57. Tartar, L.: An Introduction to Navier–Stokes Equation and Oceanography. Springer, New York (2006)
58. Temam, R.: Navier–Stokes Equations. Theory and Numerical Analysis. North-Holland, Amsterdam (1977)
59. Taylor, C., Hood, P.: A numerical solution of the Navier-Stokes equations using the finite element technique. Int. J. Comput. Fluids **1**(1), 73–100 (1973)

60. Turek, S.: Efficient Solvers for Incompressible Flow Problems. An Algorithmic and Computational Approach. Springer, Berlin (1999)
61. Weickert, J.: Navier-Stokes equations as a differential-algebraic system. Preprint SFB393/96-08, Technische Universität Chemnitz-Zwickau (1996)
62. Weickert, J.: Applications of the theory of differential-algebraic equations to partial differential equations of fluid dynamics. Ph.D. thesis, Fakultät für Mathematik, Technische Universität Chemnitz (1997)
63. Williamson, C.H.K.: Vortex dynamics in the cylinder wake. Annu. Rev. Fluid Mech. **28**(1), 477–539 (1996)
64. Zeidler, E.: Nonlinear Functional Analysis and its Applications. II/A: Linear Monotone Operators. Springer, Berlin (1990)

Index

S. Campbell et al. (eds.), *Applications of Differential-Algebraic Equations: Examples and Benchmarks*, Differential-Algebraic Equations Forum,
https://doi.org/10.1007/978-3-030-03718-5

Printed in the United States
by Baker & Taylor Publisher Services